Lecture Notes – Joint EPS-SIF International School on Energy – Course 4

Advances in Basic Energy Issues

EPJ Web of Conferences Volume 189 (2018)

Varenna, Italy
21 – 26 July 2017

Editors:

L. Cifarelli
F. Wagner

ISBN: 978-1-5108-7252-3

Printed from e-media with permission by:

Curran Associates, Inc.
57 Morehouse Lane
Red Hook, NY 12571

Some format issues inherent in the e-media version may also appear in this print version.

Printed by Curran Associates, Inc. (2018)

For additional information, please contact EDP Sciences – Web of Conferences at the address below.

EDP Sciences – Web of Conferences
17, Avenue du Hoggar
Parc d'Activité de Courtabœuf
BP 112
F-91944 Les Ulis Cedex A
France

Phone: +33 (0) 1 69 18 75 75
Fax: +33 (0) 1 69 28 84 91

contact-edps@webofconferences.org

Additional copies of this publication are available from:

Curran Associates, Inc.
57 Morehouse Lane
Red Hook, NY 12571 USA
Phone: 845-758-0400
Fax: 845-758-2633
Email: curran@proceedings.com
Web: www.proceedings.com

TABLE OF CONTENTS

Preface

In order to contribute to the training of young scientists working in the energy sector or intending to do so, the European (EPS) and the Italian (SIF) Physical Societies started the Joint EPS-SIF International School on Energy as a collaborative initiative. The Courses are foreseen to take place on a biennial basis in the beautiful venue of Villa Monastero in Varenna, Lake Como, Italy. The 1st Course was held in summer of 2012 on "*New strategies for energy generation, conversion and storage*". The 2nd Course in 2014 was devoted to an overview of "*Basic concepts and forefront ideas on energy*", covering the major scientific areas. The 3rd Course was exceptionally held in 2016 at the Ettore Majorana Foundation and Centre for Scientific Culture (EMFCSC), in Erice, Italy, in collaboration with the Materials Research Societies (MRS and EMRS), on "*Materials for energy and sustainability*". The 4th Course of the Joint EPS-SIF International School on Energy was organized in 2017 back to Varenna. The focus was on "*Advances in basic energy issues*".

The primary goal of the School is to present all research and development fields with relevance for the technologies of energy production, conversion, transmission and savings. The unique feature of the School is its multidisciplinarity and interdisciplinarity including basic and applied topics but also climate and economic aspects. This wide scope of the School is essential in order to provide the students with a global insight into the complex nature of energy supply and consumption. For this purpose, this School effectively brings together a large number of scientists working in different disciplines but all related to energy technologies with young scientists from all over the world.

The 4th Course of the School gathered 55 participants —lecturers, observers and students— of 10 different nations. The lectures were delivered by about 20 experts in various fields. The concepts and ideas involved in modern energy systems and technologies were clearly addressed by the lecturers who invested every effort to be as didactic as possible. Students and lecturers stayed together during lunch and dinner. The attractiveness of the venue helped to keep all parties on site for the whole week. In this way knowledge gaps were filled, and answers were provided to long-standing questions in the mind of the students during lively discussion-times that followed each lecture. In this way, a community formed and the basic idea of such a summer School was fully met.

The proceedings of the School, published as Lecture Notes, conserve the teaching material presented and make it available to those who did not attend the School. They serve as a reference book for both specialists working in one of the energy fields but with interests in the status of other energy-related areas, and non-technical readers who want to get a general overview on the involved concepts and techniques and their prospects. Papers are ordered according to topics: global overview of world energy resources, renewable energies (including geothermal energy), nuclear energy (fission, and fusion), energy economics, energy and grids, energy storage. These proceedings are published both on paper, in a volume of the "*Lecture Notes of the Joint EPS-SIF International School on Energy*" series, and on-line open-access in The European Physical Journal Web of Conferences (EPJ WoC).

We are especially grateful to all our distinguished colleagues who have accepted to write and timely provide their contributions for these Lecture Notes. Our thanks go as well to our many sponsors. Finally, we are grateful to our Scientific Secretary, Gianluca Alimonti, for his dedicated work concerning the programme of this Course and these proceedings, and to the staff of the Italian Physical Society for their warm and efficient hospitality in Varenna.

L. CIFARELLI and F. WAGNER

Statement of Peer review

In submitting conference proceedings to *Web of Conferences*, the editors of the proceedings certify to the Publisher that

1. They adhere to its **Policy on Publishing Integrity** in order to safeguard good scientific practice in publishing.
2. All articles have been subjected to peer review administered by the proceedings editors.
3. Reviews have been conducted by expert referees, who have been requested to provide unbiased and constructive comments aimed, whenever possible, at improving the work.
4. Proceedings editors have taken all reasonable steps to ensure the quality of the materials they publish and their decision to accept or reject a paper for publication has been based only on the merits of the work and the relevance to the journal.

Title, date and place of the conference

JOINT EPS-SIF INTERNATIONAL SCHOOL ON ENERGY
COURSE 4 - ADVANCES IN BASIC ENERGY ISSUES
21-26 JULY 2017
VILLA MONASTERO - VARENNA, LAKE COMO

Proceedings editor(s):

L. CIFARELLI AND F. WAGNER

Date and editor's signature 1/6/2018

EPJ Web of Conferences **189**, 00002 (2018) https://doi.org/10.1051/epjconf/201818900002
LNES 2017

The complex story of energy transition —an introduction

F. WAGNER(*)

Max-Planck Institut für Plasmaphysik - Garching/Greifswald, Germany

Human beings grew in numbers as any other biological system would do under favourable living conditions. The growth of mankind up to presently 7.5 bn people was enabled by the access to cheap energy. Technology made it possible to exploit fossil fuels at large scale and to transform the chemically bound energy into different forms for practical application.

Figure 1 shows the growth of the US population between 1835 and 2009. Initially, at still a small population, the renewable energy (RE) share in the energy demand was high, close to 1. In course of time, the share of nuclear and fossil energy grew at the expense of REs. This historical evolution is representative for most industrialised countries. The interrelation between energy consumption by a growing population and the population growth by energy availability is, however, not "free lunch".

Figure 2 shows the increase of world energy consumption being steeper than the growth of world population. However, about 80% of primary energy use gives rise to CO_2 emissions. Mankind is suddenly faced with the threat of climate change. In the future, the course has to be reversed back to large RE shares but under conditions where large populations, 300 Mill people in case of USA (see fig. 1), have to be safely and economically supplied by energy.

As nutrition, clean water, healthy living conditions all depend on access to energy a classical conflict emerged in the last decade between efforts to locally improve living conditions by higher energy use and the threat of thereby jeopardising the global living conditions if energy consumption causes CO_2 release in the atmosphere as is the case up

(*) E-mail: `fritz.wagner@ipp.mpg.de`

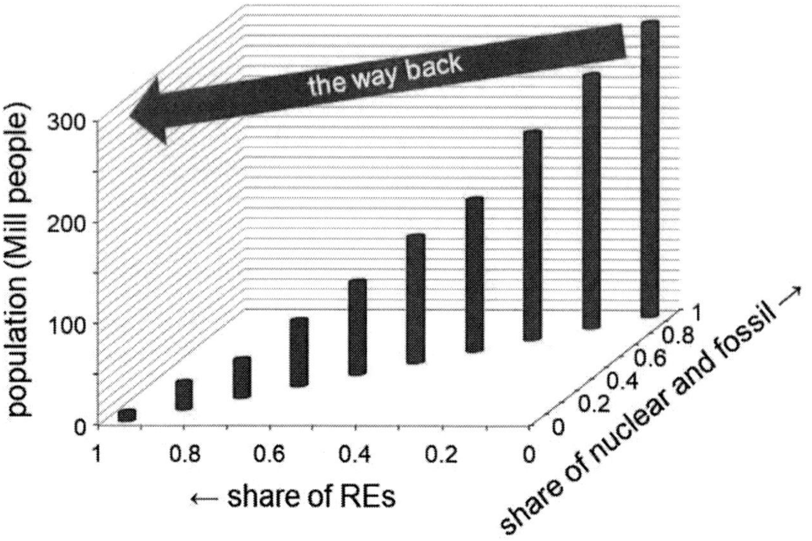

Fig. 1. – Growth of the US population between 1835 and 2009 plotted against the share of renewable, nuclear and fossil energy.

to now. Figure 3 shows the human development index HDI, based on life expectation, education, and gross domestic product (GDP) together with the "life expectation at birth" *versus* the electricity consumption per capita. There is a close correlation —more electricity, or generally energy, allows a better and healthier life. These relations drive

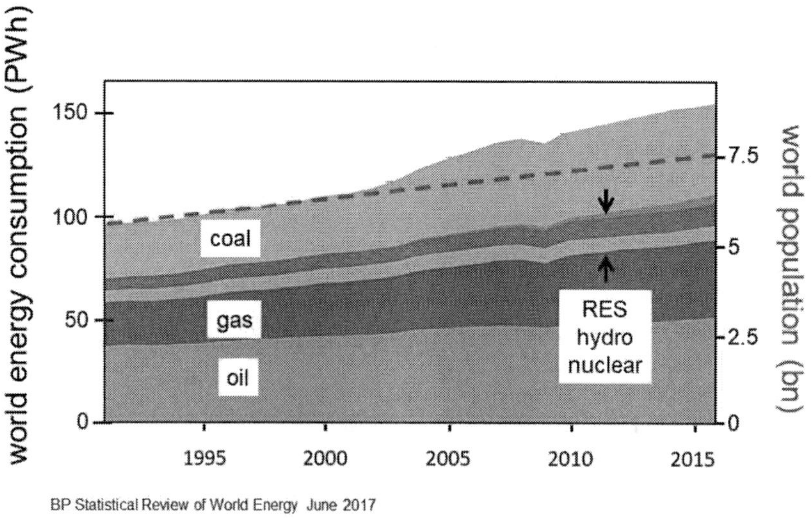

Fig. 2. – Growth of world energy consumption and of world population.

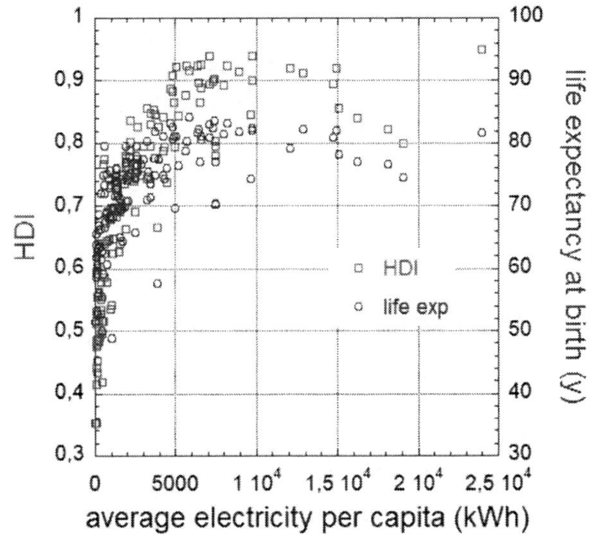

Fig. 3. – Human development index HDI, and "life expectation at birth" *versus* the electricity consumption per capita.

the world to higher energy production and consumption and it has to be expected that this development continues as governments have constitutionally the task to improve the welfare of their nations.

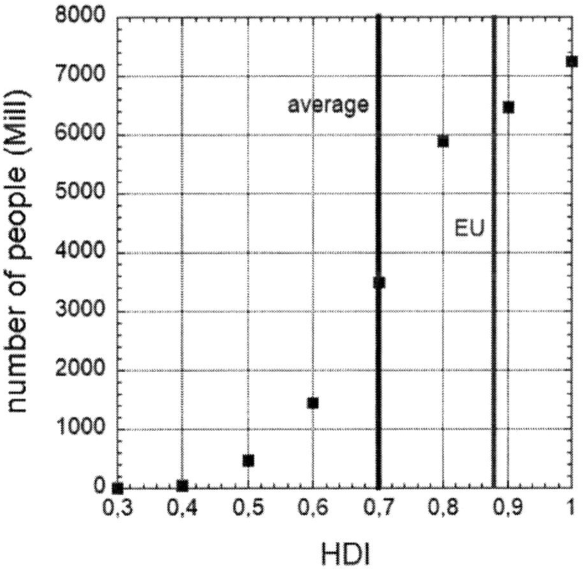

Fig. 4. – Number of people living below a specified HDI value.

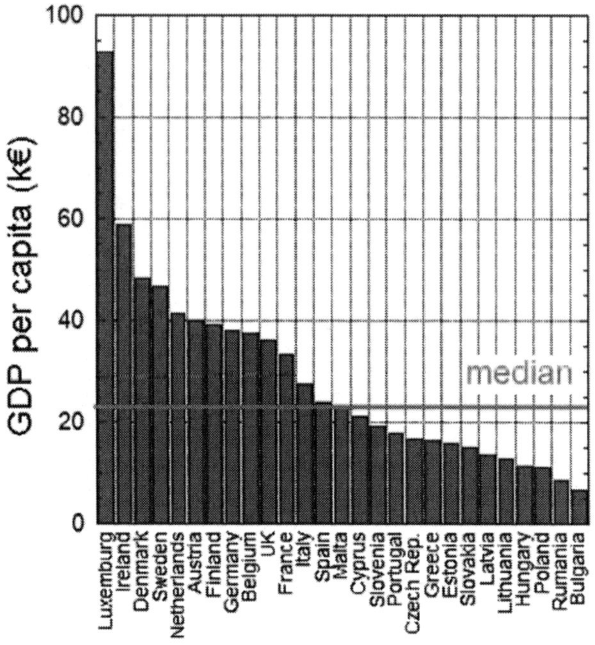

Fig. 5. – Distribution of the GDP per person in Europe.

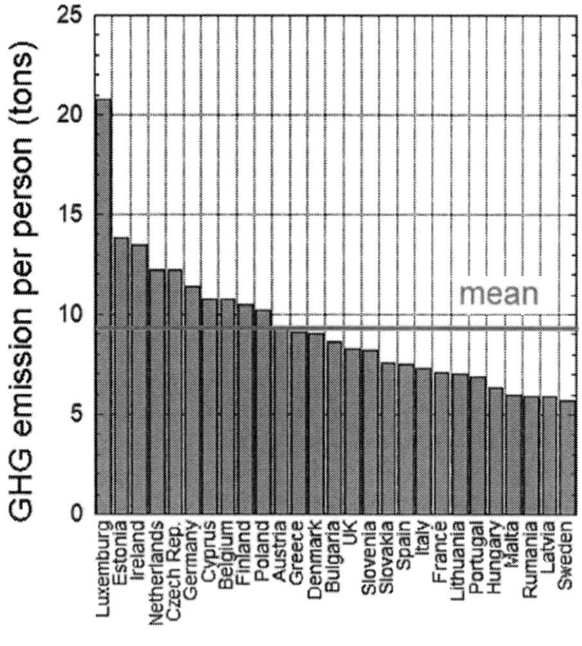

Fig. 6. – Per capita GHG emission for the different EU countries.

Figure 4 analyses the number of people on earth enjoying high HDI factors —or not, respectively. The average HDI value and the one of the EU are indicated by vertical bars. About 6.5 Mill people live under circumstances below the average standard of Europe. This aspect should be the source of much concern because tremendous societal upheavals can be expected when this distribution of welfare expands toward a more homogeneous and just distribution specifically if the future promises energy scarcity out of environmental concerns.

Europe offers unfavourable conditions for an energy transformation process toward less CO_2 emission. Figure 5 shows the distribution of the GDP per person, the economic productivity of different European countries. There is a large spread by a factor of 15 in wealth within the European countries, which can be expected to be an obstacle if the transformation process of energy technologies enforces sacrifices and abdication of accustomed life-style.

As we have seen above for the global situation, the variation in GDP in Europe reflects itself in a corresponding variation in primary energy (PE) use and in greenhouse-gas (GHG) emission, as demonstrated by fig. 6. Plotted is the per capita GHG emission for the different EU countries. In tendency, those countries with a large GDP per capita are individually also causing large GHG emissions. *E.g.*, Germany, the most populated European country, gains its wealth at the expense of a large GHG footprint. But there

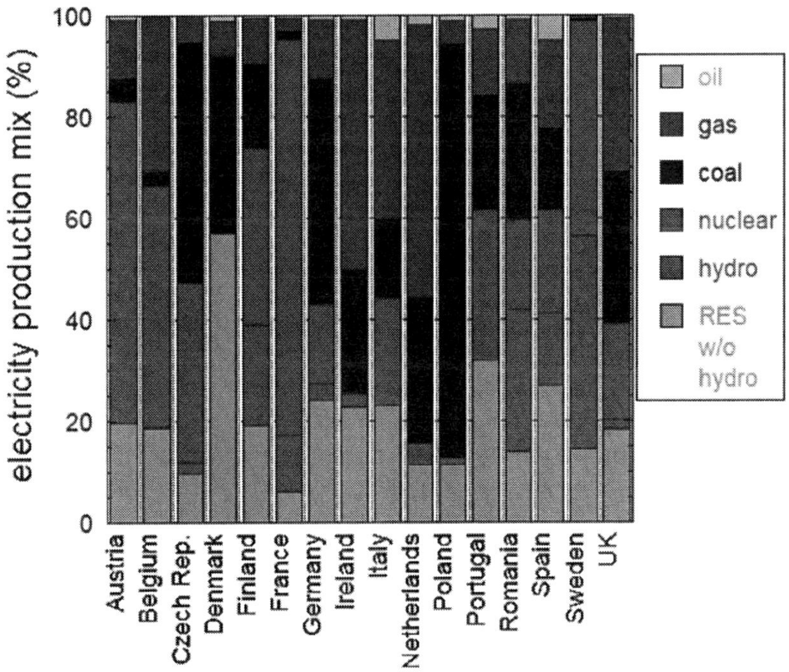

Fig. 7. – Sources of electricity production of European countries.

are deviations from this rule notably Sweden and France with high GDPs at low GHG emission. Those cases have to be studied in some detail and they might provide a specific message how to properly guide the transition.

But Europe will be an interesting test bed for the energy transition thanks to its strong national diversification in electricity production today, as fig. 7 shows. Poland is nearly 80% coal based in its electricity generation, France nearly 80% nuclear and Austria, thanks to a large hydro share, more than 80% renewable.

It is obviously easy to collect the European countries behind formal climate goals. Europe has formulated common energy and climate targets, *e.g.*, up to 2030 —to reduce GHG emission by 40% compared to 1990, to increase the REs share to 27% and to improve energy efficiency by 27%. But it will be a tremendous political task to collect these countries behind these goals as soon as the process boils down to decide upon suitable CO_2-free technologies. This book presents and discusses many technical solutions. But in principle, there are not so many and they all have or may have caveats. Most of the technologies produce electricity, which will be the primary energy of the future from which other forms of energy have to be derived —least efficient and powerful storage technologies can be developed so that electricity can serve all energy consuming sectors including mobility, room- and process heat. But this exclusiveness should not be expected because also local circumstances as the availability of geo-thermal or solar-thermal energy will play a role (or does it already now).

The future nuclear technologies are fission on the basis of fast neutrons, developed in the frame of Generation IV reactors and fusion, with ITER being the first, still experimental, fusion reactor presently realised in France. The problems of fission are well known —predominantly concerns about safety and radioactive waste. It may make a difference whether a country never accumulated fission waste because it did not use fission energy. In the other case, a storage technology has to be developed and a repository for waste has to be identified. This is unavoidable. In this case, the quantity of waste to be handled is of minor importance.

Unfortunately, fusion is not ready and will not play a role up to mid of this century where a solution for the climate problem has to be found and implemented. Fusion promises favourable supply and environmental characteristics and radioactive waste with short decay times (about 100 years). Around 2035, ITER will demonstrate the soundness of the fusion concept from the physics and partly the materials point of view.

The use of fossil fuels can in principle be continued but in the frame of carbon-capture and sequestration (CCS) where CO_2 is collected and separated and finally stored in underground cavities.

Finally, there are many renewable energy forms with proven or expected capabilities —wind and solar power and bio-mass may offer the best prospects. But each has specific characteristics and problem areas like low power density, intermittency in case of wind and photovoltaic (PV) power or the general concerns in case of biomass about nutrition and losses of bio-diversity. Material to these topics is presented in these proceedings.

Because of the environmental concerns, fossil energy supply technologies have to be replaced by new techniques without CO_2 emission in the short period of about one

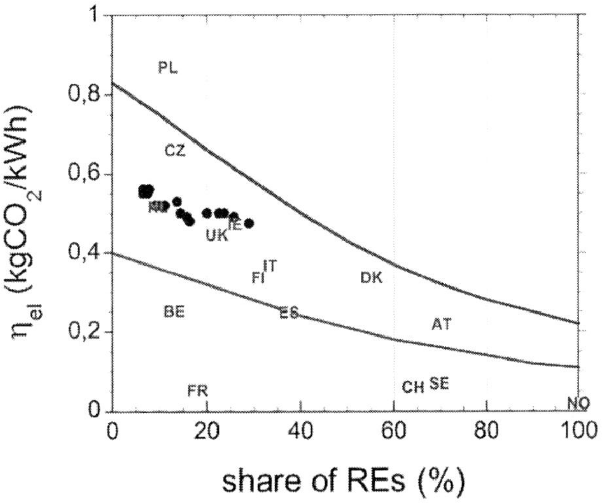

Fig. 8. – Specific CO_2 emission *versus* the share of renewable energies involved in electricity generation. Black dots: German development from 2002 to 2015. The black curve represents an upper limit —Germany's electricity is produced with the present mix of fossil fuels. The blue curve represents the case that Germany's electricity is produced by gas. The country symbols represent the situation of different European countries.

generation. But in this period the number of energy consumers will increase by further 2.5 bn aggravating the complexity of a world-wide technology change. In the past, the time scales for a technology change were specified by the resources of fossil fuel, *e.g.*, 50 years for gas and oil, about 120 years for coal whereas the shale oil and gas bonanza in the USA demonstrated that these figures are lower estimates and that the Hubbard peak has the tendency to move further into the future. The new time scales are by far more restrictive with the tendency to rather move closer. The "new" timescales are dictated by the 2-degree limit and the remaining possible CO_2 pollution. If the world would pollute the atmosphere with CO_2 as the average American does, the limit would be reached in about 8 years; it would be reached in about 75 years with the average Indian emission and with the one of the average European, it would be met in 16 years. Time is short now; nevertheless, the global development may not be encouraging. On the one hand there are pompous climate meetings, on the other hand, it was demonstrated how quickly climate policies can change in democracies, *e.g.* after an election. Also the developments in Europe are not convincing. The main and most meaningful concept —to give an increasing price to the emission of CO_2— has not successfully been implemented. Despite tremendous efforts for the energy transition in Germany where the wind and PV capacity surpasses already the peak electricity power demand, the GHG emission has not been reduced in a convincing way. Germany also demonstrates the high price tag for the technology change with the highest electricity prices in Europe. France rules out CCS by law. On the other side, coal will not disappear quickly and the last cold winter and fuel crisis in China will have left their mark. About 70% of Chinas electricity is generated by burning coal and China develops and exports coal power plants with higher efficiency and lower emission. It would be wise to use the European expertise and culture in technology development and research to contribute to all energy relevant fields, including CCS.

This applies also to research into future fission concepts. Unfortunately, fission has become a field for party politics which ignore the demonstrated capability of this technology. Figure 8 shows the specific CO_2 emission connected with electricity production first for Germany (black dots, black and blue curves) and for different European countries (country symbols). The black dots show the German development from 2002 to 2015 where a fleet of RE systems has been built up. The black curve represents an upper limit for the hypothetical case that the German electricity production happens exclusively with the present mix of fossil fuels (coal, lignite and gas). The actual data fall below this curve mostly thanks to nuclear power. The blue curve represents the other hypothetical limit that electricity in Germany is exclusively produced by gas. Of interest are the points of other countries falling below the blue limit: Norway thanks to hydro power and the other countries thanks to their hydro and nuclear power mix. France has nearly 80% electricity from nuclear power, Belgium nearly 50%, Sweden and Switzerland about 40%. If Europe is truly concerned about global warming and climate change, these figures and facts cannot be ignored. Even with 100% REs, Germany will not reach the low level of CO_2 emission of countries which involve nuclear power least a CO_2-free back-up system is available to bridge periods without wind during the night. In summary, some EU countries have produced electricity already at a low CO_2 emission level for 35 years which Germany plans to meet in 35 years.

This chapter should demonstrate that a solution of the climate problem —if possible at all for 10 bn people on earth— can only come from technology and can only be implemented by adhering to the rules and laws of economy. Therefore, research is mandatory into all energy related fields with good prospects. It might be profitable for political parties to rule out research into fields which do not agree with ideological positions but this is a dangerous route for their nations. With insufficient and wrong technology a reduction of GHG emission might nevertheless be possible by the deindustrialisation of Europe. Its effectiveness has been demonstrated by the reduction of the German GHG emission following the deindustrialisation of the former DDR starting in 1990, by the effects of the economic crises of 2008/2009, or by the creeping deindustrialisation in several European countries. This may not be the way the European young generation might favour. This book collects facts on energy technologies and issues addressing young scientist which will soon have a say when it comes to decisions on energy politics. This book should provide a sound technical basis.

EPJ Web of Conferences **189**, 00003 (2018) https://doi.org/10.1051/epjconf/201818900003
LNES 2017

Our energy future starts from actual energy limits

Gianluca Alimonti

INFN and Università di Milano - Via Celoria 16, Milano, Italy

Summary. — Since the discovery of fire, the history of human evolution reflects the history of mankind's ability to control and transform energy. Steam engine gave rise to the industrial revolution and fossil fuels entered and shaped our development: diffused and cheap energy allowed to replace human labor, and activities, impossible until that time, became common. Nowadays more than 80% of the global primary energy consumption comes from fossil fuels, but this energy system has limitations: green house gases and polluting emissions are generated by the fossil fuels combustion. Moreover fossil fuels are limited energy resources and sooner or later they will run out. It is thus important to start shaping and building a new energy system and this is what this school is about.

The history of energy, evolution and discoveries in transforming energy, is the history of mankind. This is the case starting from the Prometheus mythology who gave fire, the first energy technology that transforms chemical bonds to thermal energy, to mankind, to Newcomen&Cawley, inventors of the steam engine.

The relationship between energy and human evolution has historically been very close and it is still very important, as shown by the Human Development Index (HDI). Such an indicator has been defined by the United Nations Development Program to emphasize that people and their capabilities should be the ultimate criteria for assessing the development of a country, not economic growth alone.

The HDI is a summary measure of the average achievement in key dimensions of human development: a long and healthy life, being knowledgeable and have a decent

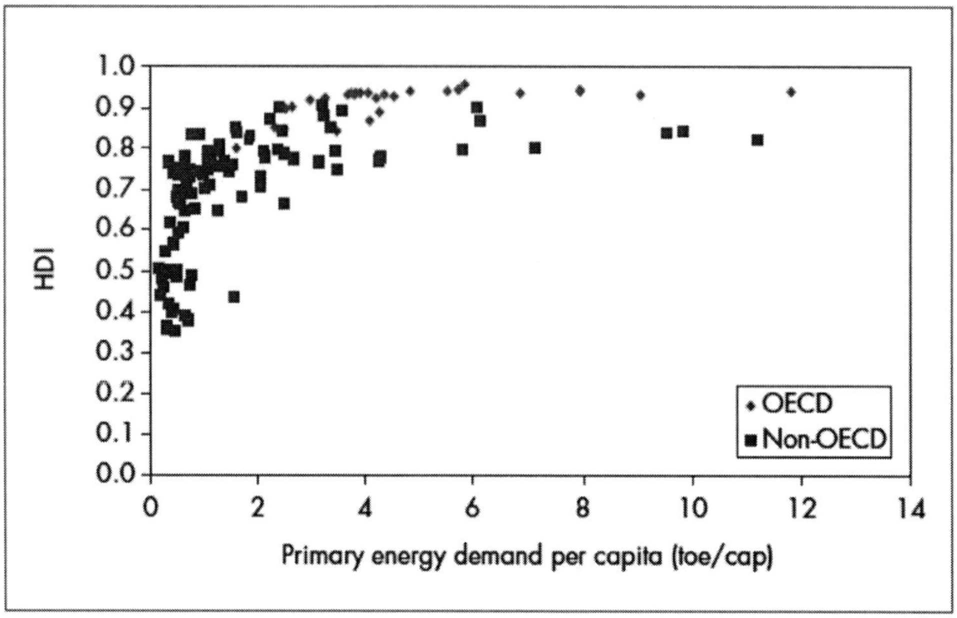

Fig. 1. – HDI and primary energy demand per capita, 2002. (Courtesy of IEA.)

standard of living. The HDI is the geometric mean of normalized indices for each of the three dimensions.

As shown in fig. 1, coming from an International Energy Agency (IEA) analysis, there is a clear relationship between HDI and primary energy demand per capita [1], with OECD countries all being in the high part of both HDI and energy use. After a steep increase, once you hit OECD country levels of per capita energy consumption, the HDI does not appear to increase much, but lower energy consumption has an impact.

As we all know, energy can neither be created nor destroyed, but only transformed into the many different thermal, mechanical and chemical forms we are used to. Depending on the form, a different unit is generally used, but with simple equivalences it is possible to go from electron volts, to joule, from calories to tons of oil equivalent([1]).

Never mix energy with power: assuming a FIAT 500 has the same weight of a Ferrari, there is not a big difference indeed, they both have the same energy when travelling at the same speed even if, given the different power they have, they may need different time to reach the same velocity.

Energy is related to power the same way length is related to velocity or the amount of water in a dam with the incoming/outgoing water flux or, for people having a more economic background, money in a bank account is related with salary and expenses!

([1]) A very useful unit converter can be found at `https://www.iea.org/statistics/resources/unitconverter/`.

Just to give few examples, we go from few watts of a modern LED light, to about 1 kW of an oven, from about 100 kW of a car up to few MW of a train.

With simple calculations, assuming we use all the energy we consume in one day with a standard diet, we can evaluate that the average power of a human body is about 100 W or that a hydroelectric plant has to convert the potential energy of 1 ton of water falling from about 367 meters (assuming an ideal conversion efficiency equal to 1) to produce just 1 kWh. Similarly, we can calculate that the energy content of 10 grams of wood would ideally be enough to lift a man by more than 100 meters and the energy content of a 100 grams chocolate bar would lift the same man by about 2000 meters!

Primary energy is the energy directly available in nature.

There are renewable primary energy sources (or simply renewable energy sources, RES) like hydroelectric plants but also ocean energy sources, wind, solar, biomass; part of the urban solid waste, as a biomass, is also considered a renewable energy source. All RES come directly or indirectly from the Sun and as such they are considered to be unlimited.

The fossil energy sources like oil, coal and natural gases like methane are the most important primary energy sources: they also come from the Sun but their generation time is much longer than the one of RES and they cannot be considered renewable energy sources. Energy from nuclear fission, not generated by breeder reactors, is in this same group of limited primary energy sources.

Some authors define a third group: the "almost unlimited" primary energy sources. Among these we can find nuclear energy generated by breeder reactors (that will be described in a dedicated paper) and, hopefully in the future, by fusion reactors. Also geothermal energy, being generated partly by the radioactive decays inside our planet and partly by the fossil heat remaining since the formation of Earth, is not, strictly speaking, a RES and it is listed among the "almost unlimited" primary energy sources. More often though, geothermal energy is considered a RES and fission, even if from breeder reactors, and fusion are considered in the group of "not renewable" primary energy sources.

Primary energy sources are generally transformed into secondary energy sources or energy carriers that are more suited to transport and/or final use. Unfortunately every time there is an energy transformation, there is an energy loss.

An example of secondary energy is electricity: it has the huge advantage of being easily transportable over great distance to the final user, who can very easily convert it with very high efficiency into mechanical energy, heat, use it as light source or in many different and common ways.

Electricity has the drawback that it cannot be stored in significant amount and for long time and therefore must be produced at the same time of the request: this is, as we will see during this school, an important limit that is causing more and more stringent problems for the grid stability. To solve such a limit, important developments on electricity storage are going on, and the grid stability problem may partly be addressed by the development of smart grids that will be presented during this week.

Let us now consider the energy to the final consumption and its link with the primary energy sources.

If we have 1 kg of dry wood having about 4 kWh of primary energy in the form of chemical bonds and we burn it in a stove to heat a room, we then have 4 kWh of energy to final consumption.

Starting again from the same primary energy in the form of dry wood, but converting it into electricity burning the wood in a thermoelectric power plant with an efficiency of 25%, we will then have just 1 kWh of electricity to the final consumption and 3 kWh of primary energy will be lost. There are thus 4 kWh of energy to the final consumption in the first example and just 1 kWh in the second example, but in both situations they derived from the same primary energy of 4 kWh. The question is: what amount of primary energy can be associated to 1 kWh produced by a hydroelectric plant? (the same question applies to the energy produced by photovoltaic or wind power).

This is not a difficult question, but the answer is not unique as can be seen comparing the world primary energy production data [2] given by the International Energy Agency (IEA) with the same data [3] coming from British Petroleum (BP): IEA is reporting a hydroelectric primary energy that, up to a few years ago, is about one third of the nuclear primary energy (both hydroelectric and nuclear produce only electricity and until the Fukushima nuclear incident they were producing about the same amount of electricity). On the other hand, BP is reporting an almost equal primary energy production for these two energy sources, as it is for the produced electricity. The reason for this difference are the different conventions for electricity.

According to IEA, figures for electricity production, trade, and final consumption are calculated using the energy content of the electricity (*i.e.* at a rate of 1 TWh = 0.086 Mtoe). Hydro-electricity production (excluding pumped storage) and electricity produced by other non-thermal means (wind, tide/wave/ocean, photovoltaic, etc.) are accounted for similarly using 1 TWh = 0.086 Mtoe. However, the primary energy equivalent of nuclear electricity is calculated from the gross generation by assuming a 33% conversion efficiency, *i.e.* 1 TWh = (0.086–0.33) Mtoe. For geothermal and solar thermal, if no country-specific information is reported, the primary energy equivalent is calculated as follows:

- 10% for geothermal electricity;
- 50% for geothermal heat;
- 33% for solar thermal electricity;
- 100% for solar thermal heat.

It is very important to understand the used conventions because they may have concrete impacts: as an example, the RES European targets are defined in terms of energy to the final consumption, while the efficiency European targets are based on primary energy consumption.

From figs. 2 and 3, taken from the IEA publication "Key World Energy Statistics" 2016 [2], it can be seen that the world primary energy supply has more than doubled

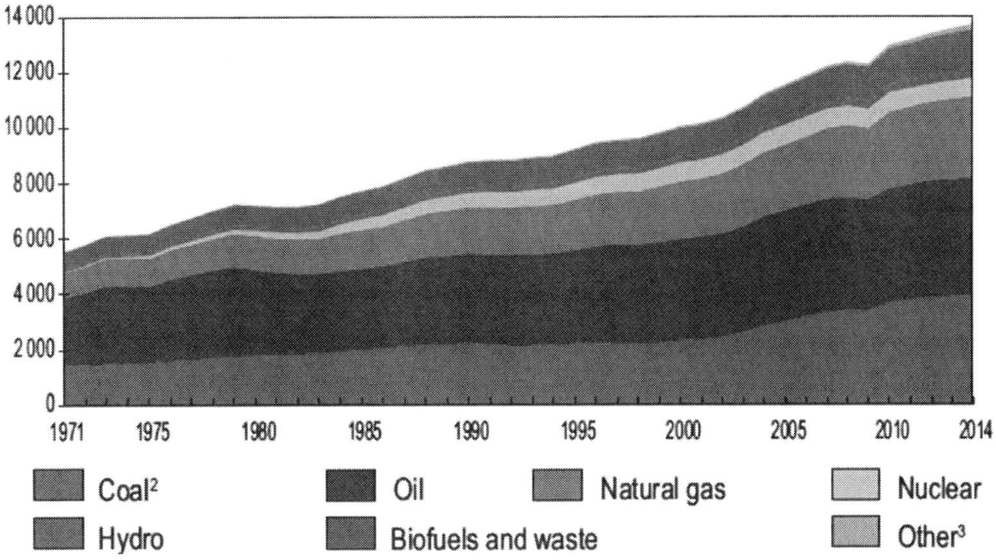

Fig. 2. – World[1] total primary energy supply (TPES) from 1971 to 2014 by fuel (Mtoe). [1]World includes international aviation and international marine bunkers. [2]In these graphs, peat and oil shale are aggregated with coal. [3]Includes geothermal, solar, wind, heat, etc. (Courtesy of IEA.)

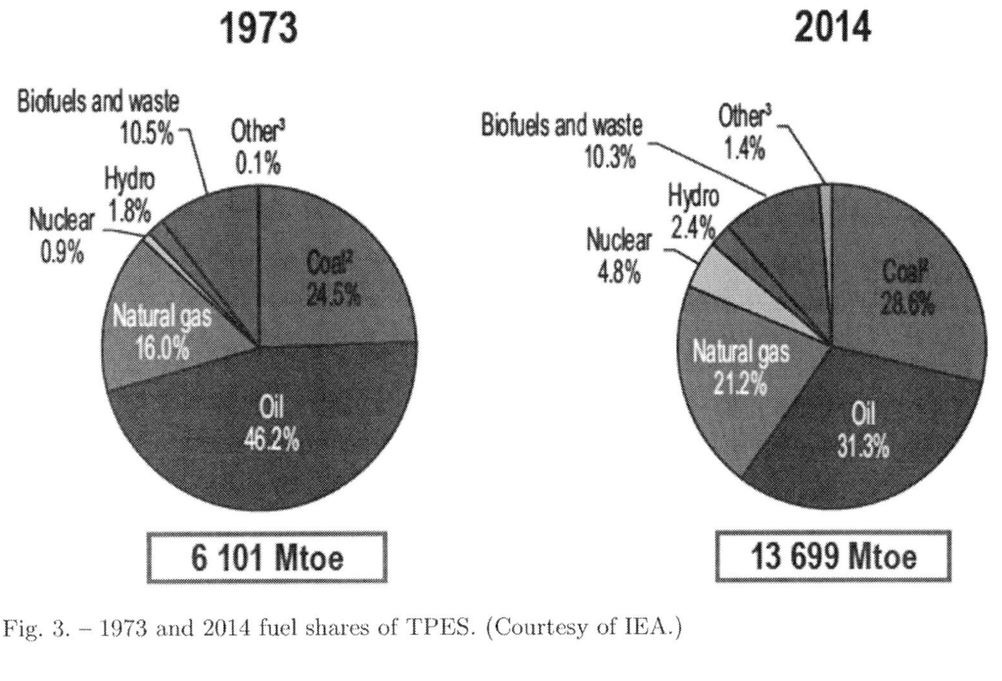

Fig. 3. – 1973 and 2014 fuel shares of TPES. (Courtesy of IEA.)

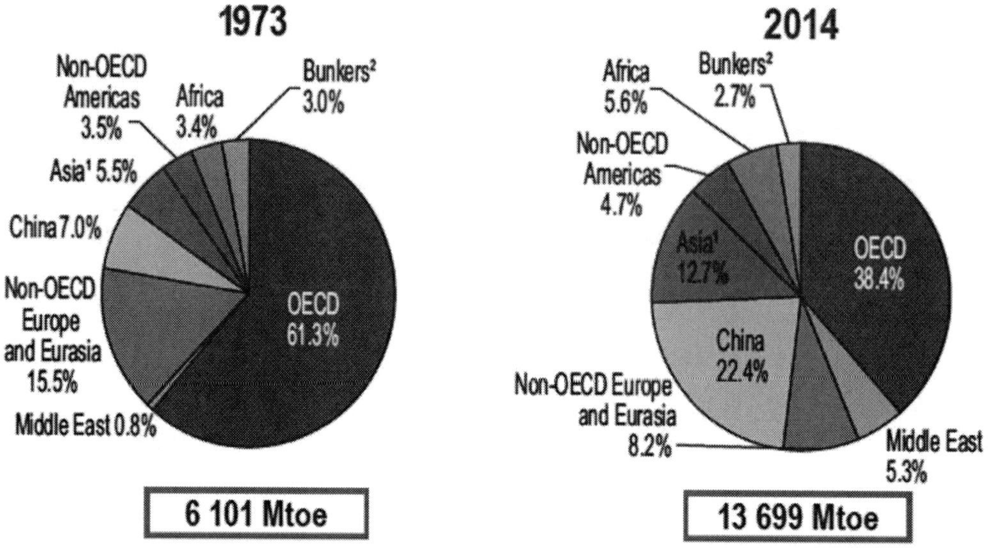

Fig. 4. – 1973 and 2014 regional shares of TPES. [1]Asia excludes China and OECD countries of Asia. [2]Includes international aviation and international marine bunkers. (Courtesy of IEA.)

during the last 43 years: more than 80% of the global supply is coming from fossil fuels. Oil is still the most important primary energy source, even if there has been an important share decrease, while gas and coal increased their share. The dip generated by the recent economic crisis with the growth going back to its "pre-crisis" trend is also shown.

Figure 4, taken from the same IEA publication [2], shows the clear geographic shift of the energy use from the OECD countries to Asia and China, separately considered, reflecting the economic growth that the world has been experiencing in recent years.

Figures 5 and 6, taken from the IEA publication "Key World Energy Statistics" 2016 [2], are related to the global electricity production: again, the dip generated by the recent economic crisis is evident with its growth going back to its "pre-crisis" trend. It is important to note that while the primary-energy demand doubled, the electricity production almost quadrupled in the same period: about two thirds of the electricity production now comes from fossil fuels, with coal, that even increased its share from 1973, being by far the most important source. Oil almost disappeared from electricity production, being now mostly dedicated to chemical industry and transportation, while natural gas almost doubled.

Hydro production decreased its share, but almost tripled its absolute production compared to 1973. Until the Fukushima accident, nuclear and hydro production were about equally sharing the remaining world electricity production; nuclear production is now about two thirds of the hydro production, with the remaining global share being mostly covered by the "new renewables" (mostly wind and solar).

The total final energy consumption in the OECD countries is almost equally divided into the three main sectors: industry, transport and "other sectors" (residential,

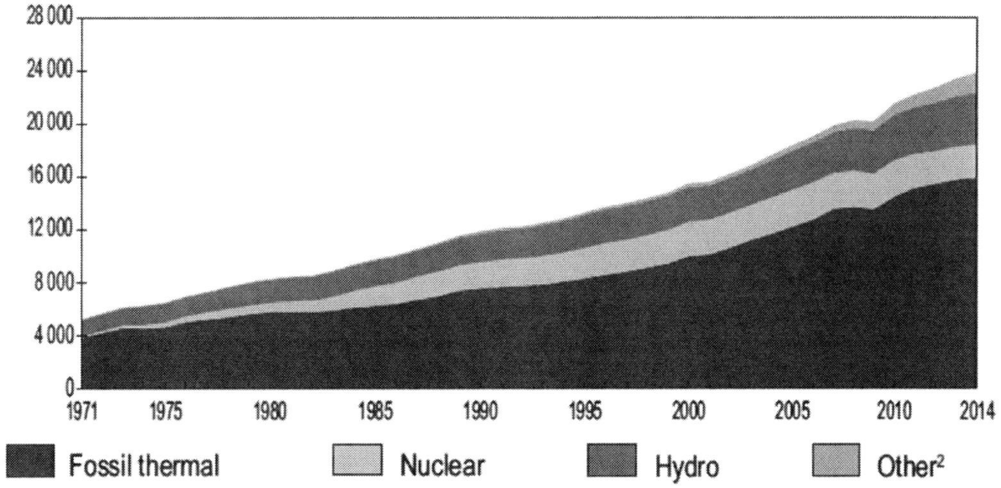

Fig. 5. – World electricity generation[1] from 1971 to 2014 by fuel (TWh). [1]Excludes electricity generation from pumped storage. [2]Includes geothermal, solar, wind, heat. (Courtesy of IEA.)

commercial and public services, agriculture/forestry, fishing and non-specified). A minimal part of the energy final consumption goes to the so-called non-energy use (fuels that are used as raw materials in the different sectors and are not consumed as a fuel or transformed into another fuel. Non-energy use also includes petrochemical feedstocks).

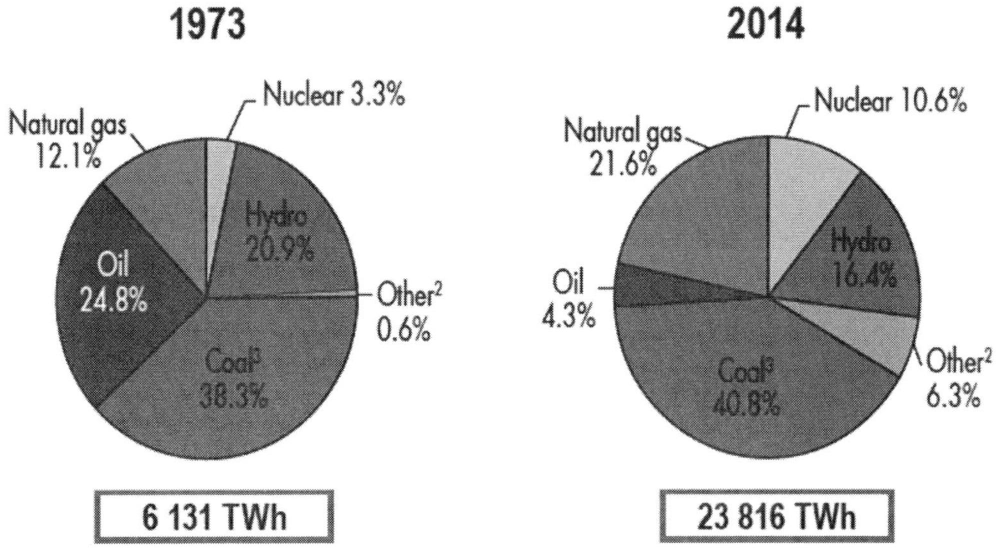

Fig. 6. – 1973 and 2014 fuel shares of electricity generation. (Courtesy of IEA.)

Though well known, it is important to underline that transportation depends worldwide on oil, with shares well over 90% everywhere.

Let us now introduce two new energy indicators: the energy intensity of a country and the electric energy intensity.

The energy intensity, Ei, is defined as the primary energy demand of a country over its Gross Domestic Product (GDP); given the economic inflation, a reference year is needed.

Ei is an indication of the energy efficiency of a nation's economy, even if it is a complex indicator and depends on several parameters:

- The higher the technological developments, the more efficient is the energy use.
- Heavy industry requires more energy.
- Colder countries require more energy for heating.
- Larger countries require more energy for transportation.

Despite the above mentioned limits, it is observed that Ei grows in emerging economies and decreases with de-industrialization and service sector (tertiary) growth.

The electric energy intensity is defined as the electricity demand of a country over its GDP. It is very similar to Ei but shows an opposite behaviour: the optimal use of electricity induces an increase in productivity and therefore energy savings in total energy.

We have been talking about global energy consumption, but we should not forget that the energy consumption is very far from being equally distributed on our planet: there are places on Earth where the equivalent energy consumption is well over 6 Mtoe per capita, while others where it is not much different from the energy consumption of Early Man.

Africa, having about 14% of the world population, has only 3% of the world electricity consumption. South Africa, having about 5% of the African population, needs about half of the total African electricity consumption. Excluding northern and southern African countries, the main energy source for the remaining African population is wood.

IEA evaluates [4] that about 1.3 billion people on Earth live without electricity and 2.7 billion live without clean cooking facilities (fuels and stoves that do not cause air pollution in houses). As we have seen, access to electricity and to energy in general, is considered to be an essential requirement for human development. IEA evaluates that investments of about 35 $ billion/year would bring energy access to the whole world population. For comparison, Germany and Italy together spend every year more than this amount of money to subsidize the RES deployment.

As we have seen, more than 80% of the world primary energy demand is covered by fossil fuels: does this energy system have any impact on the environment and on men's health? May it go on forever?

I think there are three main limits for the energy system:

- The combustion of fossil fuels generates greenhouse gases (GHG) that have consequences on the climate of our planet.

– The combustion is also generating polluting emissions like nitrogen and sulphur oxides, particulate matter, carbon monoxide and others that may have an impact on men's health and environment.

– Fossil fuels are by definition finite resources that, sooner or later, will come to an end: what will come next?

The order of this list is just for the sake of this presentation and does not necessarily reflect any priority order. Each single point will now be better specified.

The oxidation of fossil fuels C-H bonds generates water and carbon dioxide: they are both necessary elements for life on our planet but the latter is also a GHG (also water vapour is a powerful GHG).

As an example let us consider the methane combustion:

$$CH_4 + 2O_2 \rightarrow CO_2 + 2H_2O$$

Simple considerations on molecular weights of the different compounds show that out of 1 kg of methane, almost 3 kg of carbon dioxide are generated.

The average temperature of our planet is set by the equilibrium between the incoming radiative energy from the Sun and the outgoing infrared radiation: fig. 7 shows the spectra of the solar incoming energy into our planet and the infrared outgoing energy. They are very well decoupled: the atmosphere is almost transparent to the incoming energy while it is much more absorbent for the outgoing energy. As a result, the mean global temperature of the Earth is about 33 °C higher than it would be without the atmosphere (*i.e.* without the natural greenhouse gases present in the atmosphere) and this makes life possible on our planet the way we know it. More than two thirds of the total greenhouse effect on our planet (*i.e.* two thirds of the 33 °C temperature increase) are coming from the water vapour that is [5], as shown in fig. 7, the most important GHG in or planet, while carbon dioxide is contributing with about 20% of the total temperature increase.

The greenhouse radiation forcing of each single gas is well known and the green house theory tells us that a doubling of the carbon dioxide concentration in our atmosphere would give rise to a global temperature increase of about 1 °C "if the rest is kept firm". Keeping all the rest firm means that we assume there are no feedbacks on our planet.

Climate change feedbacks are important in understanding the climate evolution because feedback processes may amplify or diminish the effect of each climate forcing, as the doubling of carbon dioxide concentration in our atmosphere, and so play an important role in determining the future climate state. Feedbacks may be positive or negative: a positive feedback is, for instance, the increase of the amount of water vapour in the atmosphere with temperature, which in turn leads to further warming. A negative feedback comes from the Stefan-Boltzmann law: the amount of heat radiated from the Earth into space changes with the fourth power of the temperature of the atmosphere.

Clouds are a very important feedback and warming is expected to change their distribution and type. Clouds reflect infrared radiation back to the planet surface, having

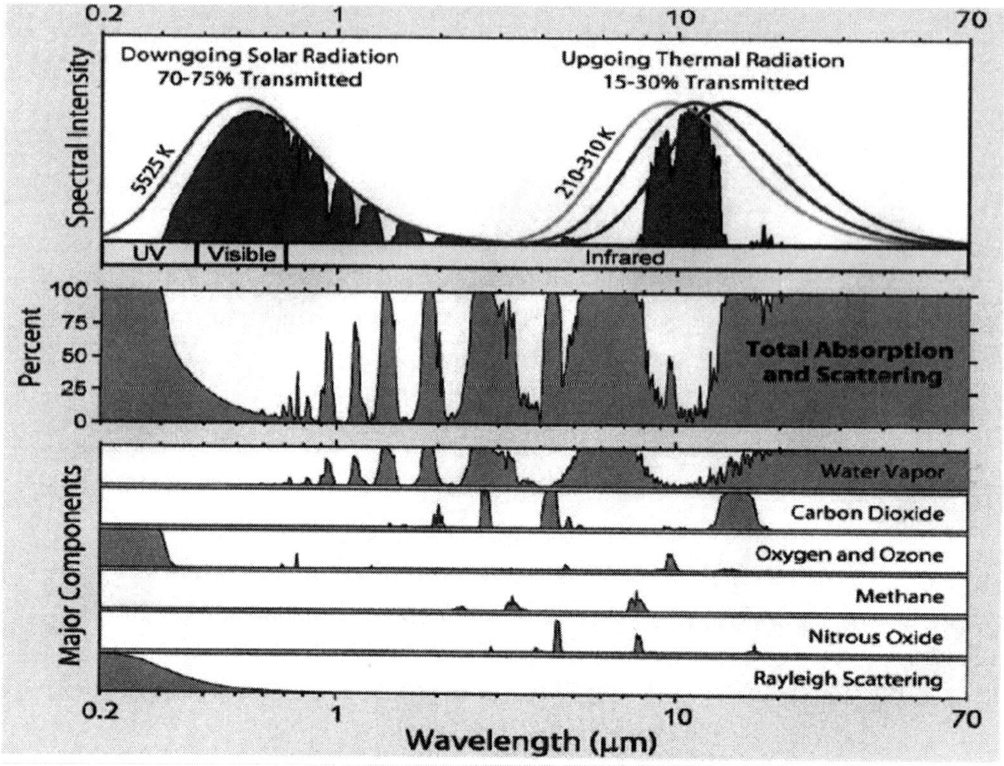

Fig. 7. – Radiation transmitted by the atmosphere.

a warming effect; at the same time clouds reflect sunlight and emit infrared radiation to space, having a cooling effect. Whether the net effect is warming or cooling depends on details such as the type and altitude of the cloud. High clouds tend to trap more heat and therefore have a positive feedback, low clouds normally reflect more sunlight having a negative feedback. According to the last IPCC report [6], cloud feedbacks continue to be the largest uncertainty in the Global Circulation Models (GCMs) that are built to simulate the complex behaviour of our atmosphere considering all possible feedbacks. According to GCMs, the response to our atmosphere to a doubling of carbon dioxide concentration would likely be a temperature increase between 1.5 °C and 4.5 °C.

The discussion about climate change is not on the well-known GHG theory but on how the complex climate system may respond to an increase of the carbon dioxide forcing considering all known feedbacks.

Let us now come to the polluting emissions generated by the fossil fuels combustion that may have a negative impact on human health and on the environment. We are talking about nitrous or sulphur oxides, generally indicated by NO_x or SO_x, Particulate Matter, indicated by PM_x, where x represents the maximum size in microns of the particulate we are referring to, ozone, carbon monoxide, CO, generated by an incomplete combustion, and other compounds.

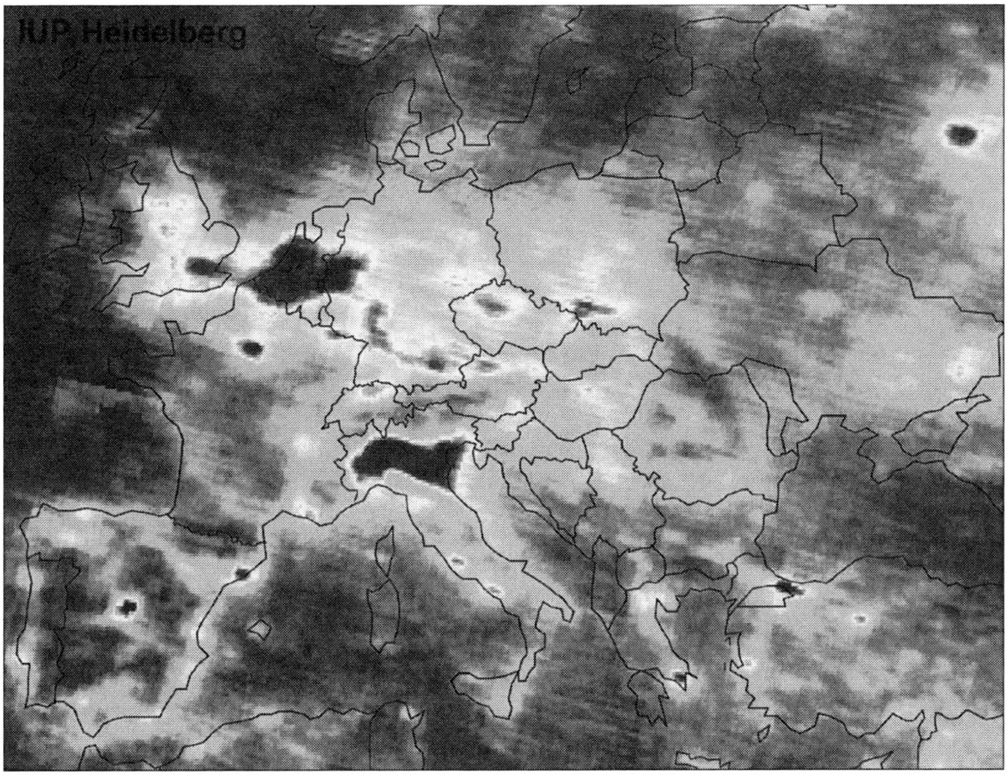

Fig. 8. – European mean tropospheric nitrogen dioxide (NO_2) vertical column density (VCD) between January 2003 and June 2004, as measured by the SCIAMACHY instrument on ESA's Envisat. Image produced by S. Beirle, U. Platt and T. Wagner of the University of Heidelberg's Institute for Environmental Physics. (Courtesy of T. Wagner.)

Figure 8 shows how the NO_2 concentration is linked to human emissions: high levels are evident in large areas like the Netherlands and the Po valley, where concentration is high also because of little exchange with the atmosphere, and over large cities like London, Paris, Madrid and, in Italy, Roma and Naples (Milan is hidden in the middle of the Po valley).

Another very evident effect of atmospheric pollution is shown in fig. 9.

Epidemiologic studies on health effects caused by these polluting emissions are not easy but the harms originated by these compounds are becoming more and more evident as shown in fig. 10 and discussed in recent studies by the World Health Organization (WHO): globally, almost 7 million deaths were attributable to the joint effects of household (HAP) and ambient air pollution (AAP) in 2012 [7].

In the next days there will be specific presentations on this subject that will show the methodology applied to populations to point out and clarify what health effects are produced by such emissions.

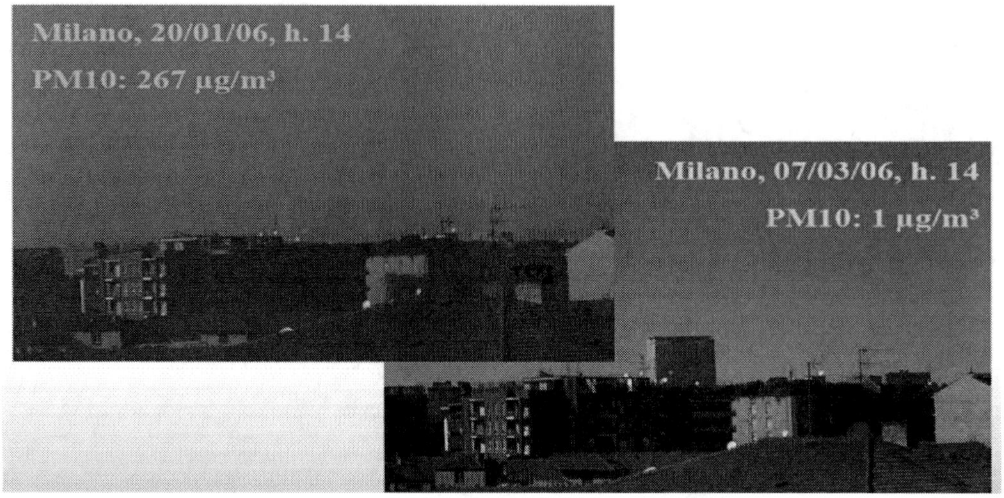

Fig. 9. – Visibility in Milan from the same observation point in days with different PM10 atmospheric concentrations. (Courtesy of Dr. Lazzarini, ARPA Lombardia.)

Last but not least: the fossil fuel sources that, as we have seen, cover nowadays more than 80% of the global primary energy demand, are limited resources and sooner or later they will run out.

Hubbert, an american geologist and geophysicist who worked at the Shell research lab in Houston, is known for his studies on the size of oil fields. He predicted that, for

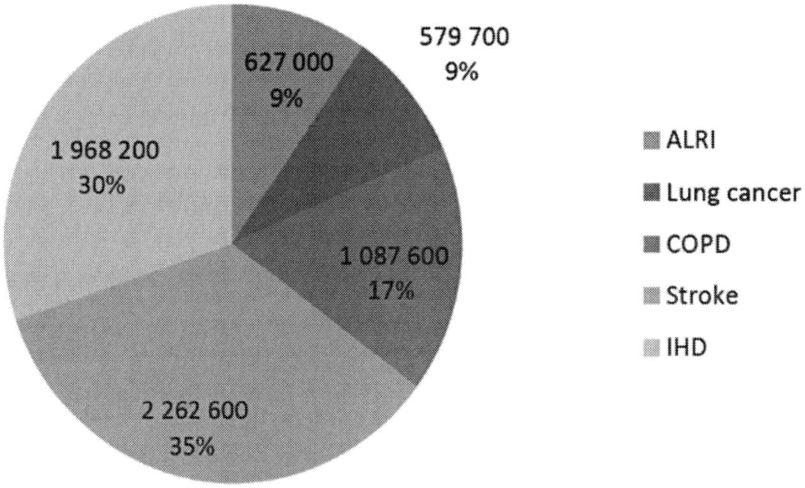

Fig. 10. – Deaths attributable to the joint effects of HAP and AAP in 2012, by disease. HAP: household air pollution; AAP: ambient air pollution; ALRI: acute lower respiratory disease; COPD: chronic obstructive pulmonary disease; IHD: ischaemic heart disease [7]. (Courtesy of WHO.)

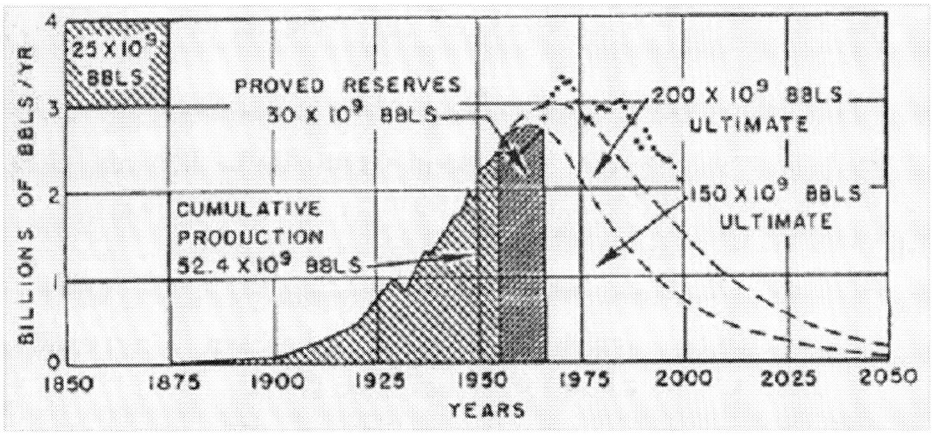

Fig. 11. – Picture taken from "Nuclear energy and the fossil fuels", M. King Hubbert, June 1956, showing the forecast of the United States overall petroleum production.

any given geographical area, from an individual oil field to the planet as a whole, the rate of petroleum production of the reserve over time would resemble a bell curve. Based on his theory, he presented a paper [8] at the 1956 meeting of the American Petroleum Institute in San Antonio, Texas, which predicted that the overall petroleum production would peak in the United States, excluding Alaska, between 1965 and 1970, as shown in fig. 11. At first his prediction received much criticism; Hubbert became famous when his prediction proved correct in 1970.

In 1974, Hubbert projected that global oil production would peak in 1995 "if current trends continue". Various subsequent predictions have been made by others as trends have fluctuated in the intervening years. Fortunately technologies evolve: new oil reserves can be found and known oil reserves may become economically exploitable. As an example, the US oil production, after the peak in the seventies and a subsequent decrease, came back with a steep rise.

The red curve of fig. 12 [9] is showing that Hubbert was correct in his projection relative to the US production excluding Alaska. When Alaska production is considered together with offshore drilling and the recent tight or the so called non-conventional production, the overall US oil production is nowadays at the same level of the peak reached in the seventies.

The message is that we are using limited resources but they will not finish tomorrow. This is a delicate subject where not only technical considerations but also economic and geopolitical considerations come in: there will be a dedicated presentation during the coming week on this issue.

As Hubbert was saying: "Energy is essential for all organic and inorganic activities: history of human evolution reflects history of mankind ability to control and transform Energy". The discovery of fire, that allowed Man to convert the chemical energy of wood into light and heat, was maybe the first and the most important discovery. Up

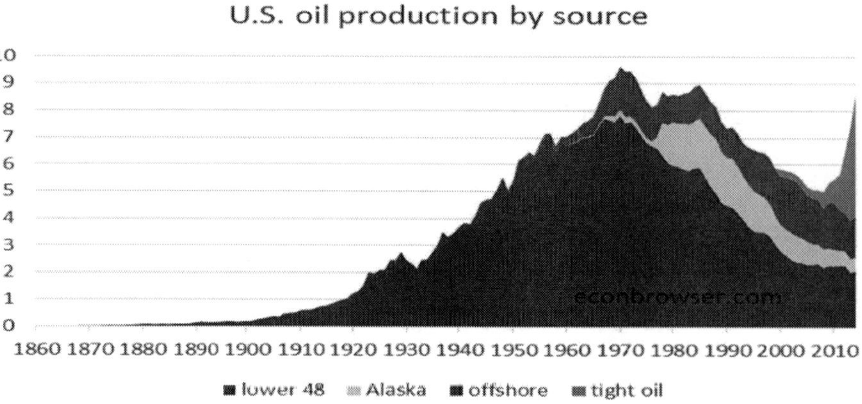

Fig. 12. – US oil production up to a few years ago. (Courtesy of Prof. Hamilton.)

to the steam engine, that initiated the industrial revolution, mankind fulfilled its energy needs thanks to agriculture and the exploitation of the original renewable sources. The industrial revolution allowed fossil fuels to strongly enter and shape our development: diffused and cheap energy allowed to replace human labor and activities impossible until that time became common.

For comparison, let us consider that during the Roman empire, a slave eating about 2000–2500 kcal/day, could produce a work output around 300–400 Wh/day, equal to about 100 grams of oil: every european citizen, who now consumes about 10 kg of oil a day, is exploiting the job of 100 to 150 slaves.

As a consequence, the energy demand increased to a level never reached before, as simply synthesized in fig. 13. We are nowadays on the fossil fuel peak: it is not clear

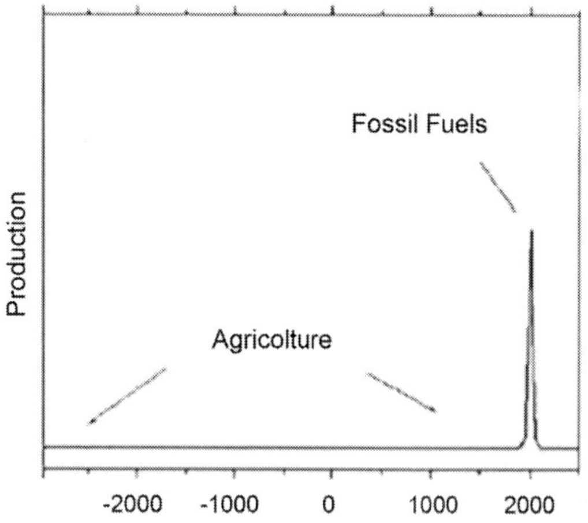

Fig. 13. – Historical mankind energy sources.

whether we are just before, exactly on, or just after the peak, but we have seen that there are important reasons to start thinking about what will come next.

This is what this school is about.

REFERENCES

[1] IEA, *World Energy Outlook* (International Energy Agency) 2004.
[2] IEA, *Key world energy statistics* (International Energy Agency) 2016.
[3] BP, *BP Statistical Review of World Energy*, June 2016.
[4] IEA, *World Energy Outlook* (International Energy Agency) 2011.
[5] Lacis A. *et al.*, *Atmospheric CO_2: Principal Control Knob Governing Earth's Temperature*, *Science*, **330** (2010) 356.
[6] Stocker T. F. *et al.* (Working Group I of the IPCC), *Climate Change 2013 - The Physical Science Basis* (Intergovernmental Panel on Climate Change IPCC) 2013.
[7] WHO, *Burden of disease from the joint effects of Household and Ambient Air Pollution for 2012* (World Health Organization) 2016.
[8] Hubbert M. K., *Nuclear Energy and the fossil fuels* publ. No. **95** (Shell Development Company, Houston) 1956.
[9] Hamilton J. D., *The Changing Face of World Oil Markets* (Department of Economics, University of California, San Diego) 2014.

EPJ Web of Conferences **189**, 00004 (2018) https://doi.org/10.1051/epjconf/201818900004
LNES 2017

Fossil energy: From conventional oil and gas to the shale revolution

Mario Marchionna

Saipem S.p.A. - Via Martiri di Cefalonia 67, 20097 San Donato Milanese (MI), Italy

Summary. — The fossil fuels have provided more than 80% of the total energy consumption for more than 100 years; although in perspective renewables are expected to be the fastest growing energy sources, it is likely that fossil fuels will dominate energy use at least through 2050. This is still due to the increased worldwide need for energy, to their superior energy intensity and reliability and to the very huge numbers that underline world exposition to fossil fuels, not easily substitutable. A few brief considerations, regarding the chain of production, transport and use of the energy carriers, make us realize that the success that oil has had as an energy source cannot be attributed only to its great availability and the relatively cheap price but also to the ease with which liquid hydrocarbon derivatives can be transported, stored and distributed for their final use. In fact, their liquid state enables their energy density per unit of volume to be optimized, and this has great advantages especially for the storage and the distribution of the carrier. This does not mean that there are not challenges and dilemmas in the exploitation of conventional reserves, such as for example the decline of great reservoirs for lacking of suitable technology and reservoir management, the increased produced water from oil reservoirs and related management and the need to cope with the issue of climate change due to the CO_2 emissions. Attention will then be paid towards unconventional resources, especially those that have lead to the "shale" revolution: the deployment of huge reserves of shale oil and gas have deeply modified the overall energy picture in the last ten years, especially in the United States. By moving towards 2040–2050, oil and gas will remain crucial energy components, maybe with some less crude-oil production but more natural gas and higher use of electricity, produced both by gas and by renewables and also pushed by the advent of electric vehicles.

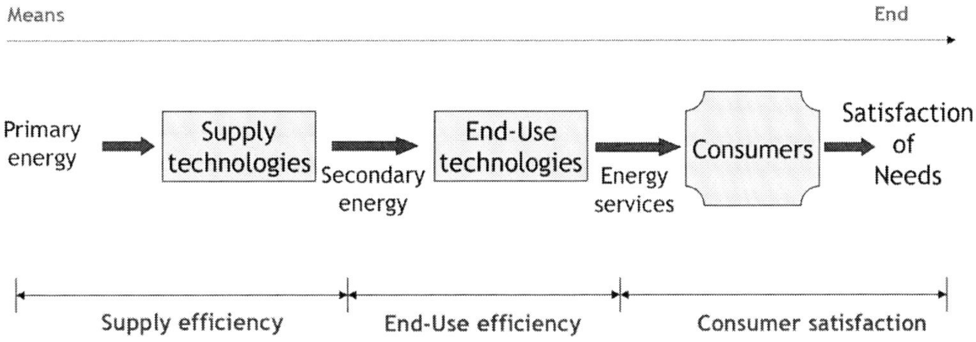

Fig. 1. – Conversion of primary energy to energy services.

1. – Introduction - from sources to market: energy carriers

Modern economic systems depend to an ever-increasing extent on the availability of energy, both for their operation and, in particular, for their development. The various forms of energy, however, are not the same, as they differ in their availability, profitability, usability and efficiency with respect to their end-use.

None of the different forms of energy available (fossil fuels such as oil, natural gas, coal, fissile materials, and renewable sources such as solar or wind power, hydropower, etc.) is capable by itself of satisfying the energy needs of the society that are linked to the main markets: the market supplying heat (or cold), mobility and transport, and ancillary services (mechanical, electronic, etc.). It is therefore necessary to introduce forms of energy that can guarantee a better link between the availability of energy sources and the particular use required: it is just the sector where energy carriers play their special role [1].

The energy carrier is a secondary form of energy which lends itself to being transported (often by means of special networks) to the place of use (fig. 1). It consists of either a transportable substance that can easily release the energy contained in it (as in the case of solid, liquid or gaseous fuels, steam, hot water, etc.), or of electricity (electric energy). The latter at present best enables its energy content, produced from the most varied primary sources, to be exploited for the needs of the end-user.

In only a few cases, it is not strictly necessary to introduce an energy carrier as the primary source is sufficiently versatile: for example, natural gas is capable of heating houses, fuelling motor vehicles, etc., but it is not always the most suitable form of energy for every use (it is not used as aircraft fuel or not yet in maritime transport, whereas it is used for road traction, but only at a very limited amount). However, there is always a natural gas treatment stage to remove not just carbon dioxide and nitrogen, but also hydrogen sulphide and the higher hydrocarbons, which are known as "natural gas liquids". It would thus be more correct to consider natural gas as an energy source, whereas methane, used commercially, should be regarded as an energy carrier.

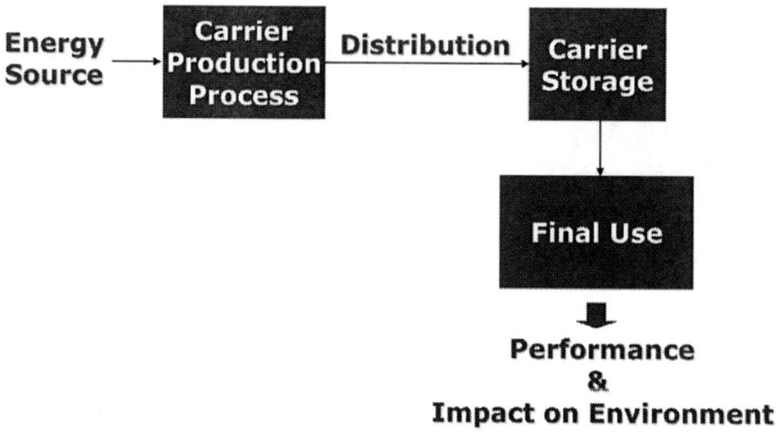

Fig. 2. – From sources to energy carriers till to the final use.

1·1. *Life cycle of the energy carrier.* – The various phases involved from generation to the end-use of an energy carrier constitute its "life cycle". The basic operations involved in the life cycle of an energy carrier are (fig. 2): a) generation using the primary fuel; b) transport (when required); c) storage (when required); d) distribution; e) end-use (with consequent impact on the environment, both in local terms in the form of emissions, and in global terms of efficiency in conserving the energy contained in the primary fuel throughout its transformation cycle).

The operations are not necessarily carried out in the aforementioned order; for example, storage could even precede transport or could be carried out at the same time as the final use (as in the case of motor traction). Linking energy sources with the market therefore entails technological and strategic choices as far as the processes for obtaining the carriers, the logistics of the sources and the end-uses are concerned.

On these grounds, there are a few considerations to be made:

- some carriers are dedicated to well-defined uses; for example, the use of gasoline and of diesel gas-oil is dominant in motor traction;
- among fossil fuels, crude oil has practically no direct use, but is converted in the refinery into the various liquid carriers that feed a relevant part of the end-uses;
- carriers are distinguished also by their different physical form: gaseous for methane and hydrogen; liquefied gas for LPG (Liquefied Petroleum Gas); liquid for all petroleum derivatives, and biofuels such as ethanol and biodiesel; solid for some other fuels; less easily definable in the case of electricity, for which the real carrier is an electric current, *i.e.* a flow of electrons.

These brief considerations, regarding the chain of production, transport and use of the energy carrier, make us realize that the success that oil has had as an energy source cannot be attributed only to its great availability and the relatively cheap price but also, and perhaps above all, to the ease with which liquid hydrocarbon derivatives can be transported, stored and distributed for their final use. In fact, their liquid state enables their energy density per unit of volume to be optimized, and this has great advantages for the storage and the distribution of the carrier. Actually, this aspect is also reflected in the higher price attributed to the "liquid" calories compared with "gaseous" and "solid" calories.

On the other side, the demand for electricity is going to and will go on increasing in the years to come, both in the industrialized countries and in the developing countries, at average rates of about 3% per annum; a high use of electricity generally allows people to live better and longer, as it happens in the most advanced countries. As already observed, electricity is the typical energy carrier and it is useful to examine its cycle that points out strengths and weaknesses.

Electricity can be obtained, both on a large scale and on a small scale, from fossil fuels (especially coal and natural gas), from fissile fuels (nuclear energy) and from renewable sources (hydroelectric, wind power, photovoltaic solar, ...); it therefore has the great advantage of being able to be produced from a wide variety of sources. Furthermore it can be put to many different uses and at very varied scales. Its environmental impact is locally very positive because, as opposed to other carriers, there is no polluting emission at the place where the electricity is used.

The development of the electricity market depends also on the development of advanced technologies of electric transmission, already now indispensable for meeting the growth in demand of the deregulated markets. Unlike transport, which is consolidated and efficient, electricity storage is a weaker point of this energy carrier. By and large, electricity, once it has reached the end-user, has to be consumed immediately. This does not mean that no storage systems exist: there are several technologies covering a broad range of applications, from those able to deal within fractions of a second with variations/interruptions of electricity (power quality) to those that enable electricity supplies to be managed according to market demand (energy management). The accumulation/storage systems are potentially useful as they can be combined with renewable sources (such as photovoltaic and wind energy) so as to mitigate their characteristics of intermittence and to ensure the recovery of the electricity produced during periods of slack demand. In recent times, continuous improvements have been achieved for storage systems based on batteries and cost is constantly decreasing along the years. However, the only energy of renewable type that has, for the moment, fully resolved this problem commercially is hydropower, as the system of pumping water back into hydroelectric basins is widespread and is used also for generating very high power.

In conclusion, electric energy, the carrier *par* excellence, is used everywhere and has zero emissions in its place of use. Storage is still its weaker point but it should be noted that every small improvement in this field could lead to an increase in its use well beyond all imagination. What is sure is that, especially nowadays, electricity is the most desired energy carrier from the public and it is likely that in the long term it will progressively substitute conventional carriers produced by fossil sources, also in the field of transportation.

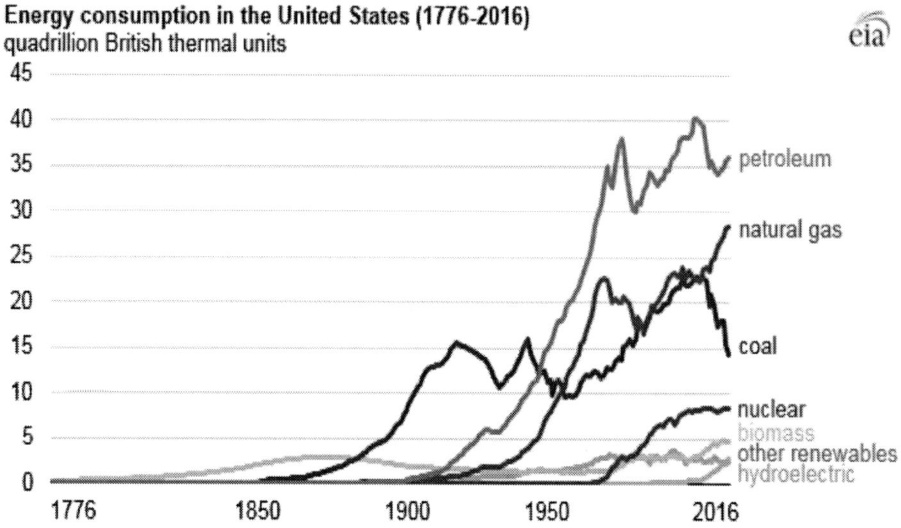

Fig. 3. – Energy consumption in the United States (1776-2016) (source EIA, 2017).

2. – Fossil fuels: conventional resources

On the other side, the present energy figures tell us another story: referring to the US (taken here as a single but very representative example) the fossil fuels have provided more than 80% of the total energy consumption for more than 100 years. According to a study of the US Energy Information Administration (EIA) [2], since 1928, when consumption of natural gas surpassed that of biomass, the three fossil fuels —petroleum, natural gas, and coal— have been the most consumed fuels in the United States. In 2016, fossil fuels accounted for 81% of the total US energy consumption (fig. 3), the lowest fossil fuel share in the past century.

In 2016, the renewable share of energy consumption in the United States was 10.5%. This was the largest renewable share since the 1930s, when the overall energy consumption was lower and the amount of biomass consumption (mainly wood) was relatively high. The greatest growth in renewables over the past decade has been in solar and wind electricity generation. Liquid biofuel consumption —more than half of which is ethanol blended into motor gasoline— has also increased in recent years, contributing, especially in the US, to the growing renewable share of total energy consumption.

In addition to the increasing share of renewables, the decline in the fossil fuel share of consumption is attributable mainly to declines in coal consumption. US coal consumption fell nearly by 9% in 2016, followed by a 14% drop in 2015. Overall, US coal consumption has declined by almost 38% since 2005. In each of the past 20 years, the power sector has accounted for more than 90% of the total US coal consumption.

Petroleum, which encompasses nearly all transportation fuels and several petroleum-based fuels used in homes, businesses, and industries, continues to be the largest source of energy consumption in the United States. Petroleum consumption has increased in each of the past four years.

Consumption of natural gas has risen in nine of the past 10 years. As recently as 2006, the United States consumed more coal than natural gas (in energy-equivalent terms), but as natural gas consumption has increased —particularly in the electric power sector— natural gas use in 2016 was about twice that of coal, contributing also to a significant reduction of CO_2 emissions [3].

Similar trends (with some difference) may be also extended to a worldwide figure. In Europe the renewables share is somewhat higher but there are not any radical differences, and the US situation can be extrapolated without significant differences to the world picture (maybe with the exception of a less pronounced switch of coal to natural gas, that is typical of the US situation).

Although in perspective renewables are expected to be the fastest growing energy sources [4], it is likely that fossil fuels will dominate energy use at least through 2050. This is still due to the increased worldwide need for energy consumption, to their superior energy intensity and reliability, to the very huge numbers that underline world exposition to fossil fuels ($> 80\%$ energy consumption) and also to the massive investments that have already been made. The mix, however, will change, according to a McKinsey study [5]:

- Natural gas, which has a lower carbon intensity than coal and petroleum, will continue to grow quickly, but the global demand for coal will likely peak around 2025. The relatively high rate of natural gas consumption growth is attributed to abundant resources and rising production — including supplies of tight gas, shale gas and coalbed methane.

- Growth in the use of oil, which is predominantly used for transport and petrochemistry (differently from all other sources that mostly produce electricity), will slow down as vehicles get more efficient [6] and more electric: here peak demand could come as soon as 2030. However, its potential decline will be very slow as the electric vehicles penetration will have to face the need of substituting a huge fleet of light duty vehicles equipped with conventional engines; presumably it will not have the same impact on heavy duty vehicles [7].

- By 2050 the McKinsey research estimates that coal will be down to just 16% of the global power generation (from 41% now) and fossil fuels to 38% (from 66% now). Overall, coal, oil and gas will continue to be 74% of the primary energy demand, down from 82% at present. After that, the rate of decline of the use of fossil fuels is likely to accelerate.

As regards oil and natural gas, impending reserves' depletion is more a myth than a reality, as proved oil reserves (only a part of resources in place) have continuously increased thanks to exploration activities, improved technology and market conditions. Furthermore the discovery of unconventional reserves (oil sands, shale oil & gas, ...) has even more improved the overall figure.

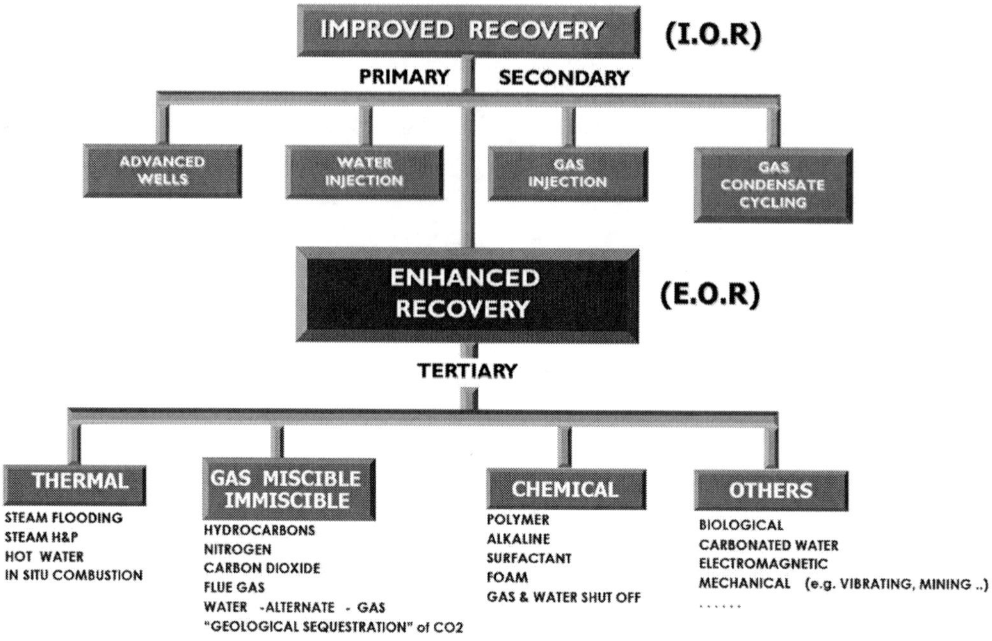

Fig. 4. – Increase of reservoir recovery factor in oil & gas fields.

However, this does not mean that there are not *challenges* and *dilemmas* in the continuous exploitation of conventional reserves, a few of them are reported in the following:

2˙1. *Decline of great reservoirs for lacking of suitable technology and reservoir management.* – This is a typical problem of old big reservoirs, which continue producing more water than oil: the first simplest remedy is just to pump even more water into the field to increase oil production, another opportunity may be the injection of natural gas to achieve the same effect but eventually many other techniques have been developed to increase the recovery factor of the reservoir, as shown in fig. 4.

2˙2. *Increasing of water production from oil reservoirs.* – Oil and natural gas companies are honing strategies for reducing production costs through effective water-management practices and wastewater disposal options [8]. Actually, Exploration and Production O&G companies are among the major water producers and the major fresh-water users, although they could consume less fresh water and do more to access alternate sources through recycling and reusing more produced wastewater.

The oil-price slump that started in late 2014 has forced producers to cut spending and reduce operating expenses, encouraging more innovative, efficient water management (fig. 5). It is expected that industry will intensify its focus on water management. In the next section on unconventional resources the same issue will be discussed again.

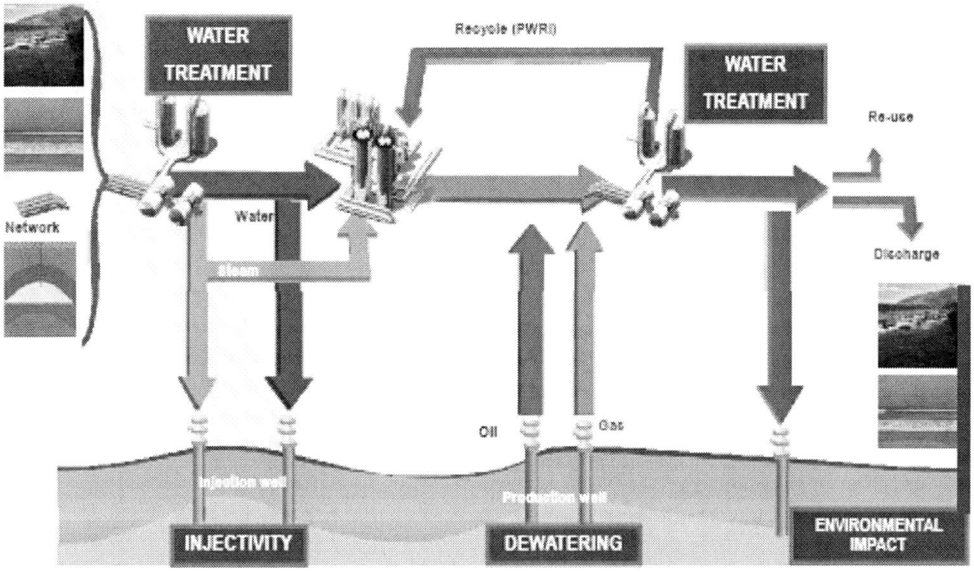

Fig. 5. – Water Management in oil & gas fields.

2˙3. *Need to cope with the issue of climate change.* – Fossil fuels (oil, gas and coal) are the most widely used energy sources today, accounting for more than 80% of worldwide energy consumption. Their undisputed leadership is however related to the problem of carbon dioxide (CO_2) emissions into the atmosphere, regarded as one of the main causes for climate change. Sixty per cent of the CO_2 produced by mankind originates from combustion of fossil fuels (both conventional and unconventional) (fig. 6), which consequently is not the only activity that produces CO_2.

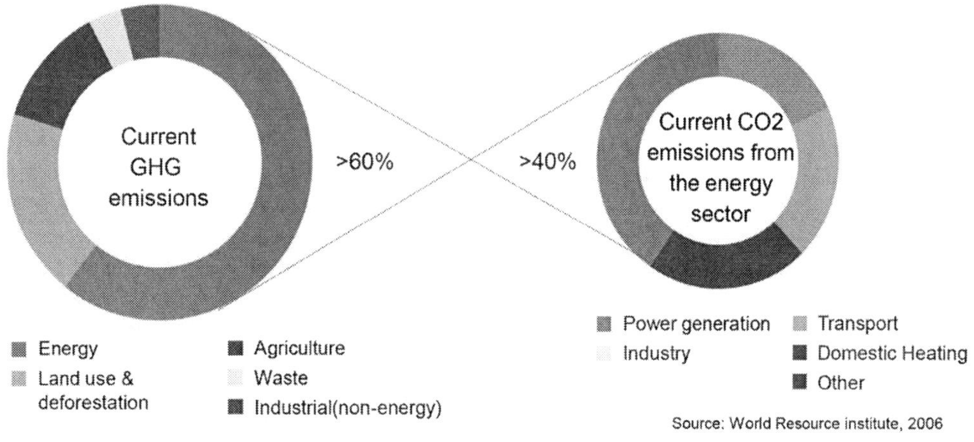

Fig. 6. – Total global man-made CO_2 emissions.

Thus the need to satisfy the growing world demand for energy, among the emerging economies in particular, goes hand in hand with the need to tackle the risks to the climate stemming from increased production of CO_2.

The first solution, and the most immediately practicable approach in the short term, is energy efficiency: *i.e.*, measures, attitudes and technologies that help reduce consumption. Another solution is to use alternative energy sources, which play a relatively modest role so far. Biomass and related materials (woody substances, various types of waste) cover about 10% of our total energy requirement, a percentage that will remain about steady. This situation seems similar in nuclear energy, which, together with hydroelectric power, covers about 8% of the total requirement. Use of other renewable energy sources (wind and solar energy) is increasing a lot, but currently satisfies only a few percent of the Earth's energy requirement.

In other words, fossil fuels will continue to dominate the world energy sector for the next two-three decades. Thus, in order to achieve a significant reduction in CO_2 emissions just over the long term direct action must be taken also on the use of these sources.

A switch between coal/oil and natural gas is beneficial to reduce CO_2 emissions, at least as regards power generation as demonstrated by the recent US experience.

In the long term, a further alternative can be represented by CO_2 Capture and Storage (CCS) technology, used to sequester CO_2 generated by fossil fuels, thereby reducing emissions into the atmosphere. The geological storage of carbon dioxide may enable real progress in the global effort to make meaningful reductions in CO_2 emissions especially from large-point source emitters such as power plants, refineries, cement plants and steel mills. Of course, CCS is not a panacea, there are several issues that need to be addressed before it can achieve a widespread application, but it does offer a tangible mean to deal with large volumes of gas emissions by using technologies already at hand, or after further improvement [9].

3. – Unconventional resources: the "Shale" revolution

As mentioned above, the deployment of huge reserves of shale oil and gas have deeply modified the overall energy picture in the last ten years. By definition, shale gas is a natural gas that is found trapped within shale formations [10]. Shale gas has become an increasingly important source of natural gas especially in the United States since the start of this century, and interest has spread to potential gas shales also in the rest of the world (although with very scarce success till now, with the only exception of Argentina). In 2000 shale gas provided only 1% of US natural gas production; by 2010 it was over 20% and the US government's Energy Information Administration predicts that by 2035, 46% of the United States' natural gas supply will come from shale gas.

Because shales ordinarily have insufficient permeability to allow significant fluid flow to a wellbore, most shales are not commercial sources of natural gas. Shale gas is one of a number of unconventional sources of natural gas; others include coalbed methane, tight sandstones, and methane hydrates. Shale gas areas are often known as resource plays (as opposed to exploration plays). The geological risk of not finding gas is low in resource plays, but the potential profits per successful well are usually also lower.

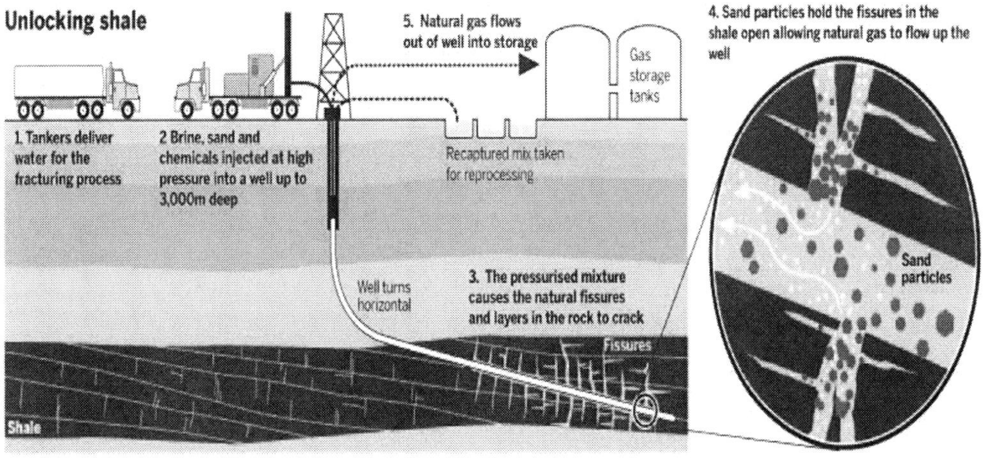

Fig. 7. – Shale gas production technology (Source: Financial Times, March 2010).

Shale has low matrix permeability, and so gas production in commercial quantities requires fractures to provide permeability. Shale gas has been produced for years from shales with natural fractures; the shale gas boom in recent years has been due to modern technology in hydraulic fracturing (fracking) to create extensive artificial fractures around wellbores.

Horizontal drilling is often used with shale gas wells (fig. 7), with lateral lengths up to 10000 feet (3000 m) within the shale, to create the maximum borehole surface area in contact with the shale.

Analogously, shale oil is an unconventional oil produced from oil shale rock fragments by pyrolysis, hydrogenation, or thermal dissolution. The term "shale oil" is also used for crude oil produced from shales of other very-low-permeability formations. However, in order to reduce the risk of confusion of shale oil produced from oil shale with crude oil in oil-bearing shales, the term "tight oil" is preferred for the latter.

Shale oil is extracted by pyrolysis, hydrogenation, or thermal dissolution of oil shale [11]. The pyrolysis of the rock is performed in a retort, situated either above ground or within the rock formation itself. As of 2008, most oil shale industries perform the shale oil extraction process after the rock is mined, crushed and transported to a retorting facility, although several experimental technologies perform the process in place (*in situ*). The temperature at which the kerogen decomposes into usable hydrocarbons varies with the time-scale of the process; in the aboveground retorting process decomposition begins at 300 °C, but proceeds more rapidly and completely at higher temperatures.

Decomposition takes place most quickly at a temperature between 480 and 520 °C. Hydrogenation and thermal dissolution (reactive fluid processes) extract the oil using hydrogen donors, solvents, or a combination of these. Thermal dissolution involves the application of solvents at elevated temperatures and pressures, increasing oil output by cracking the dissolved organic matter. Different methods produce shale oil with different properties.

Also for shale resources the same *challenges* and *dilemmas*, previously reported for conventional O&G sources, can be found: North American shale reservoirs contain billion of barrels of oil but the industry's average recovery factor is typically well below 10% compared with an average 30% for conventional reservoirs. Efforts to improve recovery from unconventional reservoirs are well under way [12].

On the other hand, as regards water management the situation is even more critical than in the case of conventional resources. Operators have always factored water handling into the cost of drilling and production operations. But the proliferation of hydraulic fracturing in North American shale plays has moved water handling from line item expense to central concern [13], making water treatment and re-use mandatory in perspective.

4. – Concluding remarks

By moving towards 2040–2050, oil and natural gas will remain crucial energy components, maybe with some less crude-oil production but more natural gas and higher use of electricity [14], produced both by natural gas and by renewables and also pushed by the advent of electric vehicles and by new perspective environmental regulations that could prohibit the use of diesel and gasoline vehicles at least in the cities [15].

However, the process will likely be very slow as the overall picture is very resilient due to the contemporary need of increasing energy consumption and to the difficulty to completely renew the worldwide vehicles fleet in less than decades.

REFERENCES

[1] Marchionna M., in *Encyclopedia of Hydrocarbons* - Treccani 3rd edition, **4.1** (Istituto dell'Enciclopedia Italiano) 2007, pp. 301–317.
[2] *Hydrocarbon Processing online*, July 3, 2017.
[3] *Hydrocarbon Processing online*, April 10, 2017.
[4] OGJ Authors, *Oil & Gas J.*, **115.10** (2017) 38.
[5] Nyquist S., in *McKinsey & Company Report,* November 2016.
[6] *Hydrocarbon Processing online*, May 23, 2017.
[7] *Hydrocarbon Processing online*, July 3, 2017.
[8] Dittrick P., *Oil & Gas J.*, **115.8** (2017) 50.
[9] Marchionna M., *La Chimica & l'Industria (Milan)*, **94.6** (2012) 100.
[10] Moniz E. J., Jacoby H. D. and Meggs A. J. M., *The Future of Natural Gas* – an interdisciplinary MIT study (MIT Press) 2011, pp. i–178.
[11] Speight J. M., *Shale Oil Production Processes* (Gulf Professional Publishing - GPP) 2013, pp. 1–192.
[12] Von Flatern R., *Upstream online*, June 9 (2017).
[13] Von Flatern R., *Upstream online*, August 9 (2017).
[14] Snow N., *Oil & Gas J. online*, October 18 (2017).
[15] *Hydrocarbon Processing online*, July 6, 2017.

EPJ Web of Conferences **189**, 00005 (2018) https://doi.org/10.1051/epjconf/201818900005
LNES 2017

Introduction to wind energy systems

HERMANN-JOSEF WAGNER(*)

Energy Systems and Energy Economics, Ruhr-University Bochum
P.O. Box 102148, 44721, Bochum, Germany

Summary. — This article presents the basic concepts of wind energy and deals with the physics and mechanics of operation. It describes the conversion of wind energy into the rotation of a turbine, and the critical parameters governing the efficiency of this conversion. After that it presents an overview of the various parts and component of windmills. The connection to the electrical grid, the world status of wind energy use for electricity production, the cost situation and research and development needs are further aspects which will be considered.

1. – Different wind turbine types: An overview

Today there are various types of wind turbines in operation, (fig. 1 gives an overview). The most common device is the horizontal axis wind turbine. This turbine consists of only a few aerodynamically optimised rotor blades, which for the purpose of regulation usually can be tumbled about their long axis (Pitch-regulation). Another cheaper way to regulate it, consists in designing the blades in such a way that the air streaming along the blades will go into turbulence at a certain speed (stall-regulation). These turbines

(*) E-mail: `lee@lee.rub.de`

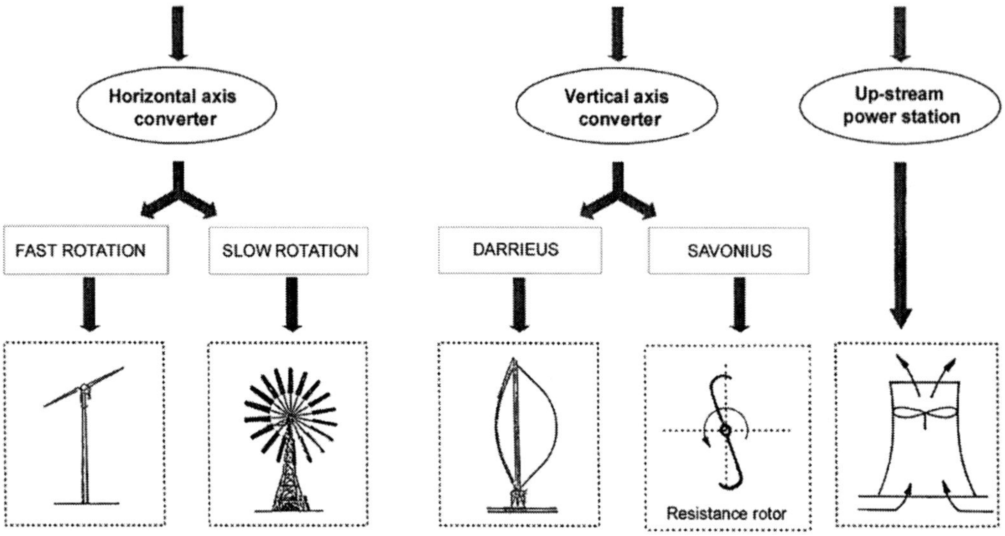

Fig. 1. – Overview of the different types of wind turbines.

can deliver power ranging from 10 kW to some MW. The largest turbine on the European market has a power of 8 MW, bigger machines are being tested. The efficiency of this type of turbine is very high. Therefore, it is solely used for electricity generation which needs "high-speed engines" to keep the gear transmission and the generator small and cheap.

Another conventional (older) type of horizontal axis rotor, is the multiblade wind turbine. It was first built about one hundred years ago. Such wind mills have a high starting torque which makes them suitable for driving mechanical water pumps. The number of rotations is low, and the blades are made from simple sheets with an easy geometry. For pumping water, a rotation regulating system is not necessary, but there is a mechanical safety system installed to protect the turbine against storm damage. By using a so-called wind-sheet in lee direction the rotor is turned in the direction of the wind. In order to increase the number of rotations, this type of turbine had been equipped with aerodynamically more efficient blades facilitating the production of electricity, here the area of a blade is smaller.

The mechanical stability of such "slow speed turbines" is very high, some have had operation periods of more than fifty years.

A third type of turbine is known as DARRIEUS; a vertical axis construction. Their advantage is that they do not depend on the direction of the wind. To start, they need the help of a generator working as motor or the help of a SAVONIUS rotor installed on top of the vertical axis. In the nineteen eighties a reasonable number of DARRIEUS turbines had been installed in California, but a further expansion into the higher power range and into the European markets has not taken place. One reason may be that they are noisier than horizontal-axis turbines. Another disadvantage is that wind velocity increases significantly with height, making horizontal axis wheels on towers more economical. Nevertheless, there are some companies producing DARRIEUS turbines in the very low power range of a few kilowatts for decentralised electricity supply in areas without electrical grids *e.g.* in rural areas of developing countries.

The SAVONIUS rotor is only used for research activities, *e.g.* as a measurement device especially for wind velocity, it is not used for power production. Therefore it will not be discussed in detail in this paper.

The last technique to be dealt with is known as Up-Stream Power Station or thermal tower. In principle, it can be regarded as a mix between a wind turbine and a solar collector. At the top of a narrow, high tower is a wind wheel on a vertical axis driven by the rising warm air. A solar collector installed around the foot of the tower heats up the air. The design of the collector is simple; a transparent plastic foil is fixed over several metres on the ground in a circle around the tower. Therefore, the station needs a lot of space and the tower has to be very high. Such a system has a very poor efficiency, only about one percent. World wide there has only been one Up-Stream Power Station built so far, it was designed by a German company. For some years it worked satisfactorily at the location of Manzarenas in Spain, but in the mid eighties it was destroyed by bad weather. This station had an electrical power of 20 kW, the tower was about 200 m high, and the collector had a diameter of approximately the same size. A second Up-Stream Power Station with an electrical performance of 200 MW was planned in Australia, but not yet realised until now. The tower height should be about 1000 m and the diameter of the collector area should be about 7000 m. No new Up-Stream Power Stations have been designed and installed so far also because of the tremendous technical progress over the last ten years of both solar farm stations and horizontal-axis wind turbines.

2. – Physical basics

2·1. *Energy content of the wind.* – The following section will be used to mathematically explain where the energy in the wind comes from and what factors it depends on.

The power is defined as

$$P = \frac{E}{t} = \frac{1}{2} \cdot A \cdot \rho_a \cdot v^3, \tag{1}$$

with E: kinetic energy, A: area, ρ_a: specific density of the air, v: wind velocity.

Therefore, it is also proportional to the cube of the wind speed, v^3.

From fig. 2, it can be seen that the power output per m^2 of the rotor blade is not linearly proportional to the wind velocity, as proven in the theory above. This means that it is more profitable to place a wind turbine in a location with occasional high winds, than in a location where there is a constant low wind speed. Measurement at different places shows that the distribution of wind velocity over the year could be approximated by a Weibull-equitation. That means that at least about 2/3 of the produced electricity will be earned by the upper third of wind velocity.

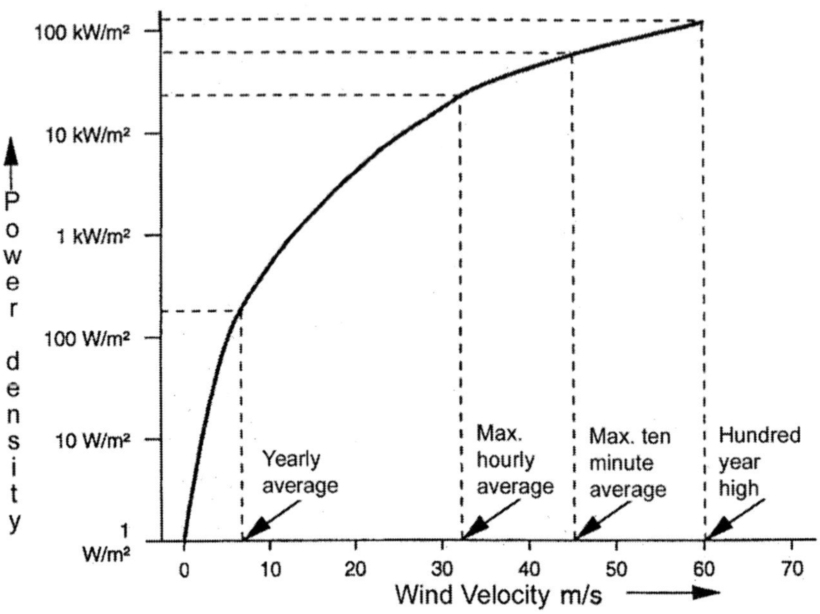

Fig. 2. – Relationship between wind velocity and power output (yearly average valid for Germany). Figure taken from [1], p. 241.

From a mechanical point of view, the power density range increases by one thousand for a wind speed change of just 10 m/s, thus producing a construction limit problem. Therefore, wind turbines are constructed to harness only the power from wind speeds in the upper regions.

2˙2. *Power coefficients*. – There is now the question of how much of the energy in the wind can be transferred to the blade as mechanical energy.

Betz' law and c_p

Betz' law states that you can only convert a maximum of 59% of the kinetic energy in the wind to mechanical energy using a wind turbine. This is because the wind on the back side of the rotor must have a high enough velocity to move away and allow more wind through the plane of the rotor.

The relationship between the power of the rotor blade P_R and the maximum power $P_{R\,\max}$ is given by the power coefficient c_p;

$$P_R = P_1 - P_2 = c_p \cdot P_{R\,\max}. \tag{2}$$

The maximum power coefficient is determined through the ratio v_2/v_1 and setting the derivative to zero.

$$c_{p\,\max} = \frac{16}{27} = 0.593 \quad \text{with } v_2 = \frac{1}{3} \cdot v_1. \tag{3}$$

Therefore, an ideal turbine will slow down the wind by 2/3 of its original speed (Betz' law).

The issues discussed in the theory can be summed up and related to the design of a wind energy turbine, by the so called Cooking recipe:

"Cooking recipe" for the design of wind turbines.

1. A high aerofoil form ratio leads to a high tip-speed ratio and therefore, a large power coefficient c_p.

⇒ Modern turbines with a good aerodynamic profile rotate quickly.

2. Simple profiles with smaller profile form ratios have a small tip-speed ratio. Therefore, the area of the rotor radius that is occupied by blades must be increased in order to increase the power coefficient.

⇒ Slow-rotating turbines have poor aerodynamic profiles and a high number of blades.

3. The profile form ratio and the tip-speed ratio have a considerably greater influence on the power coefficient than the number of blades.

⇒ The quality of an aerofoil in respect to a high-speed turbine, has an inferior significance.

3. – Technical design of wind turbines

3·1. *The design with gearbox.* – The details of a design with gearbox are shown in fig. 3. The main aspect of the classic design is the split shaft system, where the main shaft turns slowly with the rotor blades and the torque is transmitted through a gearbox to the high-speed secondary shaft that drives the few-pole pair generator.

The transmission of torque to the generator is shut off by means of a large disk brake on the main shaft. A mechanical system controls the pitch of the blades, so pitch control can also be used to stop the operation of the turbine in, *e.g.*, storm conditions. The pitch mechanism is driven by a hydraulic system, with oil as the popular medium. This system needs almost yearly maintenance and constant pressure monitoring, along with the gear box which is lubricated with oil. Due to mechanical losses in the gear box, the oil will be heated and must be cooled. For constructions without a main brake, each blade has its pitch angle controlled by a small electric motor. This is the standard of mostly all wind converters.

To reduce weight generators with permanent magnets were developed. Some producers are equipping their converters with it in the last years.

In order to reduce vibrations by high speeds in the rotating axe, generators are often designed with more than one pole pair. This allows a reduction in the rotation velocities between the gear box and the generator. The problem with this type of configurations is that the size and weight of the generator increases with the number of pole pairs.

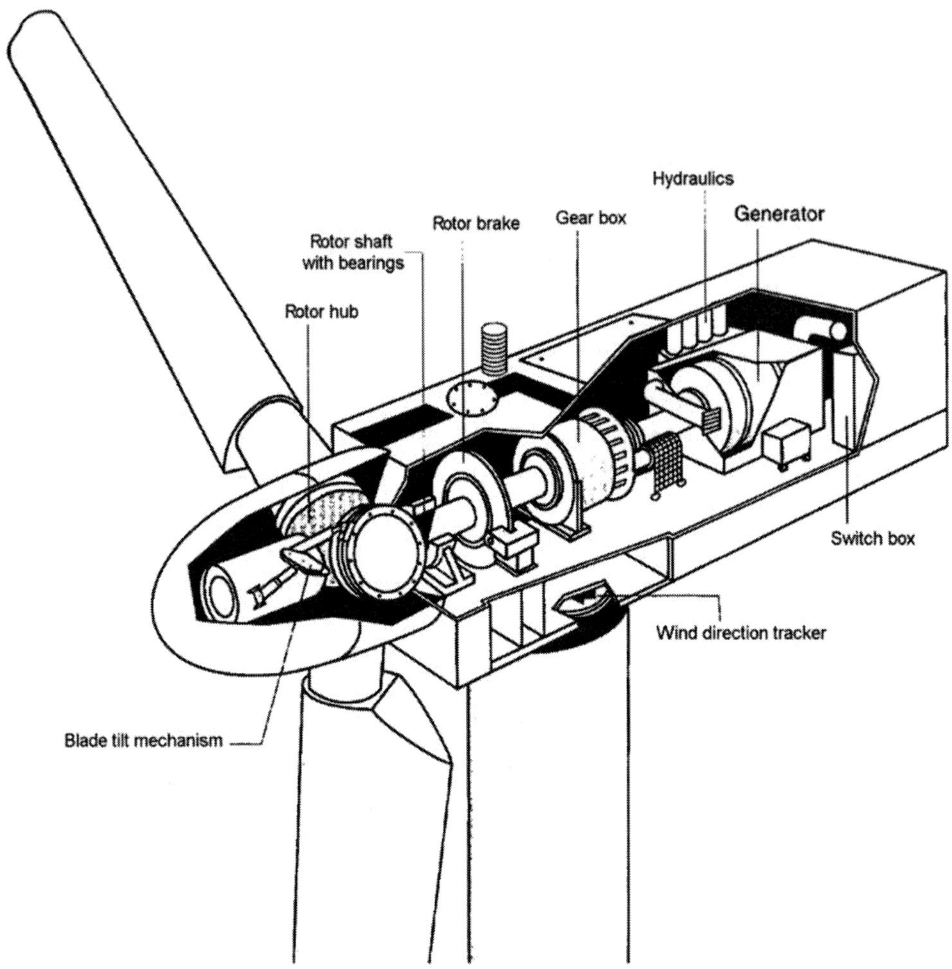

Fig. 3. – The design with gearbox.

Wind speed and direction measuring apparatuses are located at the back of the hub head. A rack-and-pinion mechanism at the join of the hub and the tower, allows the hub to be rotated in to the wind direction, and out of it in storm conditions.

3˙2. *The design without gearbox.* – Some companies *e.g.* the German company Enercon, design another turbine type, without gearbox. The scheme of such a turbine is shown in fig. 4, where the main design aspects can be clearly seen.

This design has just one stationary shaft. The rotor blades and the generator are both mounted on this shaft. The generator is in the form of a large spoked wheel with *e.g.* forty-two pole pairs, around the outer circumference and stators mounted on a stationary arm around the wheel. The wheel is fixed to the blade apparatus, so it rotates slowly with the blades. Therefore, there is no need for a gearbox, rotating shafts or a disk brake.

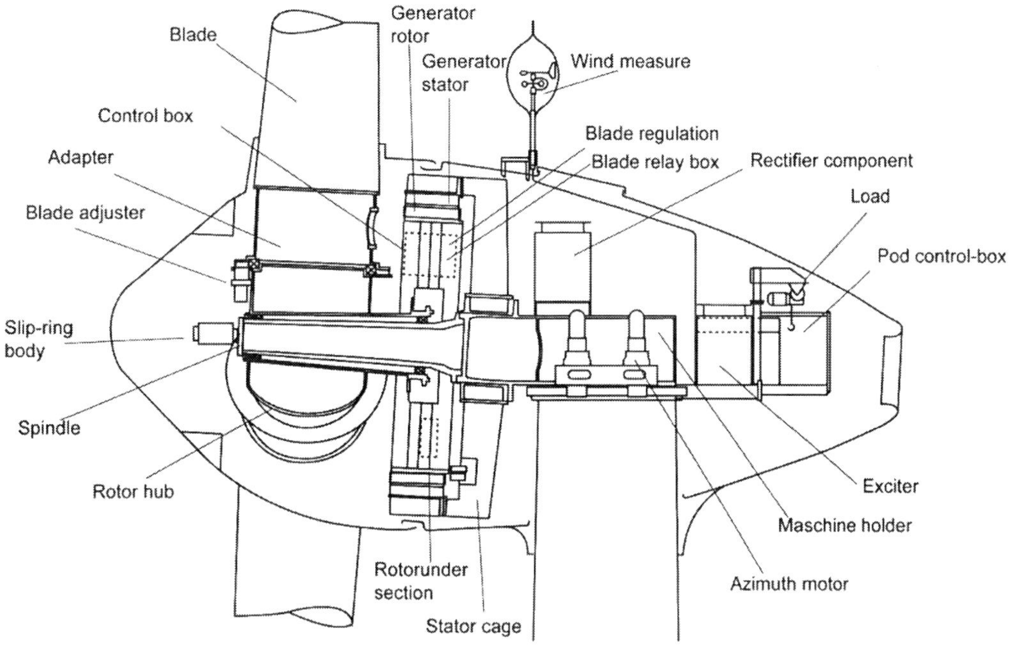

Fig. 4. – The design without gearbox (Enercon E-66).

This minimising of mechanical parts simplifies the maintenance and production of the turbine.

The whole system is automated; pitch control and hub direction are controlled by a central computer, which operates small directional motors.

3·3. *Aspects of design and development.* – There are several critical aspects of a wind turbine that need to be considered in the design phase, to ensure the turbine will be economic and durable.

The tower:

In principle, the tower needs to be as tall as possible, because the wind speed increases with height. However, the height is limited by costing issues; an increase in tower height of 10 m costs an extra fifteen thousand dollars, and a tower height of over 100 m requires an aircraft-warning beacon, which is again rather expensive.

Heat energy:

Large turbines (> 1 MW), have an average generator efficiency of 98%. Heat is also generated in the mechanical parts of the machine including the bearings and the gear box. This means that around 40 kW of power are lost during operation heating up the generator. This heat needs to be controlled to prevent damage to the machine parts.

A large fan system is mounted onto the back side of the hub of a turbine and used to draw cool air through the hub and remove this excess heat.

Control and monitoring:

The following aspects of a wind turbine need to be controlled and monitored to ensure effective operation of a wind turbine within the legal limits.

- By large turbines; vibration levels.
- Speed of rotation and the pitch angle, of the rotor blades.
- The natural wind speed and direction.
- The voltage and frequency of the electricity produced.
- The output phase angle compared to the grid phase angle.
- The consistency of the electrical power output.
- The acquisition and storage of electrical signals.
- Signal conversion equipment for the directional motors.
- Rotational speed at night, to reduce the noise levels, because the noise is proportional to the blade-tip speed to the power of six.

Mechanical stability:

The following forces affect the stability of the mechanical system:

- Gravity
- Centrifugal forces on the rotor blades
- Pressure changes on the blade due to the shadow effect the tower creates
- Stochastic power output of the turbine due to wind energy levels continually changing
- Resonance of the blades

Wind direction set-up:

A wind turbine can be designed to face in to the wind (windward), or away from it (leeward). A leeward turbine has the advantage of being self-orientating, but the disadvantage of the tower disturbing the wind velocity profile, before the wind has reached the plane of the rotor blades. The pressure and speed differences experienced by the blade as it passes the tower, result in stresses on the hub, which need to be alleviated by use of an extra mechanism in the hub to allow the rotor blades to move out of their usual plane of rotation.

TABLE I. – *The technical features of two different multi-megawatt wind turbines for onshore.*

	Enercon E-182 E3	REpower 3.2 MW
Design	without gearbox	with gearbox
Hub height	80 - 130 m (onshore)	100 - 130 m (onshore)
No. of blades	3	3
Rotor speed	6-18 rpm	6.7 -12 rpm
Rotor diameter	82 m	114 m
Material of blade	Fibreglass (reinforced epoxy)	Fibreglass (reinforced epoxy)
Blade regulation	Pitch	Pitch
Rated power	3 MW	3.2 MW
Transmission ratio of gearbox	None	approx. 99
Generator	Multi-pole	Asynchronous, few poles
Grid connection	Via frequency converters	Via frequency converters

3·4. *Technical features of two modern wind turbines.* – The largest market introduced machine up to the year 2016, is a 8 MW turbine from the company Vestas.

Table I gives an idea of the technical data of wind turbines of the 3 MW class. A design with and without gearbox have been chosen to show the different operation of their generators.

3·5. *Floating offshore windconverter.* – In addition to the fixed wind energy installations, there are floating wind converters as of a sea depth from 50 to 500 m. The wind turbine is installed via a floating platform, similar to the oil and natural gas platforms. With the help of floating offshore wind turbines, the potential of wind energy utilization increases. The platforms currently in use are the spar buoy, the Tension-leg platform (TLP), and the semi-submersible platform (see fig. 5). The spar buoy has a simple construction, since the hollow cylinder serves both as a tower and as a platform. It is weighted in the lower area and then anchored. The TLP consists of half-dive buoys, which are vertical anchored by rope at the seabed. Advantageous is the low mass of the TLP and its excellent stability. The semi-submersible platforms consist of a steel frame with vertical cylinders, which are responsible for the buoyancy and are considered buoys. This platform is also attached to the seabed by means of an anchorage. The semi-submersible platforms offer an advantage in their flexible use, in shallow as well as in deep water [2].

4. – Connection to the electrical grid

The main electrical grid has a constant frequency *e.g.* of 50 Hz or 60 Hz and a constant phase angle. Therefore, a wind turbine must produce electricity with the same constant values in order to be integrated into the main grid.

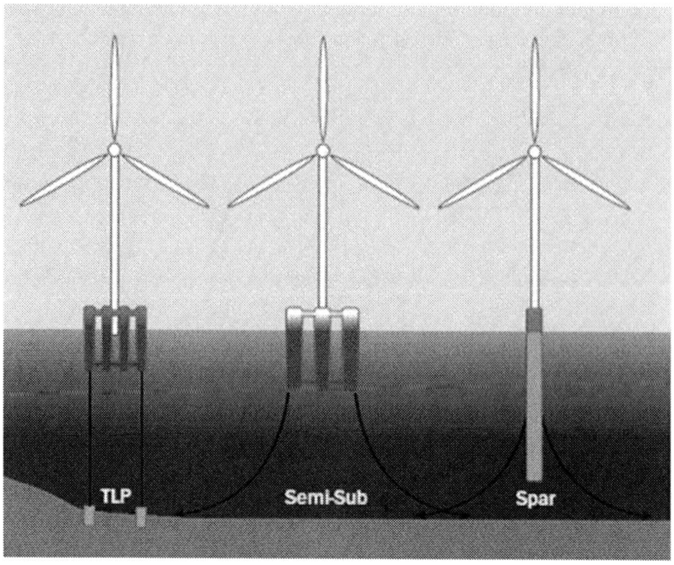

Fig. 5. – Floating offshore platforms [3].

The input energy of a wind turbine is proportional to the wind speed, but the wind speed is never constant. Each wind speed has a corresponding rotor rotation speed, at which the maximum power is produced. This maximum occurs for different wind speeds at different rates of rotation. However, the rate of rotation must be held constant in order to achieve the required constant output frequency or the wind turbine has to be connected to the grid by doubly fed asynchronous generator or by electronic frequency converters.

A small turbine can be connected directly into the grid network at 0.4 kV. When the wind turbine is integrated into the grid network, there must be no voltage change, voltage oscillation or flicker experienced in the homes on that network branch. The loss of voltage due to resistance in the cabling can be avoided by increasing the diameter of the cables. It is often required that a new network branch is constructed and linked to the transformer, in order to reduce the voltage disturbances. This increases the installation costs of the turbine.

Megawatt turbines cannot be connected to the grid at the 0.4 kV stage, but have to be connected in at 10–30 kV, which is the usual level of the city electricity share distribution. In remote areas, where a 30 kV connection is not established, the connection must be created and financed by the wind park developers. Wind parks with a lot of Megawatt turbines can also be connected into the electrical grid *e.g.* at the 110 kV level in Germany.

As mentioned earlier the maximum power output is obtained only in few hours during the year. Figure 6 shows a typical load distribution, measured in Germany. With larger wind energy installations in future this uneven distribution leads to the need of higher regulation capacities by conventional power systems.

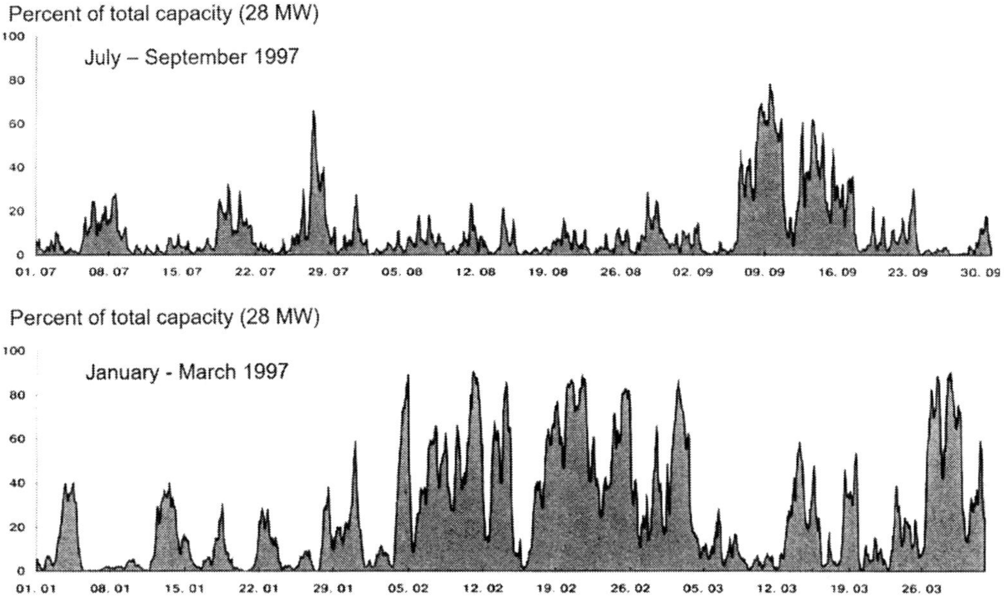

Fig. 6. – Load Distribution – Measurement (Germany).

Grid connection of offshore wind farms poses a technical and economical challenge to wind turbine and grid operators. In the initial phase, the still quite limited capacity of early pilot farms enables using a conventional three-phase AC connection to the onshore grid system which is a well-known technology and inexpensive. These wind farms were also located near to the coast, the energy losses by using AC was therefore limited.

Greater capacities and remote offshore sites make it technically difficult to connect offshore wind farms to the mainland grid by using AC undersea cables. Losses, reactive power production and limited capacity of the sea cables may become important in the future. High-voltage direct current transmission to land could be a solution but it is technically more complicated and more expensive.

In Germany there are three electrical lines in operation from offshore wind farms in the North Sea to the mainland grid. There are one line AC and two lines DC operated. To protect environment on sea several wind farms are connected to one line.

To connect the offshore wind farm to the onshore grid an internal grid is necessary. The produced power has to be fed to an offshore transformer substation. Wind turbines are connected to it via undersea cables by a voltage of about 30 kV. After stepping-up to the transmission line voltage, *e.g.* 180 kV, the power is conveyed to shore.

5. – Use of wind energy

5·1. *Worldwide status.* – In the nineteen eighties, it was the USA that took the lead in establishing wind farms. They set more than 10000 turbines into operation, each

TABLE II. – *World wind power capacities. This statistic is taken from* [2].

Land/Region	Total installed rated power until the end of 2016 [GW]
China	169
USA	82
Germany	50
India	29
Spain	23
UK	15
France	12
Canada	12
Brazil	11
Italy	9
Remaining countries	76
Total	487

generating in the range 80–200 kW. In Europe it was Denmark that was the main pioneer of wind energy. At the end of twenty-first century, Germany has taken the world lead, producing about one third of the world's wind produced electricity until 2008.

Many governments have begun to produce initiative schemes to increase the economic feasibility of wind turbines. Some initiatives used include paying more for wind produced electricity, and providing a proportion of the initial construction costs. Governments of industrial countries, or those with a high power consumption, are eager to promote wind energy, because it is environmentally clean and sustainable and limits the need for fossil fuel usage (table II).

By the end of 2016 the world total rated power was about 487 GW. This is about 1,9% of the worldwide installed capacity of power stations (water, coal, natural gas, nuclear power).

It should be taken into account that wind energy has a lower capacity factor than fossil and nuclear power stations (around 1) with around 0,2 to 0,3 for onshore wind power and up to 0,5 for offshore wind power.

5·2. *Offshore wind status.* – Due to high wind potential at Germany's coastline, wind energy should play a decisive role. Therefore, Germany's government plans to install around 6,5 GW of offshore wind parks until 2020 and 15 GW until 2030. By the end of the year 2016, the wind power installed capacity in Germany for offshore was around 4 GW and for onshore around 46 GW [4].

5·3. *Investment and operation costs.* – The costs involved in installing a wind turbine vary depending on the design, size and the chosen location of the new turbine. The infrastructure costs can be minimised by constructing wind parks, where a number of

TABLE III. – *(a) The typical investment costs of a 2 MW onshore wind power station in Germany (Amesdorf), (Status: 2013). (b) The typical operating costs of a 2 MW onshore wind power station in Germany (Wellen), (Status 2013). Data are taken from WindGuard GmbH, Germany.*

(a) Investment plan	Costs [€/kW]	
Hub height	< 120 m	> 120 m
Wind power station, transport, installation	1150	1340
Foundation	70	
Grid connection	70	
Site development (lanes)	40	
Planning, environmental measures, concession, others	190	
Total	1520	1710

(b) Operating costs: 5,1 ct/kWh (Average over 20 years operating time)	
Service, repair, others	50 %
Rent	20 %
Management (technical and business)	20%
Reserve for unforeseen events	5%
Insurance	5%

new turbines are installed on the same site. An example of the investment costs for a wind park at land in Germany is shown in tables IIIa and b.

The money invested in an average wind park on land depreciates over about a ten year period. During this period, the set-up and installation costs are high, along with the loan repayments and insurance costs. After this period, the costs then decrease. Over the next ten years, a financial return can then be made on the investment, when the price for the electricity per kilowatt hour is set at a high enough value by the government. This means it is economic to ensure the durability of the installed turbines, so maintenance costs after the ten year period are kept low and the investors can receive a good return on their investment.

The above example of investment costs for wind parks on land in Germany is non-transferable to other countries. Especially the employee's wages and the structure of financing wind farms are quite different. Nevertheless power generating by wind turbines in developed countries with existing electrical grid and sufficient installed power, costs three to four times more than power generation by conventional power stations.

Wind parks on the sea need higher specific investments, although the wind offer is better on the sea than on land. The costs of produced electricity are higher by an offshore wind park (status 2016). The expectation is that the cost could be reducing for new offshore wind parks in the next years. Many countries are supporting the installation of wind converters by funding programs.

5·4. *Environmental aspects.* – Wind energy is a renewable energy source; therefore it holds many advantages over the fossil fuels, which additionally have diminishing reserves. Wind energy is clean with regard to toxic emissions. Therefore, it does not add to Global Warming or acid rain problems.

The wind turbines can affect the environment in aesthetic and human intrusive ways. This is because they must be sited in prominent locations and through the rotation of their blades; they can produce optical distortions *i.e.* flickering shadows, and a humming noise. The land required for the sighting of a wind park can be considered large, if all the access routes are also taken into consideration. However, they very rarely require the resettlement of communities, which is a problem associated with, *e.g.*, large Hydro-Electric-Power stations. The danger to birds of the rotating blades has been debated, but it has been found that the birds change their flight paths to avoid the blades. It has also been a question whether the reduced wind speed at the ground level affects the growth of the flora. This is answered by the observation that many wind parks have animals grazing between the turbines.

5·5. *Life Cycle Analysis.* – Life Cycle Assessments (LCA) is an important tool for industry and policy makers, used to determine the actual emissions of a product or technology throughout its whole life cycle. In case of energy production systems or power plants, the following is analyzed: energy required to produce the materials and processes; emissions resulting from various processes for materials production and processes resulting into their Cumulated Energy Demand (CED) and Global Warming Potential (GWP) become important parameters when making decisions on further research, development and deployment of any technology. Results and characteristics of wind energy LCA were for the first time analyzed for the German offshore wind farm alpha ventus. Alpha ventus is a test field in the North Sea for 5 MW wind energy converters in a water depth of 30 m. Results have shown that within less than one year the energetic expenditure as well as the greenhouse gas emission of the entire life cycle of alpha ventus is amortized [5].

6. – Research and development needs

The field of wind converters is connected with high technical development activities since a lot of years. According to the first report of the project "Wind Energy Thematic Network" funded by the European Commission, some aspects the year 2012 of the R&D needs are described below, which are also relevant for:

1. *Environmental and social impacts* (*e.g.* enhancing local incentives by developing participation models)

   - methods to integrate wind turbines visually into the landscape,
   - reduction of noise impacts,
   - mitigating impacts on bird populations, habitats and flight paths,
   - turbine design regarding life cycle analysis,
   - analyse social effects like local employment, investment, taxes etc.

2. *Wind turbine and component design issues* (*e.g.* basic research in aerodynamics, structural dynamics, structural design and control)
   - new materials with higher strength like carbon fibre for the blades,
   - feasibility studies of new wind turbine concepts and innovations,
   - integration of demand-side requirements in the design of a turbine, *e.g.* electrical control system interaction with grid requirements.
3. *Testing, standardisation and certification* (*e.g.* common accepted certification procedures for wind turbines and wind farms)
   - standards for service and maintenance concepts of offshore wind converters.
4. *Grid Integration, energy systems and resource prediction* (*e.g.* forecast of wind resource)
   - development of scenarios for redesigning the grid system with high wind penetration,
   - increasing both: power quality and consistency,
   - energy management and storage systems for stand alone applications.
5. *Operation and maintenance* (*e.g.* advanced condition monitoring)
   - development of early failure detection and condition systems,
   - development in preventative maintenance,
   - standardisation of components for easy replacement,
   - certification of service and maintenance concepts,
   - cold and icing climate resource assessment.
6. *Offshore wind technology* (*e.g.* research into the control and efficiency of very large wind farms and more cost effective foundations, transport and installation techniques)
   - monitoring of environmental impacts (effects on birds, effects of noise and vibration on marine life especially by installation of wind converter etc.),
   - development of deep water foundation structures (fundamental wind turbine design research),
   - offshore meteorology,
   - special designs of systems and components for transportation, erection, access and maintenance of offshore wind turbines,
   - investigate the use of energy storage,
   - improve corrosion protecting systems regarding the offshore conditions.

TABLE IV. – *Technical developments in Wind Turbines* [6].

Generators	– Copper? Permanent Magnet? – Rare earths? – Superconductors?
Rotor Blades	– Cost reduction? – Utilization? – Active control elements in rotor blades?
Operations Monitoring	– Drive train monitoring (CMS) for onshore plants? – Tower- and foundations vibrations – Number of Sensors, also Onshore? – Periodic Inspections- Scope, Number?
Technical Uncertainties	– Handling of new developments?
Offshore Foundations technologies	– Steel? Concrete? – Gravity Foundations? – Environment, under water noise protection during the piling for the foundation?

In addition to these aspects a lot of R&D has been accomplished in parallel to operating and market introducing wind turbines, *e.g.*:

– better state of knowledge of the dynamic forces on the drive train,

– improvement of the availability of gear boxes,

– optimisation of control units and control systems,

– development of central lubrication systems.

From a technical perspective several configurations for wind technologies are under research and development in Germany (table IV).

7. – Outlook

The wind energy market has grown because of the environmental advantages of harnessing a clean and inexhaustible energy source and because of the economic incentives supplied by several governments. However, energy is required from other generation methods during the building phase of a new turbine, so in this period, Greenhouse gases and air pollution will be added to. If the life cycle of a wind turbine is looked at, more pollutants are saved during operation, than are emitted during the building phase.

A wind turbine is not a self-sustainable power station. This means that back-up power generation is needed at the times when the turbine is inoperative. This back-up is nowadays supplied by the established fossil fuel power stations. If the number of wind turbines increases, the need for extra investment in the back-up generation systems will rise, in order to maintain a stable electricity grid system. The challenges in this case are that existing power stations will deliver less electricity than before. Under this condition the electricity generation prices for new power stations are increasing due to smaller load duration time than before.

However, wind energy is still one of the most important renewable energy resources for the future, because it can be harnessed in a clean and inexhaustible manner, through the application of technically advanced and efficient machinery. These additional investments will need in some cases subsidies. For more information especially with respect to technological and ecological aspects (life cycle aspects) see [5].

REFERENCES

[1] KLEEMANN M. and MELIß M., *Regenerative Energiequellen*, 2.Auflage (Springer-Verlag, Heidelberg) 1993.
[2] www.GWEC.de.
[3] GRAZIANO L., *Literaturrecherche zum Thema schwimmende off-shore Windkraftanlagen*, Student project, March 2017.
[4] ENDER C., *Wind Energy Use in Germany*, DEWI Magazin 02/2016.
[5] WAGNER H.-J. and MATHUR J., *Introduction to Wind Energy Systems – Basics, Technology and Operation*, second edition (Springer-Verlag, Berlin) 2013, ISBN 978-3-642-32975-3.
[6] HERMANN-JOSEF WAGNER, *Erneuerbare Energien Wind und Sonne – Wo geht es technisch hin?"*, VGB PowerTech 1/2, 2015.

EPJ Web of Conferences **189**, 00006 (2018) https://doi.org/10.1051/epjconf/201818900006
LNES 2017

Biomass and biofuels

Marco Fiala(*) and Luca Nonini(**)

Department of Agricultural and Environmental Sciences - Production, Landscape, Agroenergy University of Milan - Via G. Celoria, 2, 20133 Milan, Italy

Summary. — Biomass includes all materials that contain organic carbon bound in the chemical structure of molecules, resulting from the chlorophylline photosynthesis, carried out by autotrophies organisms. Lots of biomass from agricultural, agri-food and forestry sectors can be used for energy purposes, representing an essential renewable energy source that, if appropriately managed, can help to reduce the negative environmental impacts arising from the exploitation of fossil fuels. The possibility of using biomass for a specific production process mainly depends on its physical and chemical properties. This paper is organized in two sections: in the first one, the most important biomass used worldwide for energy generation (thermal energy and/or electric energy), as well as its properties, are described. In the second one, the main biomass-to-energy processes (thermochemical and biochemical conversions) are shortly explained. Finally, some emerging techniques (such as bio-methane and bio-hydrogen production) are discussed in more detail.

(*) E-mail: marco.fiala@unimi.it
(**) E-mail: luca.nonini@unimi.it

1. – Biomass: Source of energy

1·1. *Definition of biomass.* – The term *biomass* includes all the materials that contain carbon (C) in an *organic form*. Plants in general, as autotrophies organisms(1), through *chlorophylline photosynthesis* are able to transform C from the *inorganic form* (carbon dioxide, CO_2) into an organic form, by chemical bonds with other elements, such as hydrogen and oxygen, using solar energy. The solar energy is, thus, converted into *chemical energy* able to assemble organic molecules. The *structure* of these molecules can be very simple (as in the case of glucose and other monosaccharides) or more complex (as in the case of *hemicellulose*, *cellulose* and *lignin*). The demolition of these molecules through biological or physical processes causes —closing the cycle— CO_2 regeneration, which comes back to the atmosphere. The chemical energy is thus released and can be converted into other energy types. As a result, as long as in the human time scale equilibrium is maintained between the use (through energy conversion processes) and the production (through photosynthesis), biomass is —as *renewable*— an inexhaustible energy source. Therefore, its use does not affect natural resources for future generations. Biomass used for energy purpose is considered to be "*neutral*" both in terms of the increase of atmospheric CO_2 concentration(2) and of *global warming* (caused by the *greenhouse effect* enhancement). In fact, the mass of CO_2 released during the energy conversion processes (*e.g.*, combustion) is the same which was previously absorbed by plants during the chlorophylline photosynthesis.

In general, biomass can be classified, taking into account the *source sectors*, as follows [1, 2]:

- *forestry* (wood and derived products);
- *agricultural*, resulting from both crops (by-products and energy crops) and livestock waste (manure and slurry);
- *industrial* (agri-food and industrial residues or waste, in particular);
- *urban* (Organic Fraction of Municipal Solid Waste (OFMSW) and residues derived from trees and green public areas management);
- *different* (*e.g.*, seaweed).

Biomass-to-energy transformations (or conversions) can be carried out in two different ways [1, 3]:

- *directly*, in the case of *solid (lignocellulosic) biomass*;
- *indirectly*, through transformation into a liquid or gaseous carriers (*biofuels*).

(1) Autotrophies organisms are able to feed using only simple inorganic substances, as it happens for plants, which need only CO_2 from the air, water and mineral substances that absorbed from the soil.

(2) Fossil fuels utilization —organic compounds stored for millions of years in the underground— is the main cause of the exponential increase of CO_2 concentration in the atmosphere, recorded from the early decades of 1800 (about 280 parts per million (ppm)) to nowadays (408 ppm in the 2016). These enormous quantities of emissions would not have been registered in the absence of human activities (extraction and utilization —combustion— processes).

1˙2. *Chemical and physical properties.* – In the natural state, the *Fresh (wet) Mass* (FM) of each biomass (m_{FM}, kg) consists of a *Water Mass* ($m_{\mathrm{H_2O}}$, kg) and a *Dry Mass* (m_{DM}, kg). Expressing these masses as percentages, the *Moisture Content* (MC, see below) and the *Dry Matter content*(3) (DM; %FM) can be respectively defined [4]. The possibility of using biomass in an energy conversion process rather than in another depends on its chemical and physical properties, the most important of which are i) Moisture Content and ii) carbon/nitrogen ratio (C/N).

1˙2.1. Moisture Content. In a generic biomass, *water* is contained in two forms: i) *bound*, to the molecular structure and ii) *free*, within cells and plant tissues. The mass of contained water depends on: i) climatic conditions, ii) harvesting period and iii) conservation processes. The *drying process* (both natural and artificial), aims to reduce free water, while the amount of binding water remains unaltered. The *free water content* (variable from 4–5% to 90–92%) is the parameter that, as well as affecting chemical and physical properties, and thus the energy content, mostly influences the type of supply chain where biomass can be used. The *Moisture Content* can be referred to both the m_{DM} or to the m_{FM} ($m_{\mathrm{H_2O}} + m_{\mathrm{DM}}$). In the first case we talk about *Moisture Content* calculated *on dry basis* ($\mathrm{MC_{DM}}$; %DM), while in the second case we talk about *Moisture Content* calculated on *wet basis* (MC; %FM). As a result, the water content can be calculated as follows (eq. (1)):

$$(1) \qquad \mathrm{MC_{DM}} = \frac{m_{\mathrm{H_2O}}}{m_{\mathrm{DM}}} \quad \text{and} \quad \mathrm{MC} = \frac{m_{\mathrm{H_2O}}}{m_{\mathrm{FM}}} = \frac{m_{\mathrm{H_2O}}}{m_{\mathrm{H_2O}} + m_{\mathrm{DM}}}.$$

1˙2.2. C/N ratio. C/N is the ratio between C and N *content* of the dry mass. This ratio is dimensionless and i) in biomass that derives from plants and crops in general (*e.g.*, crop residues) it indicates the *lignification degree*, while ii) in biomass that derives from animals (manure, slurry), it depends on i) *animal species*, ii) *diet*, iii) *farm characteristics* and iv) *type of waste management*.

1˙2.3. Density and bulk density. *Density* (ρ; kg/m^3) is defined as the mass contained in the unit of volume (m^3). It is important because it influences both physical-mechanical properties (size or resistance to stress) and chemical characteristics. The ρ of biomass depends on MC: the greater the MC, the higher the ρ, due to the increase of both the FM and (slightly) the volume. In addition to ρ, it is important to consider the *Bulk (apparent) Density* (γ, kg/m^3), which indicates the mass contained in the unit of volume also *considering the empty spaces* [4]. This parameter strongly influences the technical and economic aspects concerning *biomass storage, handling and transport* (table I).

1˙2.4. Heating value. Generally speaking, the *Heating Value* (HV, MJ/kg) represents the *energy released from the full combustion* of biomass and, thus, it is the fundamental

(3) Or *Total Solids* (TS, %FM).

TABLE I. – *Biomass: preparation, bulk density and main dimensions.*

Biomass	Preparation	Bulk density (γ) kg/m^3 FM	Dimensions V = volume; ϕ = diameter
Cereals straw herbaceous stem and leaves energy crops	Raw material	30–40	–
	Small bales (prismatic)	80–120 (stacked)	$V = 0.1–0.2\,m^3$
	Cylindrical bales	120–180 (stacked)	$V = 1.5–3.0\,m^3$
	Big bales (prismatic)	120–180 (stacked)	$V = 2.0–4.0\,m^3$
	Chopped	150–250 (piled)	10–250 mm
Mais stem and leaves	Raw material	50–60 (piled)	–
	Cylindrical bales	100–150 (stacked)	$V = 1.5–3.0\,m^3$
Pruning residues	Raw material	50–70	–
	Small cylindrical bales	150–210 (stacked)	$V = 0.6–0.8\,m^3$
	Chipped	200–300 (piled)	1–100 mm; $V \leq 2.5 \cdot 10^{-3}\,dm^3$
Wood	Logs	600–700 (stacked) 300–400 (piled)	300–1000 mm; $V = 1.5–15\,dm^3$
	Chipped	200–300 (piled)	1–100 mm; $V \leq 2.5 \cdot 10^{-3}\,dm^3$
Sawdust	Milling process	120–180 (piled)	1–5 mm
Mais silage	Cut	450–750 (pressed) 350–400 (piled)	10–25 mm
Pellet	Extrusion process	800–900 (piled)	$\phi < 25$ mm
Briquettes	Extrusion process	190–340 (piled)	$\phi > 25$ mm; $V = 1–1.5\,dm^3$
Olive oil residues	Compression. centrifugation process	400–500 (piled)	1–5 mm
Grape residues	Compression process	250–500 (piled)	1–5 mm
Rice/cereals husk	Separation process	130–140 (piled)	1–5 mm
Animal slurry	Collected/moved by pumps	1000 (piled)	–
Animal manure	Collected/moved by mechanical devices	500–650 (piled)	–

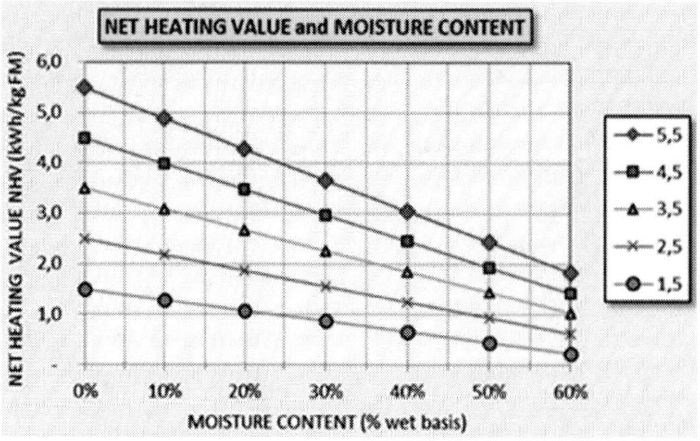

Fig. 1. – Variation of the lignocellulosic NHV *vs.* MC for different values of LHV (source: [4]).

energy parameter to characterize lignocellulosic biomass that can be used for *thermo-chemical conversions.* HV can be expressed as

- *Gross Heating Value* (GHV, MJ/kg DM), which represents the energy released from the complete combustion, at a constant pressure, of a unit of *dry lignocellulosic biomass* after cooling at 25 °C the combustion fumes and, therefore, taking into account the Thermal Energy (ET; MJ) released in the condensation of the water generated during the complete C oxidation([4]). GHV can be calculated taking into account the amount of C, H, O, N, S, as well as ash, as follows [5] (eq. (2)):

$$\text{GHV} = 85.65 + 137.04 \cdot [\text{C}] + 217.55 \cdot [\text{H}] + 62.56 \cdot [\text{N}] + 107.73 \cdot [\text{S}] + 8.04 \cdot [\text{O}] - 12.94 \cdot [\text{Ash}] \quad (\text{MJ/kg DM}) \tag{2}$$

- *Lower Heating Value* (LHV, MJ/kg DM) which is determined subtracting from GHV the heat of water condensation (in fact the vapor is usually released and lost via a chimney into the atmosphere);

- *Net Heating Value* (NHV, MJ/kg FM), which represents the ET really recoverable from the biomass, taking into account the heat for evaporating $m_{\text{H}_2\text{O}}$. Practically, *a fraction of the ET released during the full combustion process must be used to evaporate the biomass free water and, consequently, it is not available for final user* (eq. (3)):

$$\text{NHV} = \text{LHV} \cdot \left(1 - \frac{\text{MC}}{100}\right) - 2.261 \cdot \frac{\text{MC}}{100}, \tag{3}$$

where 2.261 MJ/kg (equal to 540 kcal/kg) is the heat of water vaporization.

([4]) The cumulative reaction of the complete combustion of an organic compound is $C_xH_yO_z + O_2 \rightarrow CO_2 + H_2O + \text{Energy (heat)}$.

Thus, in the definition of the energy characteristics of a lignocellulosic fresh biomass, it is necessary to know the MC value, by which to calculate the NHV (fig. 1).

1·3. *Plant derived biomass.* – This biomass essentially consists of a mixture of three organic compounds with high molecular weight: i) *cellulose* (a linear polymer of anhydrous glucose), ii) *hemicellulose* (a relatively short and branched mixed polymer consisting of sugars with 5 and 6 C atoms) and iii) *lignin* (a complex of three-dimensional polymer mainly consisting of phenols, that ensure compactness and mechanical strength of the tissues).

In addition, there may be *other compounds* (such as resins, fats, oils, waxes, starches, sugars, proteins, tannic substances, pigments, alkaloids, etc.) and, in smaller quantity, *inorganic elements* (Na, K, Mg, Ca, Cd, Zn, As, Pb, S, Cl, N, P, Si, Al, etc.). Concerning the elemental composition, the main elements are: i) C, ii) H, iii) O, iv) N and, in smaller amount, v) alkaline and heavy metals. Compared to fossil fuels, the content of C, N and S of biomass is lower but the content of O, H and ash is higher.

1·3.1. Agricultural by-products. Agricultural by-products are available in the field: i) during the crop's biological cycle (*e.g.*, pruning residues) and ii) at the harvesting operations of the main product (*e.g.*, cereals straw, corn stumps, etc.). Although by-products have some critical aspects, such as i) strong *seasonality*, ii) *harvesting period tightness*, and iii) high *dispersion on territory*, the advantage is the *great availability* [6]. In temperate climates, by-products available at the end of the crop's biological cycle have a $C/N > 50$ (lignified) and $MC < 30\%$; conversely, if their availability occurs at earlier stages of the biological cycle, C/N is much lower (< 30) and MC is higher (> 40–50%). The opportunity to recover an agricultural by-product for its valorization (raw material for industrial purposes, energy conversion, etc.) must always be carefully evaluated taking into account that nutrients and *organic matter* are removed from the soil and will not be available for the subsequent crops([5]). With the repetition of the same production cycle (*monocropping*), the by-products use should be generally avoided in all cases where their removal obliges the farmer to a subsequent soil resources reintegration (N-P-K fertilization and/or livestock waste —manure and/or slurry distribution).

1.3.1.1. Straw by-products. In this category all by-products derived from *herbaceous grain crops* are included: i) *cereal straw* (mainly from soft and hard wheat, barley), ii) *rice straw* and iii) *corn and sunflowers stalks*. This type of biomass is made up of *leaves* and *stems* that, when the main product is harvested, are left on the soil. The C/N ratio is fairly high and the MC is generally $< 20\%$. Ash content is higher than

([5]) By-products derived from herbaceous crops that supply the products during the biological crop cycle (berries, roots, tubers), very rich in N-P-K, are usually left on the soil and buried with the subsequent tillage operations. On the other hand, by-products derived from herbaceous crops that supply the product (typically, seeds) at the end of the crop cycle, with high contents of hemicellulose/lignin and slowly decomposable from soil bacteria, are harvested and utilized (*e.g.*, for energy purposes).

TABLE II. – *Agricultural by-products (for straw): yield, composition, LHV and energy equivalent.*

Straw biomass	Yield	Moisture Content (MC)	C	H	O	N	S	Cl	Ash	Lower Heating Value (LHV)	Energy equivalent
	t/ha DM	%FM	$\%_{DM}$	$\%_{DM}$	$\%_{DM}$	$\%_{DM}$	$\%_{DM}$	$\%_{DM}$	$\%_{DM}$	MJ/kg DM	toe/ha
Wheat straw	2.5–7.0	12–15	43.2	5.0	39.4	0.6	0.11	0.28	11.4	16.3–16.7	0.98–2.76
Barley straw	2.0–4.5	12–15	39.9	5.3	43.8	1.3	–	–	9.8	16.0–16.4	0.78–1.75
Rice straw	3.5–6.0	20–30	41.8	4.6	36.6	0.7	0.08	0.34	15.9	15.1–15.5	1.3–2.2
Corn stalk	4.5–7.5	40–60	43.7	5.6	43.3	0.6	0.01	0.60	6.3	16.3–16.7	1.8–3.0
Corn cob	0.7–0.8	40–50	46.6	5.9	45.5	0.5	0.01	0.21	1.4	17.4–17.8	0.29–0.34
Sunflower stalk	3.0–5.0	15–20	45.0	5.1	38.0	1.1	0.11	0.16	10.5	16.1–16.5	1.2–1.9

TABLE III. – *Agricultural by-products (for wood residues): yield, composition, LHV and energy equivalent.*

Pruning residues	Yield	Moisture Content (MC)	C	H	O	N	S	Cl	Ash	Lower Heating Value (LHV)	Energy equivalent
	t/ha DM	%FM	$\%_{DM}$	$\%_{DM}$	$\%_{DM}$	$\%_{DM}$	$\%_{DM}$	$\%_{DM}$	$\%_{DM}$	MJ/kg DM	toe/ha
Grapevine	1.5–2.0	45–55	46.8	5.8	43.5	0.8	0.03	0.09	2.9	17.6–18.0	0.64–0.85
Olive trees	1.0–1.5	35–45	50.0	6.0	41.0	0.5	0.04	0.01	2.5	18.4–18.8	0.44–0.67
Fruit trees	0.5–1.7	35–45	48.0	5.5	43.0	0.5	0.03	0.01	3.0	18.1–18.5	0.22–0.74

wood biomass; in some species (rice) the ash has a low *melting point* due to the high silica content. In the European Union (EU), the straw harvesting is generally carried out by round/square balers in: i) June-July for cereals, ii) late October for rice, and iii) September-October for maize (depending on the maturity FAO class([6])). Table II shows the main energy characteristics of this type of biomass.

([6]) The classification adopted by FAO (Food and Agriculture Organization of the United Nations) is based on the maturation days of the crop variety, assigning a number from 100 (the earliest) to 800 (the latest) and increasing the maturation days of about 5–10 in each class.

1.3.1.2. Wood by-products. This category includes pruning residues resulting from cutting operations of fruit tree crops [2]. The most important species are: i) grapevine, ii) olive trees with different breeding forms and iii) fruit trees. These residues are released on the soil during pruning operations. Their wood tissues have a medium-low C/N ratio and the MC is approximately 45–55%. Ash content, mainly due to the presence of bark, is not negligible. The recovery period is after pruning operations, that usually takes place during the vegetative period, when the main product has been already harvested([7]). Table III shows the main energy characteristics of wood by-products coming from pruning operations.

1˙3.2. Herbaceous energy-crops.

1.3.2.1. For grains. The use of cereal grains for energy production, although is not widespread in the EU, is based on the *starch* contained in the grain endosperm, as: i) fuel for heat generation (combustion process) and ii) raw material for bioethanol production (fermentation process). In Italy, corn (*Zea mais*) is the most used crop for energy production [4] because, compared to winter cereals (wheat, barley), it has a higher LHV (15.1–16.2 MJ/kg DM) and a higher grain yield (7.5–10 t/ha DM). As a fuel, corn grain has "standard" characteristics (size, γ, MC, LHV) and it is easily available. This allows the easy organization of very short supply chains. Moreover, it enables the use of products previously contaminated by mycotoxins, whose use as food is prevented (or strongly restricted). The main limits, in addition to the ethical aspects associated with the use of food as an energy source, concern i) chlorine emissions and ii) ash production (1.2% DM) with a low melting point.

1.3.2.2. For lignocellulosic biomass. In southern Europe, growing attention is focused on some herbaceous crops with a long (poliannual) biological cycle and high biomass yields. These crops are: i) *miscanthus* (*Miscanthus spp.*), ii) *giant cane* (*Arundo donax*) and iii) *sorghum* (*Sorghum vulgare*). The yield of these crops (table IV) can respectively reach 29 t/ha · year DM, 27 t/ha · year DM and 14 t/ha · year DM. These biofuels generally have a high ash content with strong level of silica (low melting point) that reduces the efficiency of the conversion process and that can be a source of pollution.

1.3.2.3. For fermentable biomass. The use of biomass derived from crops dedicated to the Anaerobic Digestion (AD) process, such as organic matrices in addition (co-digestion) to livestock waste (slurry, manure), or even as the only component of the "ration" supplied to digesters, has progressively spread [7]. Also in this case, although the use of triticale, sorghum and ryegrass silage has recently been spreading, the reference species in the areas where the concentration of AD plants in co-digestion is greater (Po Valley area) is corn (harvested at waxy maturation). Table V shows the main characteristics of the most widespread crops cultivated in Italy for fermentable biomass production.

([7]) Usually, the pruning residues remain between two rows of trees until the beginning of spring, partially drying.

TABLE IV. – *Herbaceous energy crops (for lignocellulosic biomass): yield, composition, LHV and energy equivalent.*

Lignocellulosic biomass	Yield	Moisture Content (MC)	C	H	O	N	S	Cl	Ash	Lower Heating Value (LHV)	Energy equivalent
	t/ha DM	%FM	$\%_{DM}$	$\%_{DM}$	$\%_{DM}$	$\%_{DM}$	$\%_{DM}$	$\%_{DM}$	$\%_{DM}$	MJ/kg DM	toe/ha
Miscanthus spp.	15.5–29.0	15–20	41.3	7.4	48.3	0.1	0.10	0.00	2.8	17.4–17.8	6.5–12.2
Arundo donax	17.0–27.0	50–60	41.2	7.2	46.9	0.8	0.19	0.00	3.7	17.4–17.8	7.2–11.4
Sorghum vulgare	11.0–14.0	22–26	38.8	7.3	49.5	0.6	0.09	0.00	3.7	16.6–17.0	4.4–5.6

TABLE V. – *Herbaceous energy crops (fermentable biomass): yield, composition, biogas yield,* CH_4 *content and energy equivalent.*

Fermentable biomass	Yield	Dry matter (TS)	Volatile Solids (VS)	N tot$^{(\#)}$	Biogas yield	CH_4 content	Energy equivalent
	t/ha DM	%FM	%TS	kg/t DM	m^3_N/t VS	% vol.$^{(*)}$	toe/ha
Maize silage (FAO 700)	13.0–23.0	30–35	85–90	1.1–2.0	500–600	52–57	2.6–4.6
Sorgum silage	13.0–16.0	28–35	89–93	1.4–1.9	550–650	50–55	2.8–3.5
Sugar beet	1.2–1.7	21–25	90–95	2.4–2.8	450–550	55–60	0.24–0.34

$^{(\#)}$ Organic and inorganic.
$^{(*)}$ LHV of CH_4: 31.6 MJ/m^3_N.

1.3.2.4. For oil. The oil-energy supply chain is widespread all over the world and lots of autochthonous species are currently cultivated for this purpose: in tropical areas (Southeast Asia, India, China and Africa) the most important cultivated crops are: i) *oil palm* (*Elaies guineensis*), ii) *jatropa* (*Jatropha curcas*) and iii) *pongamia* (*Pongamia pinnata*) while in temperate areas (Central-South Europe, North America) the most important cultivated crops are: i) *rapeseed* (*Brassica napus*), ii) *sunflower* (*Heliantus annuus*), iii) *castor* (*Ricinus communis*) or iv) protein-oleaginous crops, such as *soybean* (*Glicine max*). Table VI shows the main energy characteristics of herbaceous oilseeds crops compared to those of other oilseeds species.

1.3.3. Tree species energy crops. Among the sources of cultivated wood biomass, *Short Rotation Coppice* (SRC) species have represented the ones that have attracted most interest in the last decade [8]. This form of wood arboriculture is interesting because

TABLE VI. – *Herbaceous energy crops (oil): seed yield, oil content, LHV and energy equivalent.*

Oil crops	Seeds yield t/ha FM	Oil content %FM	SVO lower Heating Value (LHV) MJ/kg DM	Energy equivalent toe/ha
Helianthus annuus (sunflower)	4.0–4.5	39–40	37.1–37.5	1.3–1.4
Brassica napus (rapeseed)	4.0–4.5	40–42	36.9–37.3	1.3–1.5
Glycine max (soybean)	4.0–5.0	17–19	36.6–37.0	0.57–0.71
Elaeis guineensis (palm oil)	10.0–11.0	34–38	36.4–36.8	2.8–3.1
Jatropha spp. (jatropa)	0.2–4.0	27–29	36.5–36.9	0.04–0.88
Seaweed	50.0–100.0	65–75	44.5–44.9	3.7–7.5

of the possibility to simplify the cultivation technique (mechanization of the harvesting-transport operations). This has progressively led to the cultivation of some tree species in short cutting turn, similarly to herbaceous crops [6]. In Italy, especially in the Po Valley, the most successful cultivated species is *poplar* (*Populus spp.*([8])) [9]. The aspect that characterizes plant of poplar is the cutting turn, which defines both the number of trees per unit of areas (ha) and the type of mechanization for harvesting and transport. Currently, there are poplars with the following cutting cycles:

- *two-year* cycle (Short Rotation Forestry, SRF). The wood chips (30–40 t/ha · year FM; MC = 50–55%; LHV = 17.3–18.7 MJ/kg DM) have a wood/bark ratio higher than the annual cutting cycle crops;
- *five-year* cycle (Medium Rotation Forestry, MRF). The wood chips characteristics are similar to those of the two-year cutting cycle (25–40 t/ha · year FM; MC = 50–55%; LHV = 18.0–19.4 MJ/kg DM) but, because in this case the ratio between wood and bark is higher (because of the lower ash content), the biofuel has a higher quality ("white chip").

The soil is occupied for 10–15 years, after which the crop has to be explanted (elimination of the radical apparatus), restoring the conditions suitable for the cultivation of another crop [6].

1·3.4. Forestry wood and wood derived products. Wood-based biomass comes from: i) *primary sources* (forestry wood, directly linked to the photosynthetic activity of wood trees) and ii) *secondary sources* (wood derived products, related to industrial processes and maintenance activities, such as fruit arboriculture, public greenery, etc.) [10, 11]. Wood tissues almost entirely consist of: i) C (49–51%), ii) O (41–45%) and iii) H (5–

([8]) Abbreviation of the plural of the term "species".

7%), with small quantities of iv) N (0.05–0.4%), v) S (0.01–0.05%) and vi) other mineral elements that form the ash (0.5–1.5%). Wood biomass can be used as: i) natural material and ii) treated material, after size and shape transformations. The types currently available on the market are: i) *logs*([9]) (pieces of wood, 0.3–1 m length), ii) *wood chips*([10]), iii) *pellets*([11]) and iv) *briquettes*([12]).

1.3.4.1. Forestry wood. The main forest management tools that define the use of forest biomass for energy purposes([13]) are the Forest Management Plans (FMPs). FMPs establish the organization and the sequence of the operations for wood biomass production, yarding and transport outside the forest for different final uses (energy and/or building material). Many commercial wood assortments are available, bringing the production of waste (cuttings residues) which can reach 20–25% of the total tree mass [8]. The cutting residues are generally abandoned into the forest (or close to the working areas), causing some management problems (mainly fires and parasitic attacks). If economically sustainable([14]), they can be recovered and transformed into *wood chips* (MC = 50–55% and $\gamma = 250\text{–}350\,\text{kg/m}^3$ FM, depending on the species and the chipping techniques) [9]. Table VII shows the most important energy characteristics of forestry wood.

Normally, the firewood is reduced in pieces (diameter: 50–300 mm; length: 1 m approximately), to facilitate the operations of stacking, measurements and estimation, handling and loading/unloading from the trucks [10]. Moreover, wood in pieces is usually seasoned (about 1 year) to reduce the MC from 55–60% to 25%([15]). Chipping, applied to branches and cutting residues, greatly reduces the volume (3–4 times) and, although this is an advantage for handling-transport-storage operations, the bulk density of wood chips is always lower (2–3 times, on average) than the bulk density of round wood (at the same MC). As a result, where the operating conditions are favorable and the transport distances are rather high, it is more convenient to carry out forestry yards able both to cut the small material and to arrange the medium-large diameter biomass in round wood logs (fig. 2). The chips size depends on the type of chipper. Generally, the length is

([9]) UNI EN ISO 17225-5:2014-Part 5: Graded firewood.
([10]) UNI EN ISO 17225-4:2014-Part 4: Graded wood chips.
([11]) UNI EN ISO 17225-2:2014-Part 2: Graded wood pellets.
([12]) UNI EN ISO 17225-3:2014-Part 3: Graded wood briquettes.
([13]) The dominant forest government in hilly and medium mountainous areas is *coppice* for the production of firewood (trunks or logs).
([14]) Because for coppice the production process can be very costly (especially in the case of energy supply chain), it is important to chip all the available wood. For tall trees, which always give assortments with high commercial value, preliminary operations on young plants (whose purpose is to increase the stability of the forest to maximize the final yield) are carried out. These operations cause the production of large quantities of branches, other residues or small trees with a very low commercial value. Even in this case, the integral chipping is the only sustainable solution to obtain a lignocellulosic biofuel.
([15]) In the case of longer periods of seasoning (2 years), in our climates, MC can be reduced down to 12–15% (atmospheric dry at hygroscopic equilibrium with the air).

TABLE VII. – *Forest wood: yield, composition, LHV and energy equivalent.*

Tree species cutting period (years)	Yield	Moisture Content (MC)	C	H	O	N	S	Cl	Ash	Lower Heating Value (LHV)	Energy equivalent
	t/ha DM	%FM	$\%_{DM}$	$\%_{DM}$	$\%_{DM}$	$\%_{DM}$	$\%_{DM}$	$\%_{DM}$	$\%_{DM}$	MJ/kg DM	toe/ha
Abies alba (90–110)	190–270	40–50	49.0	6.0	44.3	0.5	0.01	0.01	0.3	18.5–18.9	85.0–120.9
Picea abies (110–140)	160–240	40–50	50.5	6.3	41.9	0.6	0.01	0.03	0.7	18.3–18.7	70.5–105.8
Pinus sylvestris (60–80)	120–145	40–50	49.6	6.5	42.0	0.8	0.01	0.01	1.1	21.7–22.1	62.6–75.7
Populs spp. (9–12)	130–170	45–55	48.5	5.9	43.7	0.5	0.01	0.01	1.4	18.0–18.4	56.5–73.9
Fagus spp. (80–100)	190–220	40–50	48.3	6.0	44.3	0.6	0.01	–	0.8	18.4–18.8	84.5–97.9
Quercus spp. (70–100)	180–220	45–55	49.5	5.4	43.1	0.4	0.01	0.04	1.6	18.1–18.5	78.8–96.3
Carpinus spp. (30–40)	55–105	45–55	47.1	6.1	44.1	0.9	0.01	0.04	1.8	17.3–17.7	23.0–43.9
Robinia pseudoacacia (30–75)	30–75	45–55	49.1	5.7	42.0	0.6	0.01	0.04	2.5	18.3–18.7	13.3–33.1
Eucalyptus spp. (60–70)	130–190	40–50	49.0	5.9	44.0	0.3	0.01	0.13	0.7	18.0–18.4	56.6–82.7

Fig. 2. – Stacked round wood logs (left) and wood chips (right) (source: `pixabay.com`).

between 15 and 50 mm, the width is about 50% of the length and the thickness is about 10–20% of the length [9].

1.3.4.2. **Wood derived products.** Pelletized lignocellulosic biofuels (*pellets*) derive from an animal feed industrial process of extrusion of fine-size materials, at high pressure (> 100 bar) through horizontal or annular grooves with special section holes. According to a current definition([16]), the pellet is a "*compressed biofuel that generally has a cylindrical form, obtained compressing finely chopped biomass with or without the use of pressing additives*". Because the composition of the raw materials can be very different([17]), the energy characteristics of the final extruded biofuels can also be variable. Pellet consists of small cylinders (diameter = 6–10 mm, length = 10–50 mm), with a high "energy density" (ρ_{MC} = 1150–1400 kg/m^3, γ = 650–700 kg/m^3, MC = 6–12%). The ash content is very low (0.3–1.0%).

Another densified fuel consists in *briquettes*, also obtained —in different shapes (cylindrical, prismatic)— by the compression of a previously dried and powdered biomass (MC = 10–15%). Briquettes production takes place by mean of alternative (piston) presses and the most commonly used row materials are: i) *sawdust*, ii) *straw*, iii) *rice husk*, iv) *nuts shells*, v) *coffee shells*, vi) *bark-free chips*, with or without additives.

1.3.4.3. **Wood residues.** Wood residues (especially if they have an *industrial origin*) and, more generally, the so-called *end-life wood*, are very heterogeneous and include packaging and containers, building materials, pallets, beams and boards for demolition and refurbishment, sawdust waste, wood and furniture industry residues, etc. Due to the long life cycle, the end-life wood residues are perfectly seasoned (MC < 20%) and, consequently, their quality as a biofuel is excellent.

1·4. *Livestock waste.* – *Livestock waste* includes i) *excreta* (feces and urine) and ii) *animal waste* (mixture of excreta, litter, water, feed residues and other animal residues). The composition of livestock waste depends on: i) animal species, ii) breeding methods, iii) feed characteristics and iv) waste management system (table VIII). In general, livestock waste can be managed as: i) *slurry* (mainly made by excreta, food residues and water) with fluid or semi-fluid consistency (DM < 15%) and moved by *pumps*, ii) *manure* (mainly made by excreta, food residues, water and litter) with semi-solid or solid consistency (DM ≥ 20%), moved by *shovels*. The energy potential of livestock waste depends on its composition, *Volatile Solids* content (VS; %DM) in particular. The waste can be used in biochemical processes to produce gaseous biofuels, such as biogas, biomethane (bio-CH_4) and bio-hydrogen (bio-H_2) with high calorific values. Slurry has the most suitable chemical composition for the AD (table IX), for which the key factors are: i) *composition of the biomass*, correlated to the degradation rate, ii) *degradability*, which has decreasing values for protein, fat, cellulose and lignin([18]) respectively, iii) toxicity

([16]) PROBIO Project, 2004.
([17]) D.L. n.152 of April 3rd, 2006.
([18]) Manure is degraded more slowly than slurry because of its higher cellulose content.

TABLE VIII. – *Breeding species and excreta daily production.*

Animal category	Weight kg	Excreta production kg/day
Calf (0–3 months)	70	4–5
Calf (3–6 months)	140	8–9
Young beefs (6–12 months)	230	20–22
Young cow (12–24 months)	390	32–36
Beefs (12 months)	400	20–25
Milk cow (12–72 months)	650	50–55
Young swine	6–30	2–3
Growing pigs	30–80	$4.0^{(\#)}$; $7.0^{(\S)}$; $14.0^{(*)}$
Fat pigs	80–160	10–13
Sows	–	8–9
Hens (10–12 months cycle on land)	1.8–2.0	$0.15\,m^3/year \cdot t$
Broilers (2 months cycle on land)	1.0–1.2	$2.0\,m^3/year \cdot t$
Turkey (4–5 months cycle on land)	7.0–7.5	$0.9\,m^3/year \cdot t$

(#) Dry feed 2.5:1.
(§) Wet feed 4.0:1.
(*) Whey feed.

TABLE IX. – *Animal waste (slurry and manure): composition, biogas yield,* CH_4 *content and energy equivalent.*

Biomass	Dry matter (TS) %FM	Volatile Solids (VS) %TS	N tot(#) kg/t TS	Biogas yield m^3_N/t VS	CH_4 content % vol.(*)	Energy equivalent toe/t FM
SLURRIES						
Milk cows	10–16	75–85	3.0–4.8	300–450	55–60	0.013–0.021
Beefs	7–10	75–85	3.8–5.3	300–450	55–60	0.009–0.013
Calves	0.6–2.9	60–75	7.4–17.7	300–450	55–60	0.001–0.003
Pigs	1.5–6.0	65–80	4.0–13.3	450–550	60–65	0.003–0.010
Hens	19–25	70–75	7.0–14.5	300–500	50–55	0.022–0.029
MANURES						
Bovine	11–25	65–85	1.2–2.8	200–300	55–60	0.009–0.020
Pig	20–28	75–90	1.8–2.0	450–550	60–65	0.039–0.055
Ovine	22–40	70–75	1.9–3.5	240–500	60–65	0.028–0.051
Broiler	60–80	75–85	4.3–6.7	400–500	50–55	0.086–0.114
Chicken (dryed)	40–80	60–70	3.4–6.4	450–550	50–55	0.052–0.103

(#) Organic and inorganic.
(*) LHV of CH_4: $31.6\,MJ/m^3_N$.

TABLE X. – *Agro-industrial residues (lignocellulosic biomass): composition, LHV and energy equivalent.*

Biomass	Moisture Content (MC) %FM	C $\%_{DM}$	H $\%_{DM}$	O $\%_{DM}$	N $\%_{DM}$	S $\%_{DM}$	Cl $\%_{DM}$	Ash $\%_{DM}$	Lower Heating Value (LHV) MJ/kg DM	Energy equivalent toe/t DM
Rice husk	10–15	41.0	4.3	35.9	0.4	0.02	0.12	18.3	15.1–15.5	0.361–0.370
Almond shells	10–20	44.3	6.0	43.0	1.2	0.02	–	5.6	18.0–18.4	0.430–0.439
Walnut shells	10–20	50.0	5.7	43.4	0.2	0.01	0.03	0.7	18.8–19.2	0.449–0.459
Peach stones	15–20	53.0	5.9	39.1	0.3	0.05	–	1.6	19.4–19.8	0.463–0.473
Plum stones	15–20	49.7	5.9	43.6	0.3	–	–	0.5	21.9–22.3	0.523–0.533
Olive stones	15–20	48.8	6.2	43.5	0.4	0.02	–	1.1	20.1–20.5	0.480–0.490
Olive oil virgin residues$^{(\#)}$	20–30	–	–	–	–	–	–	–	15.3–15.7	0.365–0.375
Olive oil virgin residues$^{(\S)}$	58–72	–	–	–	–	–	–	–	15.3–15.7	0.365–0.375
Olive oil virgin residues$^{(*)}$	48–52	–	–	–	–	–	–	–	15.3–15.7	0.365–0.375
Olive oil exhaust residues$^{(\circ)}$	10–15	43.7	5.3	37.8	–	0.64	–	12.5	14.3–14.7	0.342–0.351
Grape residues (marc)	55–65	–	–	–	–	–	–	–	6.8–7.2	0.162–0.172

$^{(\#)}$ Discontinuous oil extraction process.
$^{(\S)}$ Continuous oil extraction process.
$^{(*)}$ 3-phase oil extraction system.
$^{(\circ)}$ Oil extraction by solvent.

of some elements if they are present in high quantities (especially Na, K, Ca, Mg) and iv) presence of heavy metals (Cu, Cr, Ni, Zn, Pb) which can cause damage to bacterial metabolism. *Pig slurries*, due to the different types of animal breeding and of excreta removal system, have a rather variable DM. The use of "*fresh*" slurry, avoiding any intermediate storage (tanks or lagoons) promotes the reduction of CH_4 emissions and odors [4].

1·5. *Industrial waste.* – Large quantities of residues (or waste) obtained from food processing can be conveniently transformed into energy. This residual biomass is extremely heterogeneous (table X) but, compared to agricultural by-products, is much less spread on territory.

1·5.1. Olive oil extraction residues, rice husk and grape skin residues. *Virgin olive oil extraction residue* (pomace) is the biomass remaining after the extraction by mechanical systems, while the *exhaust olive oil extraction residue* derives from the previous one after extraction by solvents (mainly hexane). MC and composition of the *virgin residues* depend on the extraction process([19]), which may (or not) include the addition of water to the olives. Before combustion, the residues from continuous systems must be dried; the drying process is expensive but leads to a significant increase in NHV. The process for production of the *exhaust residues* starts from the virgin residues drying phase (up to MC = 8–12%), followed from the extraction by hexane. The obtained biomass (equal to 50–55% of the virgin residue mass) is granular, with excellent energy characteristics (fuel for domestic or industrial combustion plants). *Rice husk* is the residue of rice processing and it includes the external grain coatings (glumes). It represents about 20% of the whole rice grain weight. Rice husk chemical composition and quantities depend on: i) rice variety, ii) maturation degree, iii) harvesting techniques, iv) drying process. Rice husk utilization as a biofuel has some problems because of: i) high hardness, ii) strong abrasiveness, iii) high silica content, iv) low γ (= 130–140 kg/m^3) and v) high ash content (15–20%). The most important residues from oenological industry are the *grape skin residues*. They are produced by crushing and pressing the grapes together with the solid parts of the grape bunch, in quantities ranging from 15 to 30% of the grape. Depending on the degree of pressing, the grape skin residues contain 2–5% of alcohol and 2–6% of tartaric acid salts. The high MC makes them suitable for biochemical conversions, but a preventive drying process promotes their use also as biofuel.

1·5.2. Wet processing waste and urban waste. Wet processing waste, conveniently used in AD plants, mainly refers to *agri-food industry* residues (skins and waste from the processing of fruit, whey, threshing operations, vegetation waters, etc.) [8]. Because of a VS content higher than animal waste (table XI), these organic materials increase the biogas yield. In Germany and Italy there are hundreds of AD plants fed with wet processing waste, mixed with other biomass and/or animal waste (co-digestion). In wet processing waste *animal residues* (blood, fat, whey, gut, stomach contents, etc.), mainly from meat industry, are also included. These residues have a high VS and fats content (which increase the methanigenic power). However, their utilization as biomass-for-energy processes must comply with strict hygienic standards, especially in relation to the high odorous power. Therefore, these residues must be carefully treated in AD plants, adequately conceived (pre-treatment section). Lastly, there are two other types of fermentable biomass, remarkable for quantity and, above all, because of their capillary collection network already present in the territory: i) OFMSW and ii) *sewage sludge*. The OFMSW (table XII) is mainly composed of food residues (kitchen waste) that have: i) high MC (70–85%) and ii) high γ (500–700 kg/m^3). In Italy, despite the

([19]) In Italy, 46% of the olive oil production is carried out by discontinuous methods (hydraulic press), 45% by continuous or 3-step methods (centrifugal separation), with water addition. The remaining 9% is carried out with mixed methods.

TABLE XI. – *Agro-industrial residues (fermentable biomass): composition, biogas yield,* CH_4 *content and energy equivalent.*

Fermentable biomass	Dry matter (TS) %FM	Volatile Solids (VS) %TS	N tot[(#)] kg/t TS	Biogas yield m_N^3/t VS	CH_4 content % vol.[(*)]	Energy equivalent toe/t FM
Fruit juices residues	25–45	90–95	1.1–1.2	500–600	55–60	0.055–0.100
Fruits and vegetable fresh residues	5–20	80–90	3.0–5.0	350–500	50–60	0.008–0.030
Molasses	80–90	85–90	1.3–1.7	300–450	50–55	0.104–0.117
Potatoes residues	6–7	85–95	5.0–13.0	500–600	50–53	0.011–0.013
Tomatoes peels	27–35	96–97	3.1–3.2	300–400	30–55	0.029–0.038
Cereals distillation residues	6–8	83–88	6.0–10.0	400–500	50–55	0.009–0.012
Beer residues	20–25	70–80	4.0–5.0	300–400	50–55	0.021–0.026
Whey	4–7	80–92	0.7–1.0	330–400	50–55	0.005–0.009
Olive oil residues (liquid)	3.5–4	70–75	4.0–5.0	450–550	50–55	0.005–0.006
Grape residues (liquid)	10–14	85–95	nd	650–750	50–54	0.025–0.035

(#) Organic and inorganic.
(*) LHV of CH_4: 31.6 MJ/m_N^3.

TABLE XII. – *Waste materials (wet biomass): composition, biogas yield,* CH_4 *content and energy equivalent.*

Biomass	Dry matter (TS) %FM	Volatile Solids (VS) %TS	N tot[(#)] kg/t TS	Biogas yield m_N^3/t VS	CH_4 content % vol.[(*)]	Energy equivalent toe/t FM
Food residues (catering industry)	9–37	80–95	0.6–5.0	650–800	50–60	0.024–0.098
Pigs stomach content (slaughterhouses)	12–15	75–86	2.5–2.7	650–800	60–65	0.033–0.041
Bovine rumen content (slaughterhouses)	18–20	90–94	2.0–3.0	650–800	60–65	0.057–0.063
Blood (slaughterhouses)	6–20	93–95	14.7–17.0	600–850	60–65	0.019–0.064
Incubating process waste	44–48	41–45	5.0–5.5	600–800	60–65	0.063–0.068
Broken eggs	21–25	95–97	7.8–8.5	600–850	60–65	0.069–0.082
Bread processing waste	88–92	90–94	nd	740–840	50–60	0.266–0.278
Bread (out-of-date)	78–82	96–98	nd	710–810	53–57	0.239–0.282
Biscuits (out-of-date)	90–94	98–99	nd	750–850	50–54	0.279–0.291
Flours (out-of-date)	86–90	96–98	nd	705–805	51–55	0.253–0.264
Cellulose pulp	12–14	89–91	5–13	450–550	50–55	0.021–0.025
OFMSW	40–75	50–70	0.5–2.7	300–450	50–60	0.037–0.070

(#) Organic and inorganic.
(*) LHV of CH_4: 31.6 MJ/m_N^3.

strong variability among different urban areas, OFMSW represents, on average, 25–35% of the total mass of municipal solid waste([20]). The average composition, if the waste derives from dry-wet differentiated domestic collection, does not differ greatly from the collectable organic residues from selected users (such as fruit and vegetables wholesalers, fish markets, grocery stores, and refreshments). The presence of other foreign materials (plastic, glass) must be less than 5% of the total mass.

2. – Biomass-to-energy processes

2·1. *Generality.* – In the *thermochemical conversions*, where chemical bonds are broken through *heat*, lignocellulosic biomass with the following properties can be used [1,3]:

- low MC (15–40%);
- high C/N (> 30).

In the *combustion* process, lignocellulosic biomass is transformed directly into energy, while in others (*gasification, pyrolysis*) biomass-to-energy conversions occur through the intermediate generation of liquid or gaseous compounds [11]. These "energy carriers" are generally used as fuels for *Internal Combustion (IC) engines* (mainly reciprocating but also rotary), generating Mechanical Energy (EM) and, then, Electric Energy (EE) [4].

On the contrary, for *biochemical conversions* (where chemical bonds breaking occurs through the action of bacteria) biomass with the following properties can be used [1,3,7]:

- high MC (70–90%);
- low C/N (< 30).

Even in this case, biomass-to-energy conversions (usually through *fermentation*) occur through the production of liquid or gaseous intermediates, subsequently used as fuels for *IC engines*.

Lastly, there are other processes (mechanical or chemical *extraction*) through which vegetable fats (*oils*), finally used as pure (pure vegetable oil) or after a chemical modification into methyl-esters, are removed from biomass (*seeds*, but also *algae* in the near future) and used as biofuels for vehicles. Figure 3 shows the general scheme of the biomass-to-energy conversion processes.

2·2. *Thermochemical processes.* – Depending on the *amount of air* used to support the C oxidation, these processes are classified as follows [6,11,12]:

- *combustion*, performed with a mass of air greater than the stoichiometric mass required for complete oxidation;

([20]) According to ISPRA (*Urban Waste Report 2009*), the Italian production of urban solid waste in 2008 was about 32.5 Mt (541 kg/inhabitant per year, with 541, 619 and 496 kg/inhabitant per year, respectively, in North, Center and South-Islands Italian Regions). The landfill is still the most widespread processing mode, but the mass of collected urban solid waste in differentiated way is increasing, representing about 30.6% of total urban solid waste.

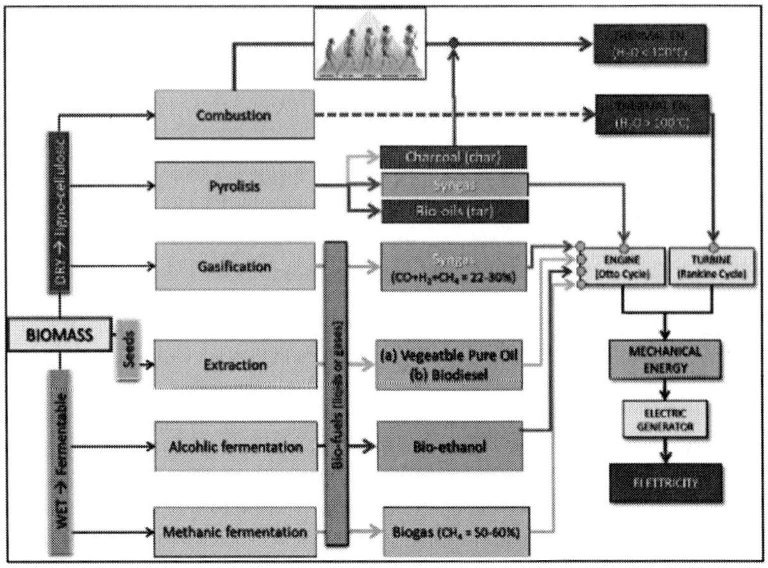

Fig. 3. – Biomass-to-energy conversion processes: general scheme (source: [4]).

– *gasification* performed with less air than the stoichiometric need;

– *pyrolysis*, performed without air.

2·2.1. Combustion. Combustion is a long sequence of chemical reactions according to which an organic substance (C) combines with an *oxidizing agent* (O_2 of the air), carrying out an exothermic redox reaction and producing heat (ET), CO_2 and H_2O. The process presupposes the simultaneous presence (fire triangle), in correct proportions, of fuel, oxidizing agent, as well as starting heat([21]). In the case of *lignocellulosic biomass* (that has a complex chemical structure if compared to some fossil gaseous hydrocarbons, such as CH_4) the process consists of the following *three phases* that, in a combustion biomass bed, *take place at the same time*:

– *drying*: the lignocellulosic mass receives heat from the surrounding environment and progressively increases its temperature, evaporating the contained free water (until 105 °C is reached). The dry biomass continues to heat, reaching very quickly the gasification-pyrolysis threshold (about 600 °C). Because the evaporation absorbs a fraction of the energy that is released by the process, to limit the process slowdown, biomass should not have MC $>$ 60–65%;

([21]) In the absence of even one of these factors, the process does not take place and, if the fuel/oxidizing agent ratio is incorrect, incomplete combustion can occur, causing the formation of still oxidizing carbonaceous compounds.

- *gasification-pyrolisis*: the heat propagates from the external to the internal layers of the biofuel particle, causing —starting from a biomass temperature of 200 °C and in the absence of oxygen— the formation of([22]): i) a *volatile* fraction (*gases* and *tar*) that rapidly leaves the ii) remaining *solid* fraction (*char*) that consists of fixed C and ash;

- *oxidation*: the gasification-pyrolisis products, coming in contact with the oxygen of the air, cause the *combustion*. A highly exothermic reaction occurs both above the char bed for volatile fraction —very intensively (*flame* formation) with emission of light radiation in the visible spectrum— and also inside the char bed for the fixed C —but much more slowly (*ember* without flame) and with light emission mainly in the infrared spectrum.

To fully oxidize the char, *primary air* is blown into the lignocellulosic biofuel bed, while *secondary air* is blown *over* the bed, to oxidize gaseous and tar compounds. If an incomplete combustion takes place, a reduction of the *energy conversion efficiency* can occur, with a significant *increase of the emission of organic carbonaceous compounds to the chimney*. The "quality" of the combustion is, therefore, strongly linked to three key-factors connected with the conversion device: i) gases residence *time*, ii) combustion room *temperature* and iii) *turbulence*. The *excess air ratio* (λ) is the ratio between the real air mass used in the combustion process and the stoichiometric air mass. For wood logs, this technical parameter ranges from 1.25 to 1.40. Generally, O_2 into dry fumes should represent 5–8% (in volume), while —at these levels of O_2— the CO_2 concentration should reach 13–16%. The combustion chamber temperature should be in the range of 850–1200 °C([23]). Fumes temperatures should remain below 150–170 °C. Moreover, the greater the uniformity of the biomass and the DM, the better the combustion results [12]. Figure 4 shows a schematic representation of the combustion process.

The type of the devices used for biomass combustion processes (*boilers* or *thermal generators*) differ in terms of i) fuel and its characteristics (size, shape and granulometric distribution), ii) thermal power (domestic or industrial purposes) and final energy use (Thermal Energy, Electric Energy), iii) complexities and technological solutions adopted (manual or automatic feeding and ash removing systems). The main characteristics of the lignocellulosic combustion devices are shown in table XIII.

2·2.2. Gasification. Gasification means incomplete oxidation of a substance in a high-temperature environment (900–1000 °C) to produce a gas (*producer gas*). As oxidant agent: i) air, ii) steam or iii) pure oxygen can be used. In the most common air gasifiers, the producer gas has a LHV = 3.9–4.3 MJ/m_N^3, while, in the less widely used oxygen

([22]) The first component that is degraded by heat is hemicellulose, followed by cellulose, and finally, by lignin.
([23]) The technical parameter is linked to the NO_x production.

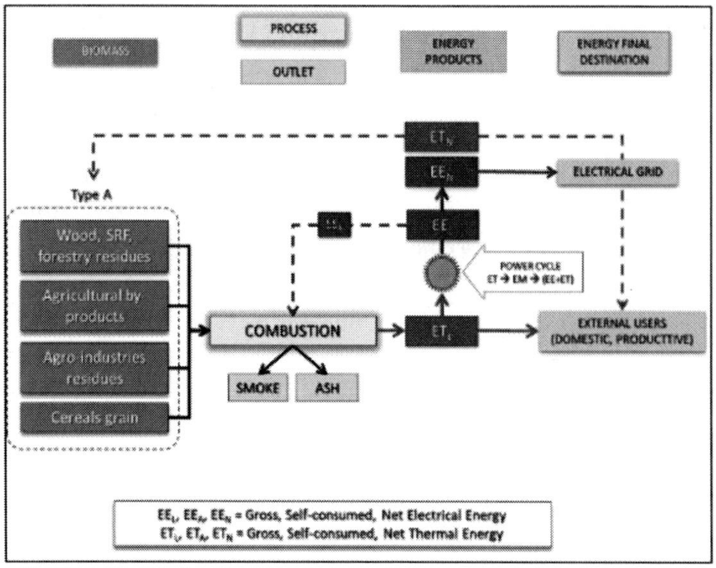

Fig. 4. – Combustion process scheme. The ET_L can be sold to neighboring external utilities through a distribution network (district heating). The EE generation (EE_L) is carried out only in high power plants (dedicated or cogenerated power cycle units) with the sale of residual heat to external utilities. A small fraction of the gross production is used for self-consumption (EE_A) while the rest (EE_N) is sold to the grid at feed-in tariff (source: [4]).

gasifier(24), the gas has a LHV = 13.0–14.0 MJ/m^3_N. The *gas cleaning phase*, that follows the gasification phase, is always particularly complex. For the gasification process, the biomass physical-chemical characteristics must be as constant as possible, especially in terms of MC and size (often related to an increase of management costs due to biomass preparation). The producer gas can be used to feed: i) gas-burners for *heat* production or, more conveniently and diffusely, ii) *IC engines* to generate EM and, thus, EE that is sold to the grid at feed-in tariff (biomass-to-electricity efficiency: 18–21%). This last solution can be also carried out by *co-generated IC engines* (*CHP, Combined Heat and Power* unit) for the simultaneous generation of electricity and heat, to use within the same process (*e.g.*, for biomass drying) and/or for external final users (fig. 5).

As an alternative to energy generation, the gasification products can be used (only for medium-high power plants) as raw materials for *liquid biofuels synthesis*. For this purpose, the producer gas is transformed into a *syngas* and then into liquid biofuels (*Biomass-To-Liquid, BTL*)(25). Depending on the applications, the requested character-

(24) For gaseous fuels, the LHV refers to the volume unit under *normal* conditions (MJ/m^3_N): anhydrous gas at $T = 0\,°C$ and atmospheric pressure $p = 760\,mm_{Hg}$ (1.013 bar). Commercially, however, reference is often made to $1\,m^3$ *standard* (m^3_S at $T = 15\,°C$ and $p = 760\,mm_{Hg}$).
(25) Fischer-Tropsch synthesis.

TABLE XIII. – *Main characteristics of the devices for lignocellulosic biomass combustion.*

Type of device	Biomass load	Power (kW)	Type of energy and destination	Biomass used	Energy efficiency (%)
Boiler with grate	Manual(#)	15–35	ET, Domestic	Wood logs	65–75
	Manual(*)	15–35	ET, Domestic	Wood logs	85–90
	Mechanical	15–35 > 100	ET, Domestic ET+EE, Products processing	Pellet	80–85
	Mechanical	15–35	ET, Domestic	Corn grain	70–75
	Mechanical	15–35	ET, Domestic	Straw residues Agricultural by-products	70–75
	Mechanical(o)	> 500	ET, District Heating ET+EE, Products processing	Wood chip Agricultural by-products	70-75
Blowup boiler	Mechanical	15–35 > 100	ET, Domestic ET+EE, Products processing	Food processing residues Processing waste	70–80

(#) Upwards flame combustion.
(*) Reverse combustion.
(o) Inclined grate.

istics of the gas mixture are variable. Because of this, the reactors and the subsequent *gas cleaning* are based on different technological solutions. Considering the high amount of impurities transported by the gas flow (especially tars and particulate), all biomass gasification plants are equipped with a *gas cleaning* and a *gas cooling* section. For low power plants ($< 300\,\mathrm{kW}$), *fixed bed gasifiers* are used (fig. 6) as well as coarse lignocellulosic biomass, while for high power plants ($> 1\,\mathrm{MW}$) *fluidized bed gasifiers* are used as well as fine or powdered lignocellulosic biomass.

2˙2.3. Pyrolysis. When biomass heating is performed in the *absence of* O_2, only chemical bonds breaking is recorded and, consequently, the formation of simple organic molecules (that at room temperature can be solid, liquid or gaseous, in relation to

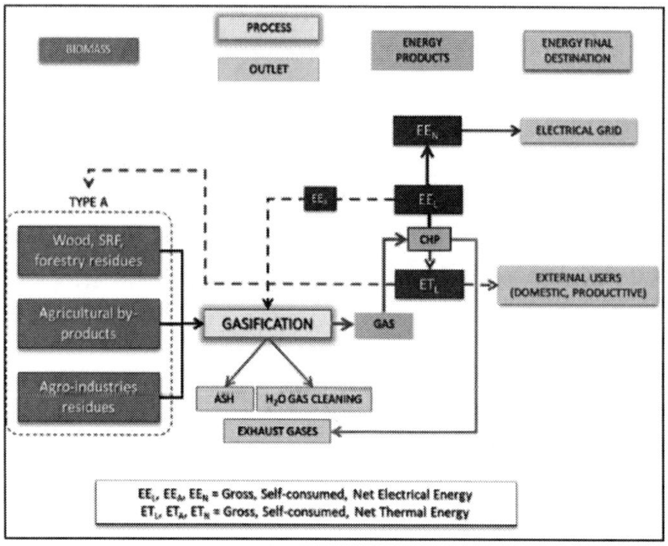

Fig. 5. – Gasification process scheme. The producer gas supplies IC engines, generating EE_L. A fraction of this EE_L is reused (EE_A) while the remaining (EE_N) is sold to the grid (feed-in tariff). Operating with CHP engines, the ET_L can be used for biomass drying or for neighboring utilities through a heating distribution grid (source: [4]).

their molecular weight) takes place. The ET required for this pyrolytic splitting is then provided —at temperatures between 400 and 800 °C: i) indirectly, through the *pyrolysis reactor wall* or ii) directly, by the recirculation of a heating agent in the bed (fig. 7). Pyrolysis products can be divided into the following:

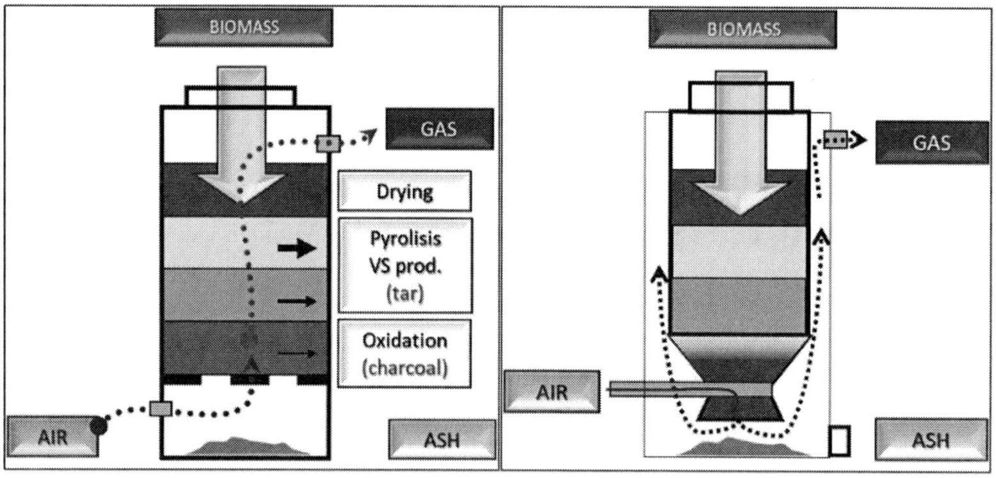

Fig. 6. – Fixed bed gasifiers: updraft type (left) and downdraft type (right) (source: [4]).

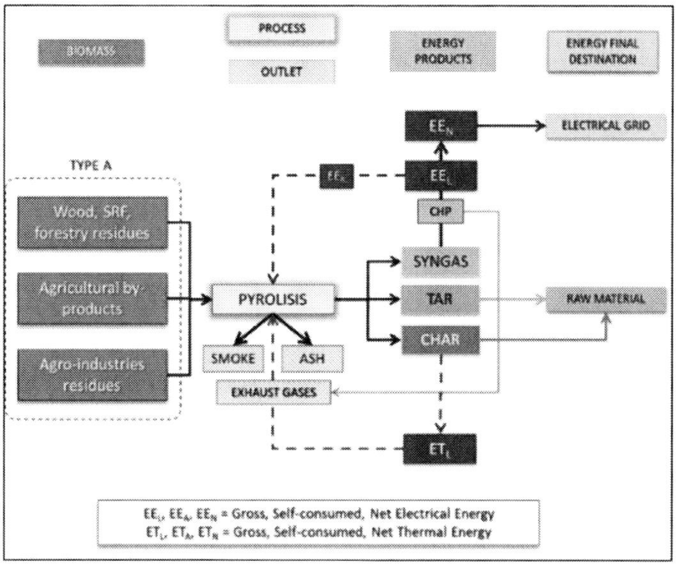

Fig. 7. – Pyrolysis process scheme. The products have different destinations: i) syngas, which is generally used to generate EE_L powering an IC engine, ii) tar (bio-oil) and iii) char which may be used as raw materials in subsequent transformation activities. Char is frequently reused as a fuel to produce the process heat (source: [4]).

– non-condensable gases (*syngas*), mainly composed of CO, H_2O, H_2, CH_4 and other light hydrocarbons. The syngas can be burnt to provide *heat* to the pyrolysis reactor or —more conveniently— to feed a *turbogas system* or a *reciprocating IC engine*, generating EM and, thus, EE;

– condensable volatile compounds (*tar*) of complex and variable composition, that consist of a mixture of heavy hydrocarbons which —when condensed— forms a black liquid fuel (*bio-oil*);

– *solid compounds* (*char*), which consist of carbonaceous residues similar to charcoal and ash. Char can be used as a fuel (also in the same pyrolytic reactor) for active carbon production or as intermediate in the chemical industry.

The way through which the pyrolysis is carried out determines different yields and proportions in the products. Generally, as time decreases and the process temperature increases, the amount of tar and syngas increases at the expense of char:

– *conventional pyrolysis*: $T < 600\,°\mathrm{C}$, permanence time $t_r = 300$–$1800\,\mathrm{s}$ and medium heat rate; the three fractions in the same proportion can be obtained;

– *fast pyrolysis*: $500 < T < 650\,°\mathrm{C}$, $t_r = 0.5$–$5.0\,\mathrm{s}$ and high heat rate; it promotes the production of the liquid fraction (tar) (70–80% of the biomass weight);

– *flash pyrolysis*: $T > 700\,^{\circ}\mathrm{C}$, $t_r < 1\,\mathrm{s}$ and very high heat rate ($1000\,^{\circ}\mathrm{C/s}$); as the previous one, it promotes the production of tar.

A well-known process of *slow pyrolysis*, carried out worldwide since ancient times (especially in the developing countries), is *carbonization* (charcoal production) that lasts for a few days and it is usually carried out outdoors with the production of stacked wood covered with soil. T is low (300–500 °C) and, thus, the final product is almost completely represented by char, because the other components are released in the atmosphere.

2˙3. *Biochemical processes and oil extraction.* – In the biochemical transformation processes (fermentations, under anaerobic conditions), the *organic molecules of the biomass are used as a food substrate by different bacteria, resulting in liquids or gaseous biofuels production* [8, 13]. These biofuels can supply heat generators to generate ET or, more widely, EM by IC engines [2]. Depending on the fermentation type (*i.e.*, on bacteria and substrate), it is possible to obtain [7]:

– *biogas*, a gaseous mixture (CH_4 based), obtained by methanic fermentation;

– *bioethanol*, an alcohol (CH_3CH_2OH), obtained by alcoholic fermentation.

A third important supply chain for biofuels production is the *extraction of oil contained in seeds of some oleaginous species* and its use as biofuel for IC engines in the form of: i) *pure vegetable oil* (Straight Vegetable Oil, SVO) or, more widely, ii) after the transformation into methyl esters (*biodiesel*) [13]. Taking into account the energy destination, biogas is mainly used for EE generation by fixed point engines, while bioethanol and biodiesel are used for EM production by mobile point engines for vehicles traction. In these big markets, the distribution and the marketing of biodiesel and bioethanol is managed by the same operators that control the fossil fuels supply and distribution. Due to the fact that the raw materials used in the "*first-generation biofuels production*" are *agricultural products* (such as cereal grains, sugar cane, sugar beet, forages, etc.) which conflict with their use in the food/feed supply chains([26]), *serious problems in terms of resources management can occur.*

2˙3.1. Methanic fermentation. Methanic fermentation (or *AD*) is a complex biological process through which the organic substance is degraded in the absence of O_2, producing a gaseous mixture with energy characteristics that are similar to natural gas. This degradation process is carried out through different stages by a *bacterial consortium*. The biogas is (in volumetric terms) mainly composed of i) CO_2 and ii) CH_4. Some gases that take part in the biogas mixture (*nitrogen, oxygen*) do not affect its energy characteristics; others, on the contrary, are very powerful (*hydrogen*), while others are

([26]) This ethical aspect, moreover, also concerns the biogas production, which has become —especially in Italy— more and more widespread thanks to the use of feed crops (corn and other cereals silages with very high biogas yields), instead of other by-products or livestock waste.

https://doi.org/10.1051/epjconf/201818900006

TABLE XIV. – *Average biogas composition.*

Gas		Content (% vol. dry gas)
Methane	CH_4	55–65
Carbon dioxide	CO_2	35–40
Nitrogen	N_2	1–2
Hydrogen	H_2	0–1
Hydrogen sulphide	H_2S	0.02–0.2
Oxygen	O_2	traces
Ammonia	$NH_4{}^+$	traces
Mercaptans	–	traces

undesirable (*ammonia and hydrogen sulphide*) because they are harmful to the process and to the environment as well (table XIV). The amount of CH_4 contained in the biogas mixture depends on the chemical characteristics of the substrate and on the conditions under which AD takes place; it generally ranges *from 50 to 80% of the total volume of the dry biogas mixture*. Consequently, the biogas LHV ranges *from 14.4 to 25.2* MJ/m_N^3 ([27]). Thanks to its high LHV, biogas —suitably treated— can easily feed *CHP units* to produce both heat (ET, partially reused for digester heating) and EE (sold to the grid at feed-in tariff). In addition to biogas, a co-product —the *digestate*— is obtained from AD; the composition and the characteristics of this co-product depend on the substrate type. In general, the digestate is rich in N (organic and mineral form) and it can be efficiently used as *agronomic fertilizer* (fig. 8).

The anaerobic bacteria that degrade the substrate usually have low growth and reaction rates; consequently, to optimize the process inside the AD digester, it is necessary *to generate* and *maintain* the *optimum environmental conditions* (T, pH, concentration of metabolites, etc.) for the *bacterial metabolism*. The whole degradation process is very complex and it is carried out by different *groups of bacteria* that implement a series of reactions, partly overlapped.

Simplifying the reactions sequence, the first two steps (that can be grouped together into a single *acidogenic phase*) act as a "preparation" for the final step (*methanigenic phase*), during which the biogas production takes place [2]:

- *hydrolysis*: hydrolytic bacteria cause the degradation of the complex organic compounds (*proteins, fats, carbohydrates*) present in the substrate, with the formation of simple and soluble compounds (*amino acids, fatty acids* and *monosaccharides*, respectively);

- *acidification and acetogenesis:* at the same time as hydrolysis, fermentative acidogenic bacteria oxidize the simple organic substrates to obtain pyruvate, which

([27]) The LHV of CH_4 is 31.6 MJ/m_N^3 (7565 $kcal/m_N^3$).

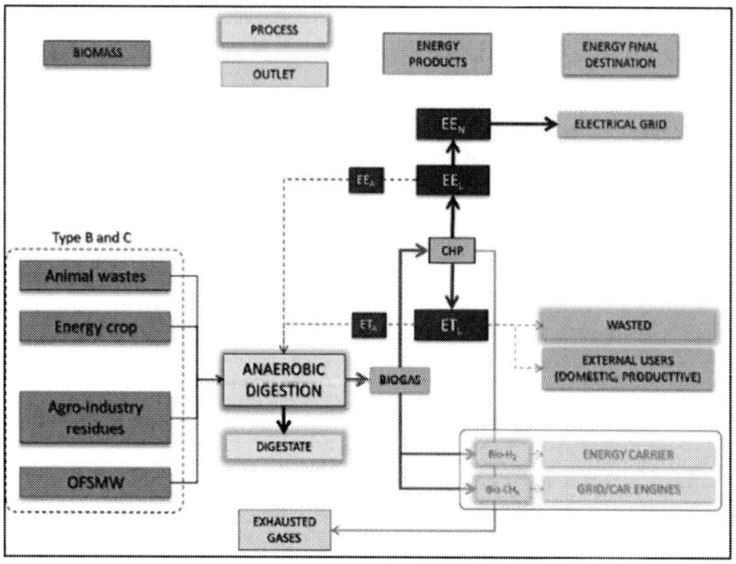

Fig. 8. – AD scheme. Biogas normally supplies a CHP unit with an electric generator, producing $EE_L + ET_L$. A small fraction of EE_L is reused (EE_A for self-consumption) while the remaining (EE_N) is sold to the grid. Also a fraction of ET_L is reused for substrate heating (ET_A) while the remaining fraction can be sold to external final users through a distribution network (but in many cases the residual heat is inefficiently dissipated). The digestate (co-product) is used for agronomic purposes. Bio-CH_4 can be produced purifying the biogas and concentrating the CH_4; it can be used as a fuel for vehicles or pumped into the natural gas distribution grid. Also the bio-H_2 production —prior to biogas production— is a plant variant (but it is still at experimental level) the purpose of which is to explore the potential of H_2 as energy carrier (source: [4]).

is subsequently transformed into ketones, alcohols and *Volatile Fatty Acids* (VFA; acids with a short chain, such as *propionic* and *butyric acids*); starting from VFA, acetogenic bacteria produce *acetic acid*, *formic acid*, CO_2 and H_2. NH_3 is also produced by the fermentation of the amino acids;

– *methanogenesis*: represents the conclusion of the anaerobic trophic chain, as CH_4 is the only non-reactive compound in the whole process.

The AD takes place in closed and isolated structures (usually cylindrical), called *digesters* (or digestion reactors). The biomass fed into digester (only one or —more frequently— a mix of different organic matrices) is heated and maintained inside the digester under:

– *mesophilic conditions (35–45 °C)*;

– *thermophilic conditions (50–55 °C)*.

The ET needed for biomass heating is recovered from the cogeneration unit (CHP; IC engine + electric generator) powered by the biogas. In general, increasing the process

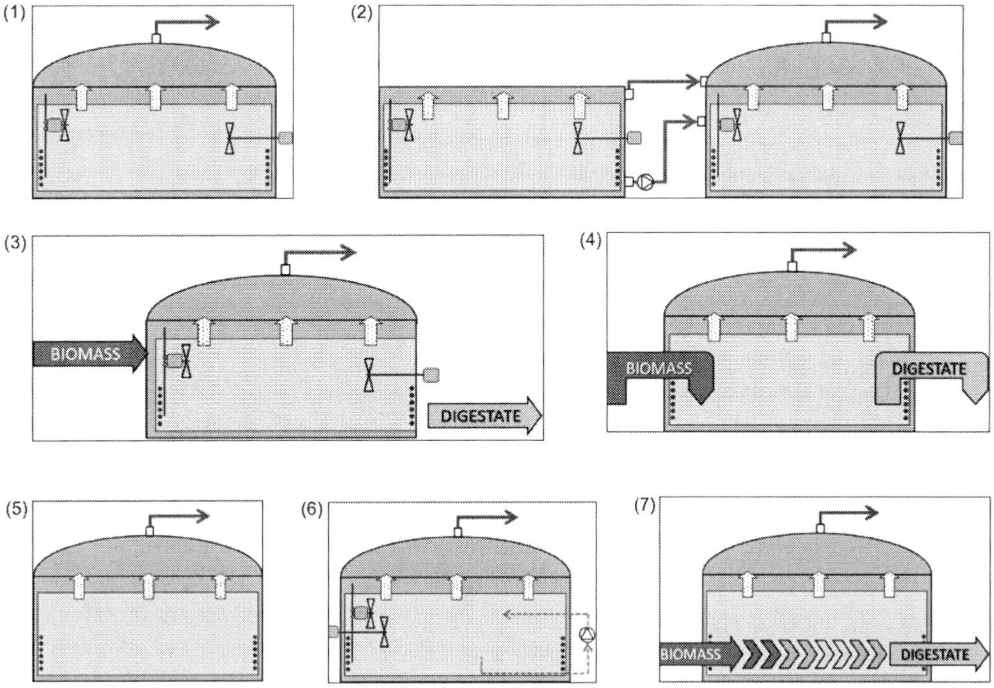

Fig. 9. – AD plant types (from left to right and from top to bottom): (1) single-stage plant, mechanical mixing; (2) two-stage plant, mechanical mixing; (3) continuous load system; (4) discontinuous load system (batch); (5) not mixed plant; (6) mixed plant (Completely Stirred Tank Reactor, CSTR); (7) plug-flow plant.

temperature, the mass of digested substrate increases and, consequently —at the same flow (m^3/day) of fresh substrate put into the digester— the *Hydraulic Retention Time* (HRT, days) decreases, reducing the digester volume (and investment costs of the AD plant). The *technical-operational criteria* for splitting out the AD plants are (fig. 9):

- *TS content* of the inlet substrate:
  - *wet digestion:* $4\% < TS \leq 10\%$; it is the most widespread AD solution in Italy, operating with animal slurry only or —very often— in *co-digestion* conditions (slurry mixed with other organic matrices such as cereals silages, food processing residues or other organic residues);
  - *semi-dry digestion:* $10\% < TS \leq 20\%$;
  - *dry digestion:* $20\% < TS \leq 35\%$; it is generally applied for particular organic substances or for biogas plant fed with OFMSW.
- *plant layout (biological phases)*:
  - *single-stage* AD plants, in which the acidogenic and methanigenic phases occur simultaneously in a single *digester* (very common and easy-management solution, low investment costs);

  - *two-stage* AD plants, in which the acidogenic phase takes place in a *1st stage digester* and methanigenic phase in a 2nd *digester* (more efficient and sophisticated solution, high investment costs).

- *digester loading mode*:

  - *discontinuous load* (batch mode): only one substrate load is performed when the process starts, and the digestate is extracted from the digester once the HRT ends; with this (uncommon) solution, biogas flux over the time is not steady (there is a strong initial increase, a very short-term stability and a rapid decline) and some technical layout arrangements are necessary;
  - *continuous load:* high-frequency (12–48 times per day) fresh substrate loads and simultaneous discharges of an equal volume of digestate are carried out; this solution is currently the best choice to guarantee: i) high-automation in process management and ii) constant biogas production over the time (m^3_N/h).

- *substrate mixing inside digester*:

  - *not-mixed digester*;
  - digester equipped with *mixing devices* (currently, propellers devices submerged in the substrate represent the most widespread solution in wet digestion and in single-stage AD plants);
  - digester equipped with *piston systems*: the lateral inlet of the substrate generates a push along the longitudinal axis of the digester (that usually has a parallelepiped shape), moving forward the substrate that is already in it and releasing the digestate on the opposite side.

Co-digestion consists in the digestion, at the same time, of i) *livestock waste* (pigs, cows and poultry waste, mainly), ii) *energy crops* (maize and other cereals silage, sorghum, loin, etc.) and iii) *different organic residues or waste* (fig. 10). The main advantage of the co-digestion, compared with the use of animal waste only, is the big increase in the organic matter input and, thus, in the significant growth in the *biogas yield* (energy production).

Moreover, the possibility of using different types of biomass in an AD plant can promote an *integrated management of the resources* (at both farm and landscape level), allowing —in the case of recovery and reuse of processing residues— the rational disposal/valorization of many materials. From this point of view, residues from *vegetable processing* (potatoes, onions, legumes and fruit, etc.), which are usually clean organic materials, are of great interest and represent excellent matrices for the AD plants. In Italy, co-digestion was mostly based on the use of *corn silage* (or silage of triticale, barley, wheat or sorghum), in AD plants with a "standard" electric power just below of 1 MW

https://doi.org/10.1051/epjconf/201818900006

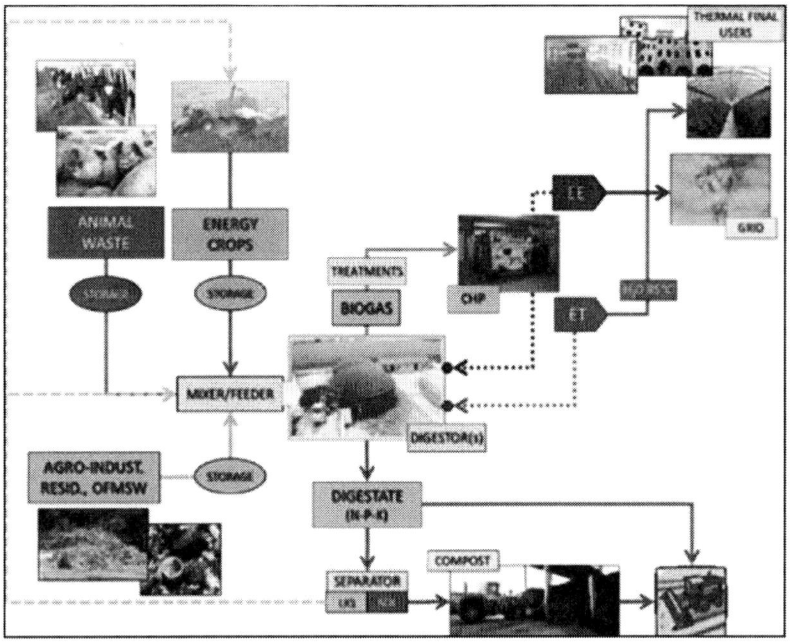

Fig. 10. – Co-digestion supply chain scheme. Biogas production aims to generate EE. Heat should also be used but final users must be very close to the AD plant (source: [4]).

(see footnote([28])) and with an energy share from energy crops often largely prevalent on animal waste rate. Under these conditions, attention has to be paid to the massive amounts of *digestate* available at the farm level which: i) has to be stocked and disposed of in accordance with the current regulation([29]), ii) generates a *significant increase in the farm N concentration*. This latter issue is a problem of not easy solution in those areas (*vulnerable zones*) where the above-mentioned regulation([30]) defines a limit about the N rate (kg/ha), coming from livestock waste, that can be spread over the time.

2·3.2. Bio-CH_4. *Bio*-CH_4 originates from biogas subjected to an *upgrading process* (CO_2 removal, mainly) in order to generate a gas with a CH_4 concentration of *95–98%*

([28]) This electric power represented —up to 31.12.2012— the upper limit for the application (15 years) of the all-inclusive feed-in-tariff (TO) equal to 280 €/MWh. For AD plants put into operation after this date, the value of the TO has been reduced (TO = 210–220 €/MWh, for a 20-year period) and the upper electric power has been set to 300 kW (`http://www.gse.it/it/Qualifiche%20e%20certificati/DM%2023%20giugno%202013/Pagine/default.aspx`).

([29]) The need to find new digestate distribution areas in accordance to the N loading limits imposed by the regulation is indicated (*Il Sistema agro-forestale della Lombardia*, Report 2011) as one of the causes of the significant increase in the land rentals in agricultural areas of Lombardy where the number of AD in co-digestion is higher than in other Italian regions.

([30]) MIPAF Ordinance of April 7th, 2006.

in volume, to be used as *biofuel*: i) for vehicles engines (cars, trucks and tractors, in particular) or ii) to supply the natural gas distribution grid. The cleaning and upgrading processes consist of the following steps: i) purification, ii) dehydration, iii) desulphurization([31]), iv) undesirable components removal. The sequence of these phases depends on the adopted *technology*: i) *physical* (such as oscillating pressure adsorption, washing with pressure water, washing with organic solvents and separation with membranes) or ii) *chemical* (such as washing with mono-ethyl-amine, MEA).

The introduction of bio-CH_4 into the distribution network allows —in addition to a reduction in the AD plant investments costs (biogas storage systems and the cogeneration unit are unnecessary)— a more flexible and efficient biogas utilization [4]. Bio-CH_4 utilization is widespread in northern Europe (especially in Germany, Austria, Sweden and Denmark), while in Italy —despite the National Energy Plan objectives and a general law about it([32])— protocols, technical standards and, above all, the TO for the "non-electric biogas" (€/m_N^3), have been approved only recently. The *advantages* of bio-CH_4 networking are not negligible. In particular, they are linked to the chance of using this biofuel in different (and even far) locations than the production sites, with the possibilities of: i) use it in more powerful cogeneration units located close to heat final users (global efficiency improving), ii) use it as a fuel for urban vehicles. The *disadvantages* are essentially related to the relatively high energy demand for bio-CH_4 pressurization; in particular, compression is advantageous only for medium-pressure pipes ($1 < p \leq 12\,\text{bar}$), while for high-pressure pipes ($p > 12\,\text{bar}$) it is much more expensive. Finally, before entering the distribution network (and also before its use in service stations as a fuel), bio-CH_4 has to undergo a purification process, achieving the full compatibility with natural gas characteristics, established by the current legislation.

2·3.3. Bio-H_2. H_2 represents the simplest, lightest, and more abundant chemical element in the Universe, making up, in combination form, about the 75% of all matter. However, in free form, H_2 represents only 0.07% of the atmosphere and 0.14% of the Earth's surface.

Currently, the most attractive application for pure H_2 is the use as *energy carrier* in *fixed and mobile conversions*. In the simplest case, H_2 combustion generates an exothermic reaction producing, in addition to ET, only water vapor. The environmental repercussion is only the *considerable amount of* NO_x generated inside the combustion chamber of the Otto cycle engines, caused by the *high operating temperatures* (2000–3000 °C). This is why the most important vehicle manufacturers have started studies concerning the development of engines derived from the Diesel cycle. Recent studies have shown that the adoption of *special engines* (Homogeneous Compression Charge Ignition, HCCI; $\eta = 45\%$), avoids the problem, maintaining considerably lower operating temperatures.

([31]) Attention has to be paid to this phase, especially when the biogas is produced from AD of cows waste, because the production of a mixture rich in H_2S can occur.
([32]) Legislative Decree No. 28 of March 3rd, 2011 defines, in general terms, the characteristics that bio-CH_4 must have to be introduced into the natural gas distribution network.

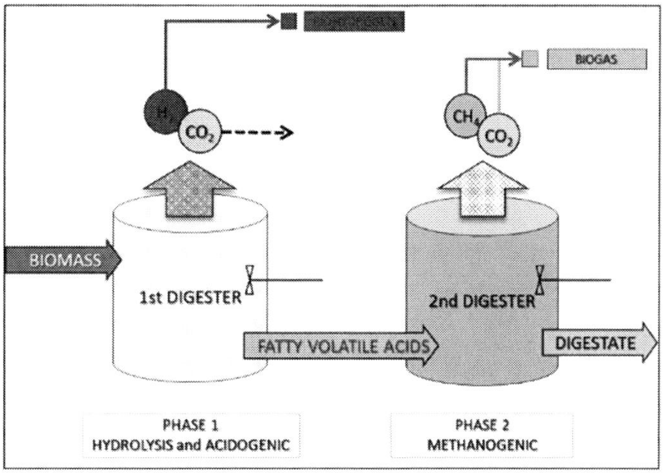

Fig. 11. – Bio-H_2 from biomass: two-stage AD plant with H_2 production (source: [4]).

H_2 is also strictly linked to the *Fuel Cell technology* [13], which —unlike the common IC engines— allows the use of H_2 to produce *EE* under isobaric and isotermic conditions *without any combustion reaction*. Since electrolytic H_2 production is very expensive, many alternatives have been exploited and they are still in an experimental phase. At the same time, many other researches have been focused on improving technologies to store and transport this very peculiar gas. Concerning the new solutions to produce H_2, a special *two-stage AD* seems to be very promising. This technique is based on the physical separation of hydrolysis (acidogenic and acetogenic) phase from the methanigenic one [4]. Practically, there are two different reactors operating in series (fig. 11). By the modulation of the HRT, it is possible to optimize the metabolic processes of VFA and H_2 production (bio-H_2) in the *first reactor*, thus stripping out the H_2 and moving the VFA into the *second reactor*, where biogas rich in bio-CH_4 is produced.

The interest in this solution is also justified by the fact that, under the same conditions, the two-stage process is —compared to a traditional single-stage AD— able to produce more energy (*higher conversion efficiency*). During recent experimental tests([33]), a two-stage laboratory scale plant, fed by a mixture of slurry and OFSMW (65% and 35% on the total mass, respectively) and operating with a HRT of 2 days in the first digester, showed that the specific bio-H_2 production (ratio between the volume of bio-H_2 produced and the volume of the reactor) was about *3.2–3.3* dm^3/dm^3 *of reactor per day*. In the extracted gas mixture from the first digester no CH_4 was present and the average content of H_2 was 42% vol. (H_2 yield: 125–130 dm^3/kg VS). Under this optimal conditions, comparing the total efficiency between the single- and the two-stage solution (9.4 and 15.8 MJ/kg VS, respectively), the conversion efficiency of the second was higher.

([33]) *Produzione di bio-idrogeno ed energia rinnovabile da residui agro-zootecnici (AgrIdEn)* Project, Regione Lombardia.

Appendix A.

List of symbols and acronyms

Name	Symbols and acronyms	Unit of measure
All-inclusive feed-in-tariff	TO	€
Anaerobic Digestion	AD	–
Bio-hydrogen	Bio-H_2	–
Biomass-To-Liquid	BTL	–
Bio-methane	Bio-CH_4	–
Bulk (apparent) Density	γ	kg/m^3
Carbon/Nitrogen ratio	C/N	–
Combined Heat and Power unit	CHP	–
Completely Stirred Tank Reactor	CSTR	–
Density	ρ	kg/m^3
Dry Mass	m_{DM}	kg
Dry Matter content or Total Solids	DM or TS	%FM
Electric Energy	EE	MJ
European Union	EU	–
Excess air ratio	λ	–
Food and Agriculture Organization of the United Nations	FAO	–
Forest Management Plan	FMP	–
Fresh (wet) mass	m_{FM}	kg
Gross Electric Energy	$\mathrm{EE_L}$	MJ
Gross Heating Value	GHV	MJ/kg DM
Gross Thermal Energy	$\mathrm{ET_L}$	MJ
Homogeneous Compression Charge Ignition	HCCI	–
Hydraulic Retention Time	HRT	days
Internal Combustion	IC	–
Lower Heating Value	LHV	MJ/kg DM
Mechanical Energy	EM	MJ

Medium Rotation Forestry	MRF	–
Moisture Content (dry basis)	MC_{DM}	%DM
Moisture Content (wet basis)	MC	%FM
Mono-Ethyl-Amine	MEA	–
Net Electric Energy	EE_N	MJ
Net Heating Value	NHV	MJ/kg FM
Net Thermal Energy	ET_N	MJ
Organic Fraction of Municipal Solid Waste	OFMSW	–
Permanence time	t_r	s
Self-consumed Electric Energy	EE_A	MJ
Self-consumed Thermal Energy	ET_A	MJ
Short Rotation Coppice	SRC	–
Short Rotation Forestry	SRF	–
Straight Vegetable Oil	SVO	–
Thermal Energy	ET	MJ
Volatile Fatty Acids	VFA	–
Volatile Solids	VS	%TS
Water Mass	m_{H_2O}	kg

REFERENCES

[1] Fidanza A., Leonardi M., Baldissara B., Ciorba U., Gaeta M., Gracceva F., Tommasino M. C., Franci T., Zanchini E., Coletta G. and Palma D., *Le Fonti Rinnovabili. Ricerca e innovazione per un futuro low-carbon* (Ufficio Studi ENEA) 2010, pp. 1–241.

[2] Gelleti R. *et al.*, *Energia dalle biomasse. Le tecnologie, i vantaggi per i processi produttivi, i valori economici e ambientali* (Consorzio per l'AREA di ricerca scientifica e tecnologica di Trieste, AREA Science Park) 2006, pp. 1–244.

[3] AA.VV., *Biomasse ad uso energetico* (Università Politecnica Marche-Regione MarcheASSAM) 2010, pp. 1–91.

[4] Fiala M., *Energia da biomasse agricole. Produzione e utilizzo* (Maggioli Editore) 2012, pp. 1–437, ISBN 978-88-387-7090-5.

[5] Grabosky M. S. and Bain R., *A Survey of Biomass Gasification*, Vol. **II** – *Principles of Gasification*, 1979, SERI/TR-33-239, Vol. **II** of **III**, UC category: UC - 61. Prepared under Task No. 3322, Solar Energy Research Institute 1536 Cole Boulevard, Golden, Colorado 80401. A division of Midwest Research Institute. Prepared for the U.S. Department of Energy. Contract No. EG 77 C 01 4042, pp. 1–243.

[6] Fiala M., *Combustibili solidi di origine agricola: le biomasse legnose*, I Georgofili – Quaderni 2007-I, Produzione di Energia da Fonti Biologiche Rinnovabili, a cura dell'Accademia dei Georgofili (Edizioni Polistampa) 2007, pp. 13–54.

[7] AA.VV., *La filiera biogas* (Università Politecnica Marche-Regione MarcheASSAM) 2010, pp. 1–50.

[8] Energy Strategy Group, *ES - Biomass Energy Report. Il business delle biomasse e dei biocarburanti nel sistema industriale italiano*, Anno 2009, Collana Quaderni AIP (AIP) 2010, pp. 1–210.

[9] Spinelli R. and Hartsough B., *Indagine sulla cippatura in Italia* (Contributi Scientifico-Pratici, XLI, Consiglio Nazionale delle Ricerche, Istituto per la Ricerca sul Legno) 2001, pp. 1–120.

[10] Caserini S., Fraccaroli A., Monguzzi A., Moretti M. and Angelino E., *Stima dei consumi di legna da ardere per riscaldamento ed uso domestico in Italia* (Agenzia Nazionale per la Protezione dell'Ambiente e per i Servizi Tecnici (APAT) nell'ambito della Convenzione con ARPA Lombardia, approvata tramite decreto No. 517 del 12 luglio 2006) 2008, pp. 1–60.

[11] Mezzalira G. *et al.*, *Come produrre energia dal legno*, Quaderno ARSIA Regione Toscana 3/2003, 2005, pp. 1–86.

[12] Chiaramonti D. and Fantozzi F., *Processi di conversione energetica delle biomasse di tipo termochimico*, I Georgofili – Quaderni 2007-I, Produzione di Energia da Fonti Biologiche Rinnovabili, a cura dell'Accademia dei Georgofili (Edizioni Polistampa) 2007, pp. 71–126.

[13] Cenčič V., Jodice R., Migliardi D., Picco D., Pin M., Tomasinsig E., Zanotto G., Radu R., Taccani R., Chindaruska S., Barbieri S., Bianco D., Michelutti R. and Monaco B., *I Biocarburanti. Le filiere produttive, le tecnologie, i vantaggi ambientali e le prospettive di diffusione* (Consorzio per l'AREA di ricerca scientifica e tecnologica di Trieste, AREA Science Park) 2007, pp. 1–216.

Geothermal energy

A. Manzella
Institute of Geosciences and Earth Resources - Pisa, Italy

These lecture notes were originally published in *EPJ Web of Conferences*, **148** (2017) DOI: 10.1051/epjconf/201714800012 as a contribution to the Proceedings of the 5th course of the MRS-EMRS "Materials for Energy and Sustainability" and 3rd course of the Joint EPS-SIF International School on Energy held in 2016. They are here reproduced as a good summary of the material presented in July 2017 during the course "Advances in Basic Energy Issues".

EPJ Web of Conferences **189**, 00008 (2018) https://doi.org/10.1051/epjconf/201818900008
LNES 2017

Concept and status of Concentrating Solar Power systems

R. PITZ-PAAL(*)

DLR Institute of Solar Research - Linder Höhe, 51147 Cologne, Germany

Summary. — Development of Concentrating Solar Power (CSP) systems has started about 40 years ago. A first commercial implementation was performed between 1985 and 1991 in California. However, a drop in gas prices caused a longer period without further deployment. It was overcome in 2007 when new incentive schemes for renewables in Spain and the US enabled a commercial restart. In 2017, almost 100 commercial CSP plants with more than 5 GW are installed worldwide. This paper describes the physical background of CSP technology, its technical characteristics and concepts. Furthermore, it discusses system performances, cost structures and the expected advancement.

1. – Introduction

Concentrating Solar Thermal Power (CSP) systems use high-temperature heat from concentrating solar collectors to generate power in a conventional power cycle instead of — or in addition to — burning fuel. The thermal energy accumulated by the concentrating collectors can be stored at low cost in a thermal energy storage (*e.g.* a tank

(*) E-mail: robert.pitz-paal@dlr.de

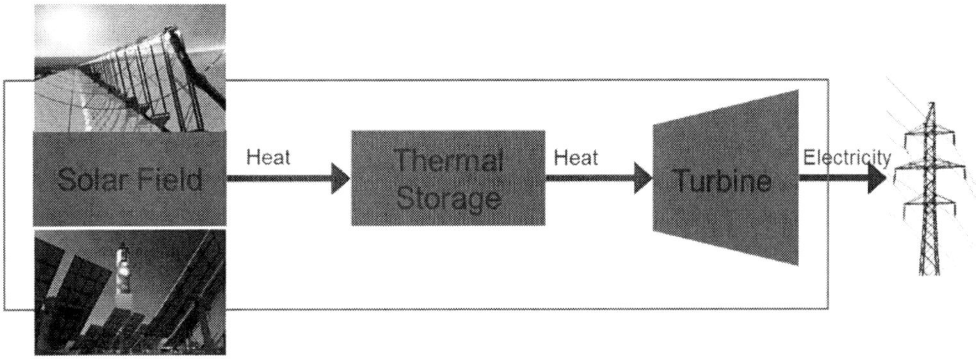

Fig. 1. – The concept of Concentrating Solar Power Systems (CSP) [source DLR].

with molten salt) before it is provided to a conventional power cycle (fig. 1). Thus, electricity production can be shifted to periods of high load demand and the technology is capable to complement the production from wind and photovoltaic as depicted in fig. 2. Firm supply can be guaranteed if additional fuel (*e.g.* from biomass) is used during the uncommon periods when the thermal storage is empty.

2. – Technical background of CSP

2·1. *Solar resource.* – Only direct radiation can be concentrated in optical systems (fig. 3). In order to achieve significant concentration factors, Sun-tracking is required, adding complexity to the system. Therefore, CSP is most suitable for centralized power production, where maintenance can be performed efficiently, and in areas with high, non-volatile direct solar radiation levels. The concentration of sunlight is achieved by mirrors directing the sunlight onto a heat exchanger (receiver/absorber), transferring the absorbed energy to a heat transfer fluid.

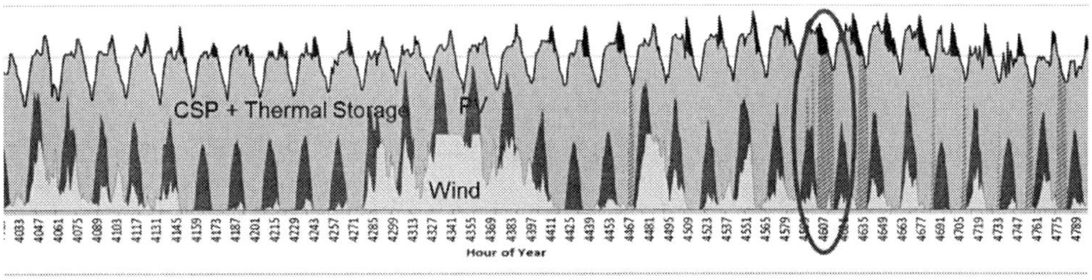

Fig. 2. – Electric load matched by a mix of wind, PV and CSP with thermal storage: During periods indicated in gray co-combustion in CSP systems is required to match the load [source DLR].

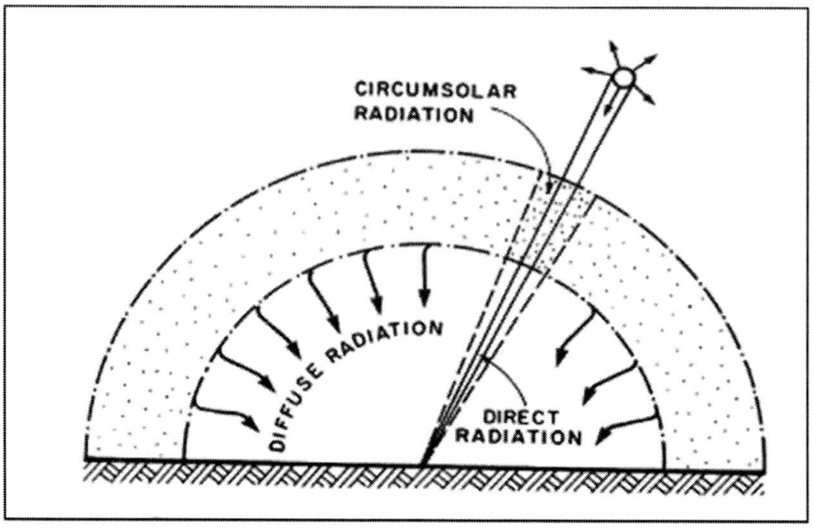

Fig. 3. – Definition of direct, diffuse and circumsolar radiation. Only direct irradicance can be used for CSP [1].

2·2. *Concentration of sunlight.* – Radiation power (unit: Watt) that is emitted from a point-like source is diluted with increasing distance from its source. This means that the power density E (unit: $\mathrm{W/m^2}$) is reduced with increasing distance, because the emitted power is distributed over a larger surface. Concentration of radiations aims to increase the power density E of the radiant energy for optimized utilization.

A generic concentrator (fig. 4) consists of a concentrator entrance aperture area A (unit: $\mathrm{m^2}$) in which the radiant energy enters and an exit aperture A' from which the radiation energy leaves. Assuming no losses in the concentrator, the law of conservation of energy leads to the concentration factor C. C is the ratio of the outgoing to the incoming energy density and can be described as the inverse ratio of the respective aperture areas:

$$C = E'/E = A/A'. \tag{1}$$

As can be derived from complex thermodynamic considerations, ideal concentrators conserve a quantity called entendue. In this case, entendue is the product of aperture area and sinus squared of the radiation cone's half angle θ:

$$A \sin^2 \theta = A' \sin^2 \theta'. \tag{2}$$

From this law, the concentration ratio of an ideal concentrator results as

$$C = A/A' = \sin^2 \theta' / \sin^2 \theta. \tag{3}$$

For a given opening angle θ, the maximum concentration is achieved if $\sin^2 \theta' = 1$ ($\theta' = 90°$).

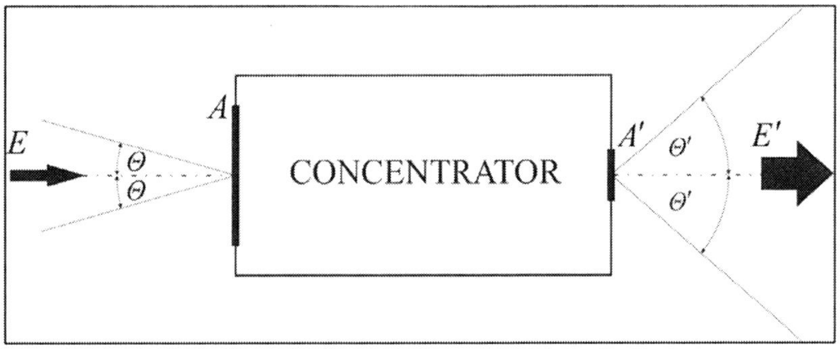

Fig. 4. – Schematic of a solar concentrator [1].

Thus, for the Sun's half angle of 4.653 mrad a maximum concentration ratio of $1/\sin^2$ (4.653 mrad) ≈ 46200 can be achieved.

In practice, imaging mirror concentrators rather than lens concentrators are applied as primary concentrators for solar high-temperature applications. Imaging mirror concentrators have better outdoor durability and lower specific costs. Their design generally approximates the parabola shape in a continuous or segmented way (fig. 5). The image of the Sun as generated by such a mirror reflector is blurred by the optical imperfections of the parabola concentrator concept itself (aberration) and by imperfect surface characteristics. The resulting image consists of superimposed images of individual concentrator segments having elliptical shapes with diameters depending on the opening angle of the Sun's disk α_D, the distance between reflector and focal point r_r, and on the opening angle of the parabola Ψ.

The concentration factor of the parabola is determined by the largest of all superimposed images A_{im} that comes from the rim of the parabola:

$$C = \frac{A_{ap}}{A_{im}} \tag{4}$$

with

$$A_{ap} = \pi r_r^2 \sin^2 \psi \quad \text{and} \quad A_{im} = \frac{\pi}{4} \frac{r_r^2 \alpha_D^2}{\cos^2 \psi}$$

leads to

$$C = \frac{4}{\alpha_D^2} \sin^2 \psi \cos^2 \psi = 46200 \sin^2 \psi \cos^2 \psi. \tag{5}$$

The maximum concentration achievable by an ideal parabola does not reach the theoretical limit of 46200, but only 1/4 of it at an opening angle of the parabola Ψ of 45°:

$$C_{\max} = 46200 \cdot 0.5 \cdot 0.5 = 11550. \tag{6}$$

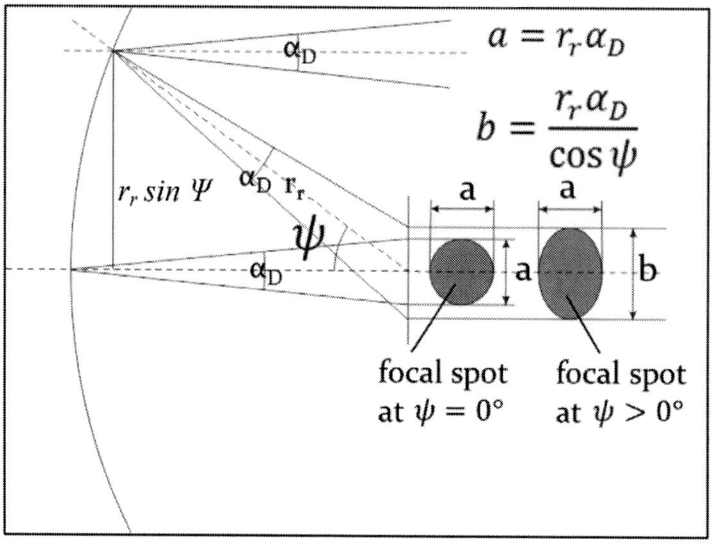

Fig. 5. – Size of the focal spot of an ideal parabola [source DLR].

The superposition of the images can often be well approximated by a Gaussian distribution. Thus, the energy density at the exit aperture of a parabola concentrator is not constant, but varies from a peak value to zero at an infinite distance. For practical application it is reasonable to use only the central part of the Gaussian profile and discard the rest. The amount of discarded energy is often an outcome of an economic optimization process with typical values ranging from 5% to 10%.

2·3. *Conversion of concentrated sunlight to heat.* – The device used in high-temperature solar concentrators for the conversion of concentrated solar radiation to heat is called receiver. It is designed to absorb the concentrated solar radiation and to transfer as much energy as possible to a heat transfer fluid. Losses have several origins: The absorbing surface may not be completely black, the surface emits thermal radiation to the environment (due to its elevated temperature) and convection as well as conduction occur. The thermal efficiency of the receiver η_{th} is defined by the ratio of the useable heat per second to the incoming solar radiation in the aperture. As shown in fig. 6, η_{th} is decreasing with increasing temperatures due to higher heat losses. On the other hand, higher concentration factors lead to higher efficiencies and both convection and conduction losses are of minor importance at high concentration factors.

2·4. *Thermodynamic cycles.* – In high-temperature solar concentrator systems, the absorbed energy is often converted into mechanical energy and subsequently to electricity in a power cycle. A part of that cycle is an engine that continuously transfers heat to mechanical energy. Steam or gas turbine cycles are common examples used in fossil power plants to generate electricity from the heat input provided by the combustion of

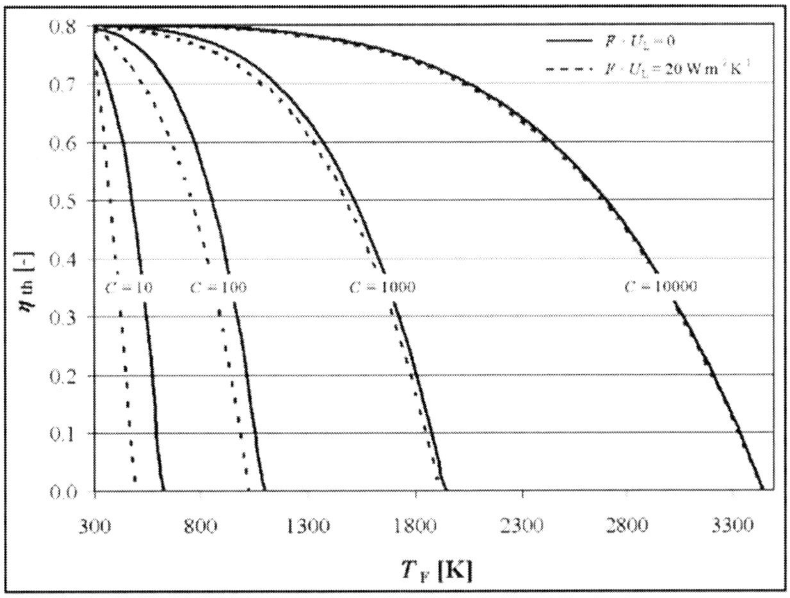

Fig. 6. – Efficiency of a solar receiver as a function of the fluid temperature, the concentration ratio C and the effective convective heat loss coefficient FU_L [1].

fuels (fig. 7a). However, thermodynamic principles forbid that all heat input is completely transferred to electricity, because a fraction of it must be removed at a lower temperature level.

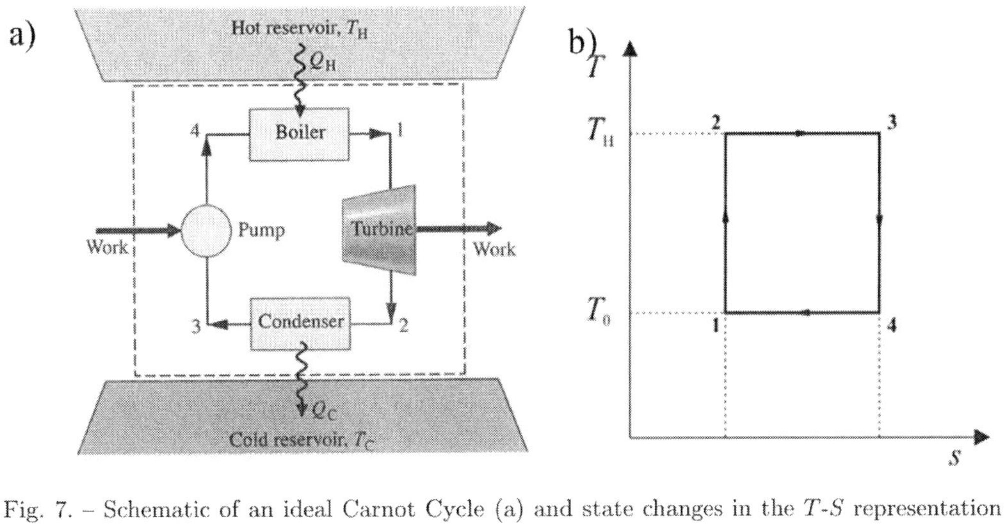

Fig. 7. – Schematic of an ideal Carnot Cycle (a) and state changes in the T-S representation (b) [source DLR].

In each generic power cycle, a working fluid undergoes a number of changes in state. In order to force the system from one state to the next, heat or work must be delivered to or extracted from the fluid. Depending on the kind of working fluid and the state changes, many different power cycles can be set up. An example is the Carnot cycle, an idealized reversible process that consists of two isothermal ($2 \to 3$ and $4 \to 1$) and two isentropic ($1 \to 2$ and $3 \to 4$) changes in state. In fig. 7b, the corresponding T/S-representation is shown, where S is the specific entropy of the fluid.

The fluid receives heat at the upper temperature T_{H} and at the lower temperature T_0 heat is released. The cycle efficiency η_{C} describing the fraction of heat input Q_{in} that is converted to work W is independent of the working fluid and can be calculated to be

$$\eta_{\mathrm{C}} = \frac{W}{Q_{\mathrm{in}}} = 1 - \frac{T_0}{T_{\mathrm{H}}}. \tag{7}$$

Today, there are no power cycles specifically developed for high-temperature solar concentrating systems but conventional fossil fuel driven power generation systems are adapted to solar applications. The most relevant ones are steam turbine cycles, gas turbine cycles and Stirling engines. Currently, steam cycles are the most common choice in commercial CSP projects. They are suited to power levels beyond 10 MW and temperatures of up to 600 °C. Steam cycles can be coupled to parabolic trough, linear Fresnel and central receiver systems. Stirling engines are used for small power levels (up to some $10\,\mathrm{kW_{el}}$). Gas turbines offer the potential to exploit higher temperatures than steam cycles (up to 1200 °C) and cover a wide range of capacities from some hundred $\mathrm{kW_{el}}$ to some $10\,\mathrm{MW_{el}}$. At high power levels, gas turbine cycles may be combined with steam cycles to highly efficient combined cycle systems promising to produce the same power output with a 25% smaller solar collector field. Up to now, solar gas turbines have been used in experimental facilities only.

2·5. *Thermal energy storage.* – Like domestic hot water systems, CSP systems have the important advantage of optional thermal energy storage systems (*e.g.* tanks with molten salt), allowing the operation of the plant to continue during transient clouds or after sunset. Thereby, a predictable power supply to the electricity grid can be achieved. In contrast to other renewable systems with electric storage, for which the inclusion of storage capacity always leads to far higher investments and higher electricity prices, CSP systems with storage are potentially cheaper than CSP systems without storage. This becomes clear when comparing a solar power plant without storage of, *e.g.*, $100\,\mathrm{MW_{el}}$ capacity, operated approx. 2000 equivalent full-load hours per year at a typical site, to a system with half the capacity ($50\,\mathrm{MW_{el}}$) but the same size solar field and a suitable thermal energy storage (fig. 8). In this case, the smaller power block is used for 4000 equivalent full load hours so that both systems can produce the same amount of electricity per year. Assuming low storage costs, the investment in the second system could potentially be lower than in the first.

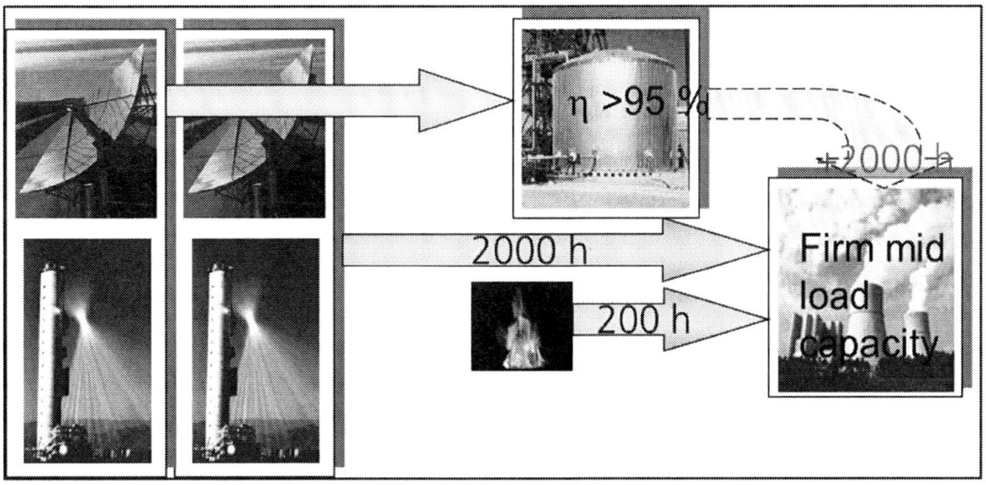

Fig. 8. – Integration of a thermal storage in a CSP system may lead to lower specific electricity cost (see text) [source DLR].

2·6. *Optimizing CSP-systems.* – According to the principles of thermodynamics, power cycles convert heat to mechanical energy more efficiently at higher temperatures. However, the collector efficiency drops at higher absorber temperatures due to higher heat losses. Consequently, for any given concentration factor there is an optimum operation temperature at which the highest conversion efficiency from solar energy to work is achieved. With rising concentration higher optimum efficiencies are achievable. Figure 9 illustrates this characteristic assuming an ideal solar concentrator combined with an ideal Carnot power cycle. If the spectral absorption characteristics of the absorber are perfectly tailored to maximise absorption in the solar spectrum, but avoid thermal radiation losses in the infrared part of the spectrum (selective absorber), additional efficiency gains can be expected in particular at lower concentration factors.

In practice, the optimum operation temperatures will be lower than these theoretical figures, because power cycles with Carnot performance and ideal absorbers do not exist. Furthermore, the impact of frequent operation under part-load conditions throughout the year on the efficiency of the system has to be considered.

3. – Technical concepts

CSP systems can be distinguished by the arrangement of their concentrator mirrors: Line focusing systems like parabolic troughs or linear Fresnel systems (fig. 10, left and middle) only require single axis tracking in order to concentrate the solar radiation onto an absorber tube. Concentration factors of up to 100 can be achieved in practice. Point focusing systems like central receiver systems (fig. 10 right) can achieve concentration factors of several 1000 at the expense of two-axis tracking. They use a large number of

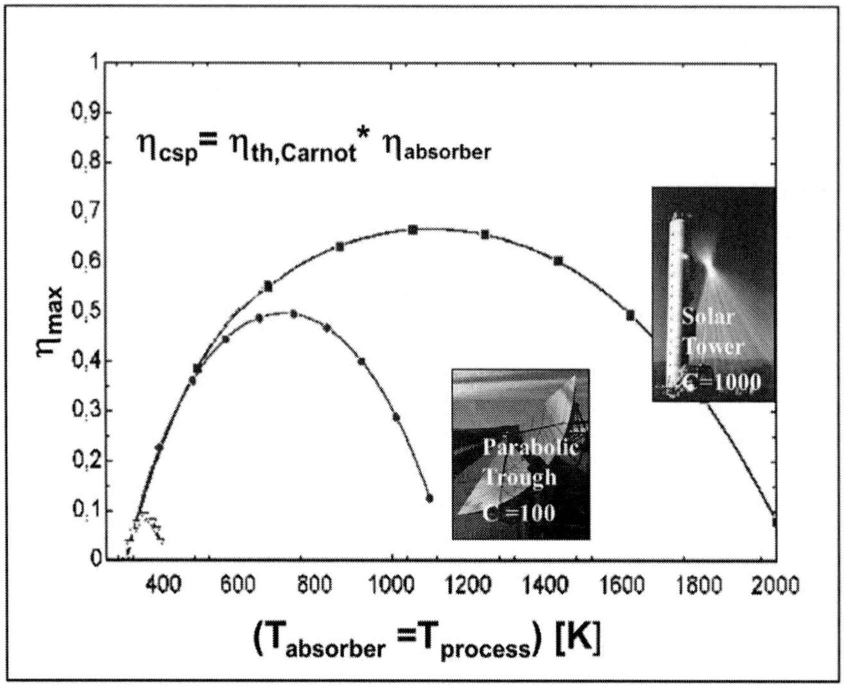

Fig. 9. – System efficiency as a function of upper fluid temperature for different solar concentrating systems [source DLR].

Fig. 10. – Schematic of a parabolic trough collector (left), linear Fresnel (center) and solar tower system (right) [source DLR].

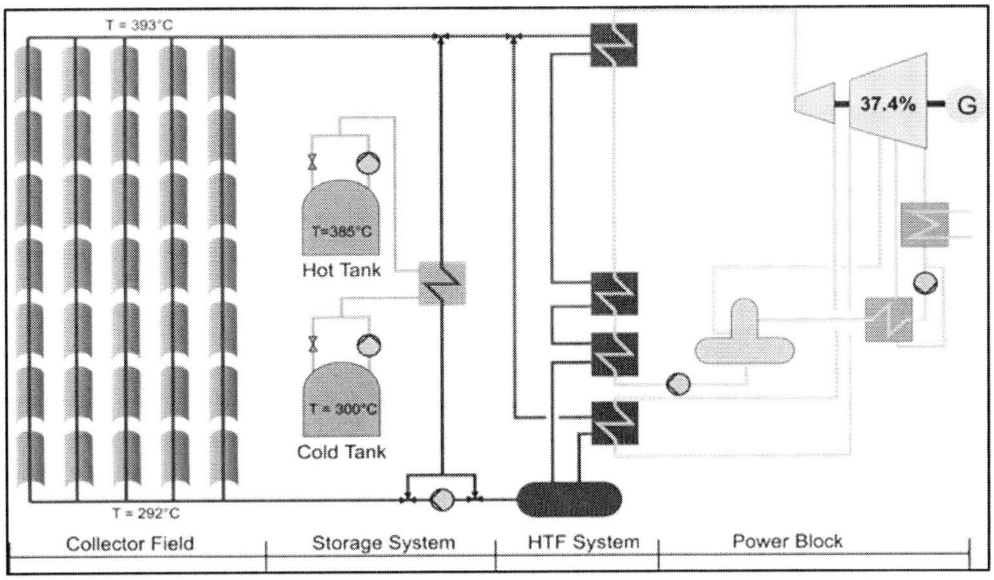

Fig. 11. – Schematic of a system based on parabolic trough collector using thermal oil as heat transfer fluid and 2 tank molten salt storage system [2].

individually tracking heliostats to concentrate the solar radiation onto a receiver located on the top of a central tower.

3·1. *Line focusing systems.* – Most of today's commercial CSP power plants ($\Sigma >$ $3\,GW_e$) are based on the design of the ANDASOL parabolic trough power plant (fig. 11) put in operation in Spain in November 2008. The plant design integrates three different fluid cycles: a eutectic mixture of 73.5% diphenyl oxide and 26.5% biphenyl is used as heat transfer fluid in the collector field. It is operated between 290 °C and 390 °C (the upper temperature limit of this thermal oil) and provides the collected energy through heat exchangers either to the thermal storage cycle or to the power cycle. The power cycle uses water steam like in conventional power plants whereas a eutectic mixture of KNO_3 and $NaNO_3$ acts as a storage medium at almost the same temperature range as the collector fluid. The mixture is chosen as it is about 4 times cheaper than thermal oil. The salt is contained in two tanks, one on a lower temperature of about 290 °C and, after being heated up in the heat exchanger, in a hot tank at almost 390 °C. The salt mixture material is available extensively in nature and used also as fertilizer with a capacity of many 100 of thousands tons per year. As it freezes at 238 °C, it needs to be kept significantly above that temperature during the lifetime of the power plant. The storage density of such a system is approximately $78\,kWh_{th}/m^3$ or $42\,kWh_{th}/t$. To provide a $100\,MW_e$ trough power plant with a 6 hour of storage capacity about 38000 t of salt are required. The largest commercial installation that uses such a storage system was put in operation in October 2013. It is the $280\,MW_e$ Solana plant in Arizona (US) equipped with a 6 hours storage equivalent to 4400 MWh_{th} of storage capacity distributed in 6 pairs of tanks of hot and cold salt each about 38 m in diameter and 14 m high. The heat losses of such tanks are significantly below 3% of the energy throughput and even two weeks of interruption does not require active heating for freezing protection of the storage tanks. Heat exchangers, pumps, valves, and pipes need to be equipped with electrical trace heating to avoid freezing problems during start-up. Such a power plant concept can achieve annual net solar-to-electric efficiency figures of up to 15%.

This concept has a number of limitations in terms of power plant efficiency and storage density that led to the development of alternative designs. First, the synthetic oil heat transfer is rather expensive and limited to approximately 400 °C. This confines the potential life steam conditions in the power block and thus the efficiency of the power cycle to values below those of fossil fuel driven steam power plants. Moreover, it limits the amount of heat that can be stored in the salt as it depends on the temperature difference between hot and cold tank. Steam, also used in the power cycle, has been investigated as an alternative heat transfer fluid. However, the integration into a storage system is difficult as the phase change characteristics of the steam during evaporation leads to a temperature profile that does not correspond to the sensible heating of the salt mixture. Using salt not only as a storage medium, but also as heat transfer fluid may overcome many of the limitations as the operation temperature of the salt can be extended to 550 °C which is a more typical operation temperature of conventional steam cycle power plants. Such a replacement would also avoid the expensive thermal oil cycle. The challenge, however, is to avoid freezing of the salt in the extended collector field under any circumstances.

If such a concept is applied to a parabolic trough system (fig. 12) the molten salt needs to be circulated through the whole collector field (> 100 km of piping) and a complete draining of the systems (*e.g.* overnight) does not seem to be an attractive option for freezing protection. In order to prevent the salt from freezing in the absorber tubes, a continuous recirculation, using salt from the cold tank and redirecting it back to the cold tank, poses a solution. Any drop in temperature of the cold tank will be compensated by small amounts of solar-heated molten salt from the hot tank. Once the hot tank is empty the heat will be provided by the auxiliary heater. Typically the auxiliary heater is not used after a sunny day. As molten salt always circulates through the solar field, it requires energy for pumping. The additionally installed electric heating devices in the solar field are only installed for emergencies and will not be used in a regular operation of the plant.

3·2. *Solar tower systems.* – In solar tower systems, a field of individually tracked mirrors (heliostats) concentrates the solar radiation on a heat exchanger located on top of a central tower (fig. 10 right). In such a point-focusing system, higher concentration factors ($\approx$ 1000) can be archived compared to parabolic trough technology ($\approx$ 100) that generates a line-focus. The collection of the energy is not performed by an extended piping circuit of more than 100 km length (including the parabolic trough receiver tubes) like in a parabolic trough field but rather by transferring the radiation energy as light

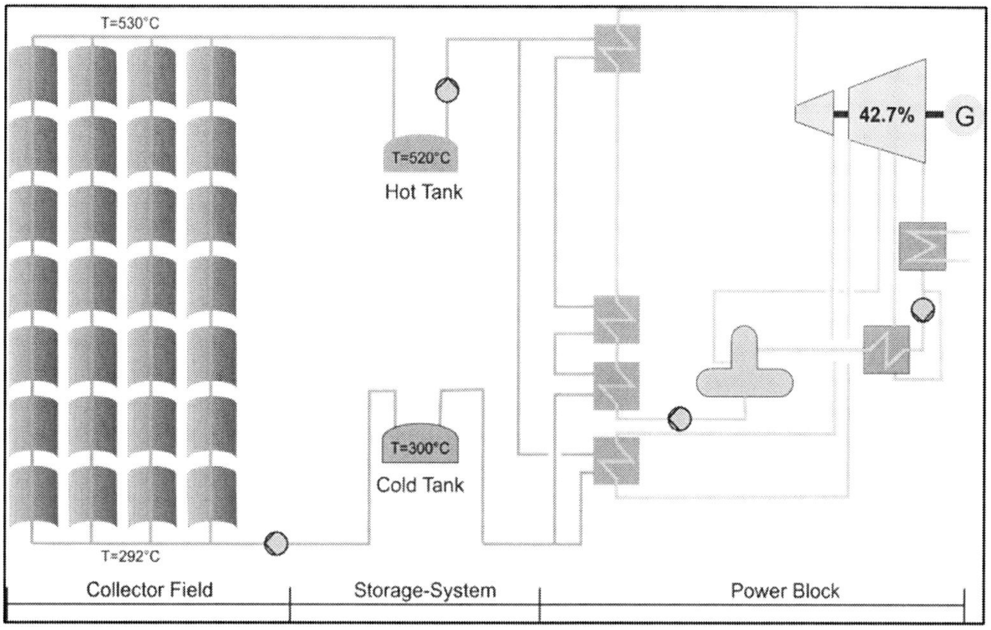

Fig. 12. – Schematic of a system based on parabolic trough collector using molten salt as heat transfer fluid and a 2 tank molten salt storage system [2].

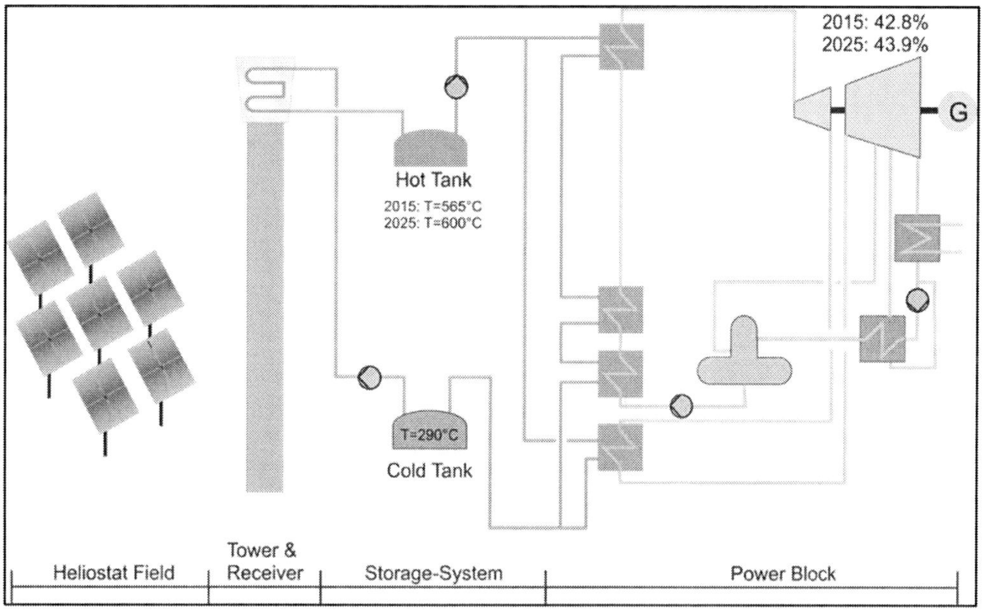

Fig. 13. – Schematic of a solar tower system using molten salt as heat transfer fluid and a two-tank molten salt storage system [2].

TABLE I. – *Data for reference system concepts to evaluate performances and costs for different site conditions* [3].

Design parameters in 2015	Unit	Parabolic Trough	Solar Tower
Site		Ouarzazate, Morocco	
Direct normal irradiation (DNI)	[kWh/(m^2 · a)]	2017 / 2558 / 2935	
Solar collector / heliostat		Ultimate Trough®	Stellio®
Heat transfer fluid (HTF)		thermal oil	molten salt
Storage medium		molten salt	molten salt
Maximum HTF temperature	[°C]	393	565
Thermal energy storage capacity (full load hours)	[h]	7.5	9[(a)]
Gross electrical output	[MW]	160	150

[(a)] Result of LCOE optimisation.

to a central location. This limits the piping length to a couple of 100 m necessary to transport the heat transfer fluid from the top of the tower to the storage and steam generator system located on the ground (fig. 13). This configuration has the benefit that gravity supports the draining of the complete salt fluid back to the storage tanks during times when operation is not scheduled (*e.g.* overnight). This reduces the auxiliary electric power requirements to heat the pipes in order to avoid freezing, in the case that molten salt is used as heat transfer and storage fluid. However, such a design also shows some significant drawbacks: A liquid salt column of 100 m equals 19 bar pressure. As the tower in a commercial solar system is typically more than 150 m tall and the pressure in the storage tanks at the ground level should be kept close to atmospheric pressure to limit construction costs, the salt circulation system cannot be designed as a closed circuit in which only friction losses need to be overcome by the pump. In an open cycle, the energy required to lift the salt to the tower top needs to be invested.

4. – Performance

In fig. 14, the annual energy flow in a CSP line focusing system is depicted qualitatively. Energy flows in tower systems differ only slightly. Optical losses associated with the solar field are caused by geometrical imperfections, surface impairments (*e.g.* dust) and extinction (aerosols). Thermal losses are linked to heat losses of high-temperature energy flows and to auxiliary warming-up effects. Power block losses are due to thermodynamic constraints that require heat removal of the cycle at low temperatures.

Performance and cost figures of CSP systems depend strongly on the power plant designs with respect to its size, its capacity factor (storage) as well as plant location

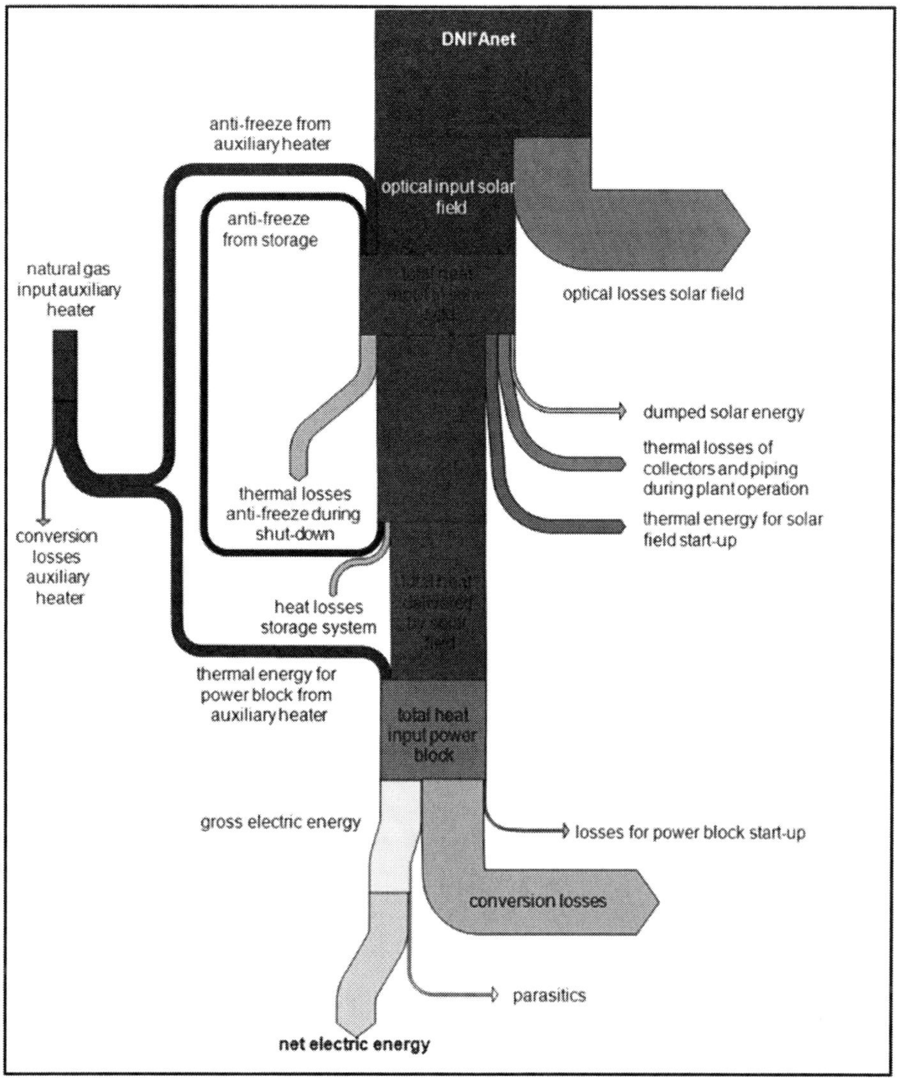

Fig. 14. – Sankey diagram for energy flows in a molten salt line focusing system [3].

and solar resource. In order to discuss this effect, the boundary conditions of different solar reference plant concepts investigated by IRENA [3] are summarized in table I. The concepts comprise of trough and tower technology with a certain capacity and storage size investigated for three different solar conditions that are representative for typical CSP markets.

Performance calculations of those concepts have been conducted with the numerical tool Greenius ® [3] and are shown for trough systems in table II and for tower systems in table III. Advancements in technology until 2025 are considered (*e.g.* replacing thermal

TABLE II. – *Electric production and system performance figures of a* 160 MW_e *parabolic trough system for state of the art technology (thermal oil heat transfer fluid) and estimations for 2025 (molten salt heat transfer fluid)* [3].

	Net electrical output [GWh / a]			Net plant efficiency [%]		Capacity factor [%]	
	2015	2025	Variation	2015	2025	2015	2025
DNI 2000	526.6	551.3	+4.7%	15.2%	17.5%	37.6%	39.3%
DNI 2550	575.9	624.6	+8.4%	14.9%	16.6%	41.1%	44.6%
DNI 2900	639.7	683.6	+6.9%	15.1%	16.9%	45.6%	48.8%

TABLE III. – *Electric production and system performance figures of a solar tower reference system for state of the art technology and estimations for 2025* [3].

	Net electrical output [GWh / a]			Net plant efficiency [%]		Capacity factor [%]	
	2015	2025	Variation	2015	2025	2015	2025
DNI 2000	574.0	610.0	+6.2%	14.4%	17.1%	43.7%	46.4%
DNI 2550	598.4	644.0	+7.6%	15.5%	18.3%	45.5%	49.0%
DNI 2900	656.6	684.4	+4.2%	15.6%	18.2%	50.0%	52.1%

oil by molten salt in parabolic trough in 2025) by different sets of technical parameters. The capacity factor shown here is defined as the annual electricity energy provided by the power plant divided by the nominal capacity multiplied by 8760 hours. Tower systems are expected to have slightly higher efficiencies compared to parabolic trough systems and are capable to achieve higher capacity factors.

5. – Cost

Based on information of commercially built power plants on investments of the different components and subsystems as well as on typical surcharges for engineering, procurement, construction (EPC), and owner's cost for the project development, land preparation and other factors, investments cost for the defined trough and tower reference concepts were calculated as shown on the left in fig. 15 and fig. 16. Technology advancements and other factors that will be discussed later are expected to lead to a cost reduction until 2025 as presented on the right-hand side of the figures mentioned. The solar concentrator (that consists of a solar field in the case of the parabolic trough and of a heliostat field, a tower and a receiver for the tower system) has the major cost share

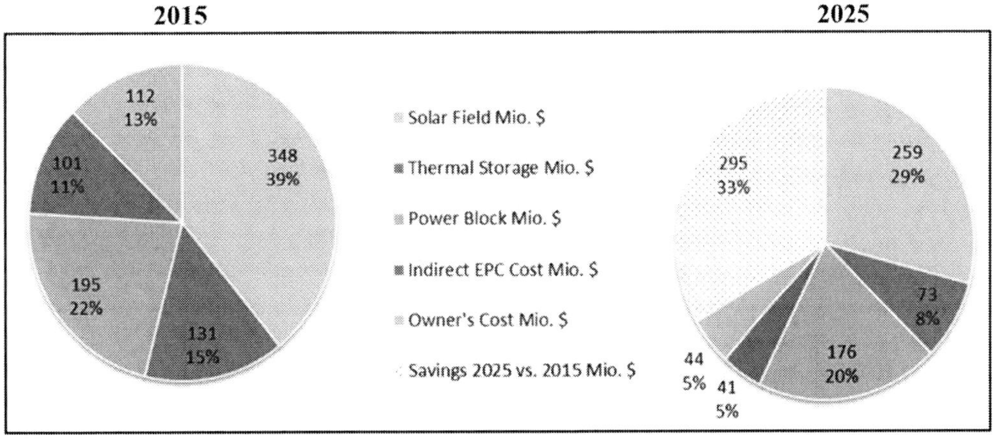

Fig. 15. – Comparison of CAPEX structure in 2015 and 2025 for 160 MW parabolic trough power plant in Mio US$ [3].

followed by the power block. Today, high surcharges in EPC and owners cost exist due the early market situation of the technology.

In order to compare the technologies, levelised cost of electricity (LCOE) are calculated using the following equation:

$$\text{(8)} \qquad \text{LCOE} = \frac{C_{invest} \cdot (f_{annuity} + f_{ins.,ann.}) + C_{O\&M,ann.} + C_{elec} + C_{fuel}}{E_{el,net,ann.} \cdot f_{avail,plant}},$$

with C_{invest} being the capital investment, $C_{O\&M,ann}$ the annual operation and maintenance costs, C_{elec} the annual electricity costs (if any), C_{fuel} the annual fuel costs (if any),

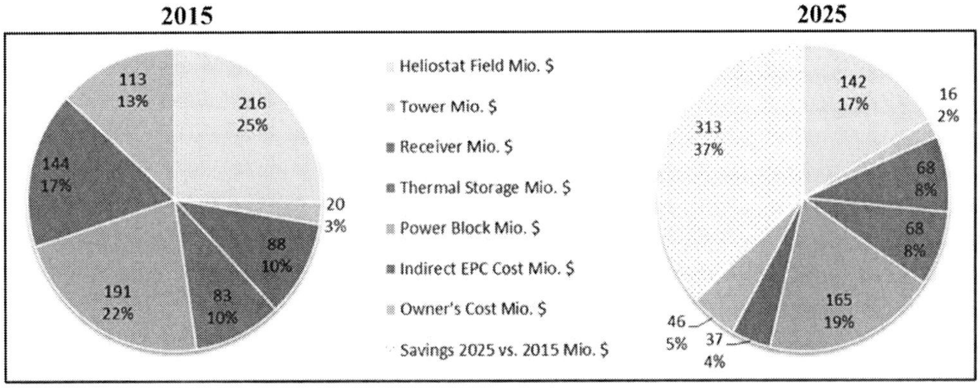

Fig. 16. – Comparison of CAPEX structure in 2015 and 2025 for 150 MW solar tower power plant in Mio US$ [3].

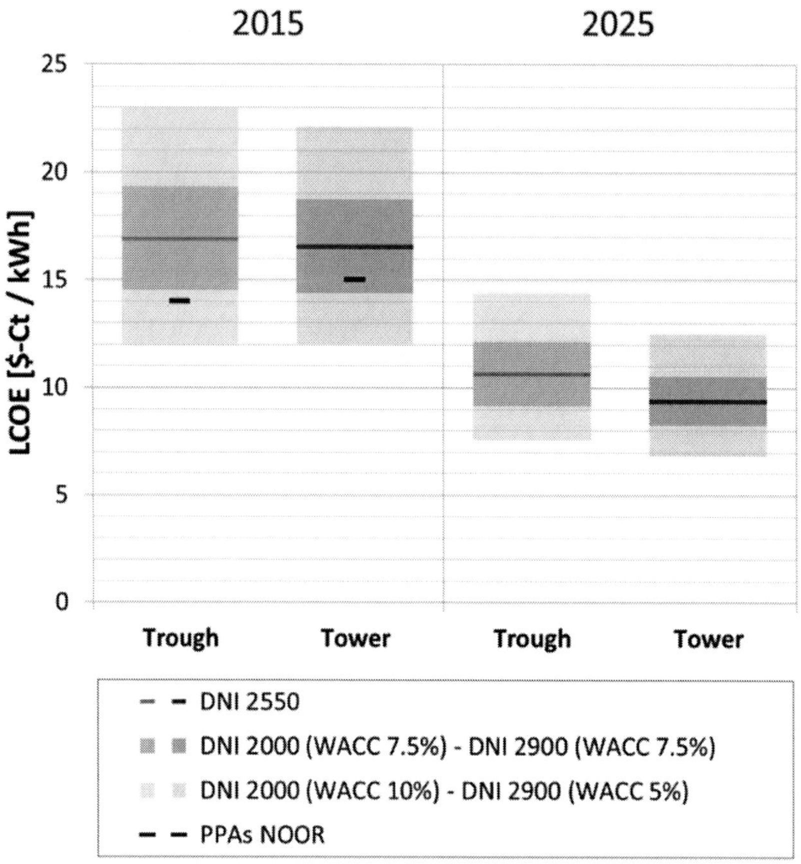

Fig. 17. – LCOE in 2015 and anticipated by experts for 2025 for parabolic trough and tower technology, different solar resource and financing conditions [3].

$E_{el,net,ann}$ the annual net electricity production, $f_{annuity}$ the weighted average cost of capital factor, $f_{ins,ann}$ the fraction of capital costs needed to cover the annual insurance costs and $f_{avail,plant}$ as the average annual plant availability.

The LCOE in US \$cents/kWh is compared in fig. 17 between trough and tower systems for the year 2015 and 2025. The colored margin shows the impact of a variation in the solar resource and in different financing conditions expressed by the factor weighted average cost of capital (WACC $= f_{annuity} + f_{ins,ann}$). In addition, electricity prices from power purchase agreements announced in 2015 of a trough and a tower plant to be built in Morocco at around 2500 $\mathrm{kWh/m^2 \cdot a}$ (Noor II &III) are indicated in the picture. They are below the average expected value of the model calculation because financing costs of those projects where lower than assumed in the model calculations due to the availability of soft loans provided by development banks.

Cost of troughs and towers are very similar and are expected to be reduced significantly over the next decade. Trough technology may keep up with the tower system if it is possible to replace the thermal oil heat transfer fluid by molten salt as it is state of the art in tower technology already today. In 2015 CSP electricity prices range today between 12 and 20 \$cents/kWh depending on financing and site conditions. In 2017 new CSP tower plants were offered in at rates of 7.3 \$cents/kWh for a project in UAE with a DNI in the range of 2200 kWh/m^{2}aA mojor reason for this are the low financing cost that could be achieved in this project. In Chile under the world's best solar resource of more than 3000 kWh/m^{2}a a price of less than 7 \$cents/kWh was offered in the same year. These figures are surprisingly low and were anticipated by most experts only to be achievable in 2025. With this, CSP is approaching a level where first unsubsidized niche markets come into reach. This would help to accelerate the market growth of this technology and enable further cost reduction through learning.

5˙1. *Cost drivers.* – As shown in the previous section, both parabolic trough and tower technology offer significant cost reduction potentials. However, the cost drivers for both technologies are quite different.

In general, LCOE reductions can be achieved either by improving the efficiency of the plant resulting in a higher electricity output or by reducing the costs (CAPEX and/or OPEX). Performance improvements as well as cost reductions contribute to the expected LCOE reductions considered by experts to be reachable until 2025. Figure 18 gives an overview of the main cost drivers for both technologies.

For trough technology, two major cost drivers can be identified. The increase of the temperature level enabled by the switchover to molten salt as HTF leads to a higher cycle efficiency of the power block and lower investment costs for the thermal storage due to the size reduction by more than 50% and the omission of a heat exchanger. Both effects together account for 1.9 \$cents/kWh of LCOE reduction. The second important cost driver is the reduction of the solar field costs which is closely related to the usage of a trough collector with wider aperture and thus less collector units. Together both factors lead to LCOE reductions by 1.75 \$cents/kWh. The residual LCOE reduction is mainly achieved by the reduction of indirect EPC and owner's cost which are assumed to reduce proportional to the direct EPC cost.

For tower technology, the major cost driver is the gain of experience. Since many of the EPC contractors and project developers built their first tower projects in 2015, risk margins are still high, the commissioning phase is extended and the operational costs are higher (at least in the first years of operation) compared to mature trough technology. The lack of experience in 2015 is represented by three factors that are altered compared to the values for troughs. The plant availability is assumed to be only 0.93 in 2015 instead of 0.99 in order to account for longer outages due to unscheduled additional maintenance and replacement of broken components. Extended commissioning phases with replacement of broken components could be seen for several tower projects built until now. These additional maintenance and commissioning efforts are also responsible for the assumption of increased O&M costs (2% of CAPEX in 2015 instead of 1% in 2025). Indirect EPC costs are expected to be 9%-points higher in 2015 due to additional risk surcharges within the supply chain. The mentioned three parameters are only altered

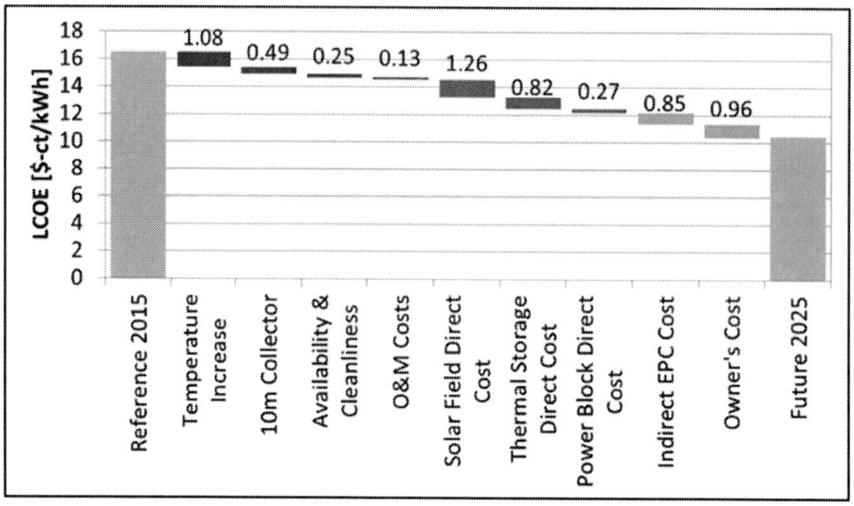

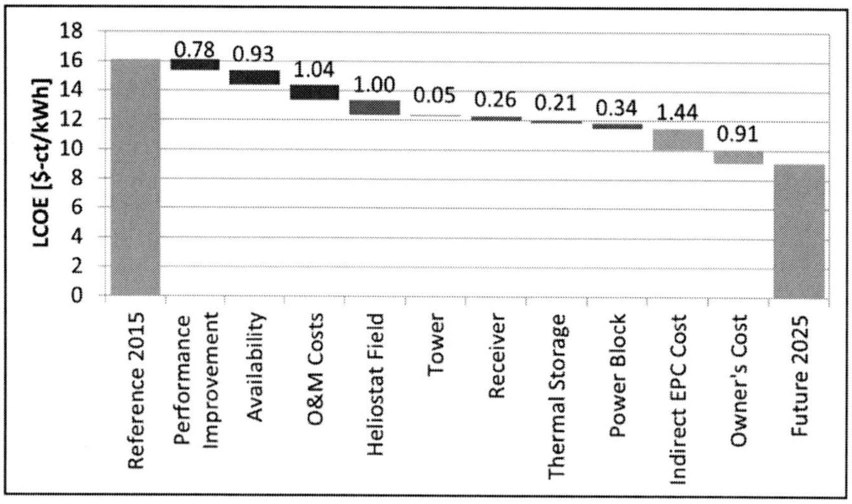

Fig. 18. – Contributing factors to the LCOE reduction from 2015 to 2025 for trough (top) and tower (bottom) systems: Performance improvement (violet), OPEX reduction (red), direct CAPEX reduction (dark green) and indirect CAPEX reduction (light green) [3].

for the year 2015. In 2025 the percentages for all indirect costs and availability are the same as for troughs which reduces the LCOE by 3.4 $cents/kWh. With the exception of the heliostat field, the reduction of direct EPC costs is marginal. Performance improvements excluding availability have about half as much influence for tower technology (0.78 $cents/kWh) as for trough technology (1.08 + 0.49 = 1.57 $cents/kWh).

6. – Summary and conclusion

In solar thermal power systems, concentrating solar collectors provide high temperatures to a power engine to generate mechanical energy that is converted to electricity using a generator. It benefits from a mature power plant technology that has been optimized for almost a century using heat based on fossil or nuclear resources. Today, several options to concentrate direct sunlight are pursued. In linear concentrators like the parabolic trough or linear Fresnel systems, a line focus is achieved by one-axis tracking of 1d curved reflector segments. In point focusing systems like towers systems, a point focus is generated by 2d curved reflector segments and two-axis tracking. All options can be combined with thermal energy storage to provide energy on demand. Thus, technology concepts that are based on molten salt both as a storage and as a heat transfer fluid are in the focus of the current deployment. A significant cost reduction noticed in recent bids for commercial projects brings CSP power plants at the edge of full competitiveness. Further cost reductions appear to be feasible for both, trough and tower systems. Both technologies bear advantages and drawbacks that do not allow a clear preference for one technology today.

REFERENCES

The work has been compiled partly from previous publications of the author:

[1] Pitz-Paal Robert, *"High Temperature Solar Concentrators"*, in *Solar Energy Conversion and Photoenergy Systems*, edited by Blanco Galvez Julian and Malato Rodriguez Sixto (Eolss Publishers, Oxford, UK) 2007.

[2] Pitz-Paal Robert, Giuliano Stefano and Wittmann Michael, *"Concepts for Cost Reduction in CSP Power Plants"*, in *Energie Erzeugung, Netze, Nutzung - Vorträge auf der DPG Frühjahrstagung*, Berlin 16.-18.3.2015 (Deutsche Physikalische Gesellschaft, Berlin) 2015, pp. 42–45. ISBN 978-3-9811161-7-5.

[3] Dieckmann Simon, Dersch Jürgen, Lüpfert Eckhard, Pitz-Paal Robert, Giuliano Stefano and Puppe Michael, *Assessment of the Cost Reduction Potential for Concentrating Solar Power Until 2025*, study performed by the German Aerospace Center (DLR) for the International Renewable Energy Agency (IRENA) 2016.

EPJ Web of Conferences **189**, 00009 (2018) https://doi.org/10.1051/epjconf/201818900009
LNES 2017

Energy storage

T. M. Taylor
CERN - Geneva, Switzerland

Summary. — The world is moving in the direction of using electricity derived from renewable sources. With the mass production of devices to enable the production of energy from such sources (wind, solar and tides), peak power capacity is competitive with that of deriving the electricity from fossil fuel. The problem is that the profile of power production does not match that of the requirements. This problem is not new, but it has become more acute: due to the extremely peaky nature of the power from the new sources, unless there is efficient energy storage, it is necessary to maintain the capacity to generate power to cover the full load in certain climatic conditions. While nuclear power is clean and would be ideal for satisfying the requirements of base load, the capital cost of nuclear plants (which should include that of decommissioning) makes the real cost of such power more expensive than that which can now obtained from renewables (and from burning gas or coal, of course). Ever since there have been electrical grids, companies have looked for ways to efficiently store energy so that power can be delivered on demand with minimum capital investment. A wide range of technologies is employed, and is being developed, so that everyday needs can be met. Energy storage is resource neutral: it allows us to use electricity from any power source more efficiently. However, it involves considerable capital investment, and can occupy a lot of space. The idea of moving towards reliance on renewable sources of energy has been taken up by a large segment of the population but there is technically still a lot to do, especially with regard to the essential corollary of energy storage. In addition to the challenge of producing the means to store energy, many other issues must be addressed. These include maintainability and longevity of the equipment (both for producing the energy and storing it). In this report we explore and compare the different methods available for storing energy and investigate trends.

1. – Fundamentals

There are several ways to store energy. When applied to electrical energy, they must be chosen to take into account the *location* and *cost* of the storage, and its *required duration.*

– potential energy $m \cdot g \cdot h$,

– kinetic energy $1/2\, m \cdot v^2$,

– magnetic energy $\int_\tau (B^2/2\mu_0)\mathrm{d}\tau = 1/2\, L \cdot I^2$,

– capacitive energy $1/2\, C \cdot V^2$,

– chemical energy reversible reaction,

where m is mass (kg), h is height (m), v velocity (m/s), L inductance (H), I current (A), C capacitance (F), V voltage (V), and g is the Earth's gravitational constant (~ 9.81 N/kg) and μ_0 the permeability of free space ($4\pi\, 10^{-7}$ H/m). $\int_\tau$ represents the volume integral.

In order to appraise and compare systems, it is necessary to understand the concepts of *entropy* and *exergy*, and when evaluating costs to take into account that of ancillary equipment and the value of occupied space. Associated with each method there is the *energy density* and the *cycle efficiency.* Processes that involve the conversion of heat into motive power or electricity are subject to *Carnot efficiency.* Reference [1] is a good modern source book on the subject of storing energy, with special reference to renewable energy.

2. – Potential energy

Pumped water is the traditional way of storing electrical energy (PHS or PHES), particularly in places where hydroelectric power is significant (*e.g.* the Alps). When power is cheap, pump water into the elevated lake (see fig. 1). For example, a mountain lake of 10 km^2, 1 km above another lake, will store about 10^{11} J ($= 10^2$ GJ) per m of depth. As 1 MWh = 3.6 GJ, this is equivalent to 27.8 MWh, and assuming a roundtrip efficiency of 75% (industry standard) this gives about 100 MWh per 5 m depth of water. At this scale the energy storage is clearly in the suppliers' hands.

Old mines are sometimes used for pumped sub-surface energy storage. Even in places without high mountains, large dams and suitable unused mines, this method is quite popular for storing electrical energy (on a smaller scale) by utilities, but is not appropriate for private use. One could consider pumping water from a well to a container at the surface, and returning the water to the well via turbines. This could possibly be considered for remote locations, but care would have to be taken to avoid contaminating the water table. For private use, it has also been suggested to use surplus power to raise a mass of say 20 t to a height of say 15 m, with appropriate gearing to a generator, but the result is not very attractive except perhaps where space is abundant because the

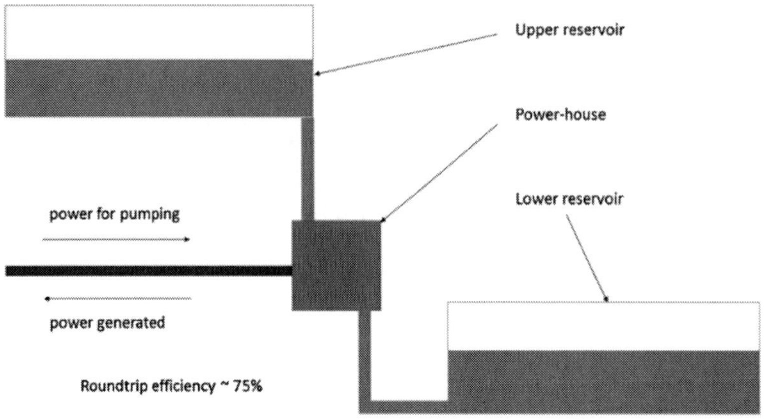

Fig. 1. – Schematic of pumped water energy storage.

energy stored would be about that one could store in 2 car batteries! Hardly appealing, except in very particular circumstances. . . .

Pumped storage is presently by far the largest form of energy storage capacity for grids, and accounts for 95% of storage installations worldwide, with a capacity of about 300 GWh, and a generating capacity of > 150 GW. There is a wealth of knowledge about the associated engineering [2].

An idea being pursued is that of an *underwater reservoir* where energy is stored by pumping water out of a spherical shell and recovered by letting the pressure of the surrounding water drive the turbine as it flows into the sphere. The advantages of this method are that only cables are required (no pipes) and space is not a problem. A trial has been successfully undertaken in Lake Constance [3], and it is confidently predicted that if multiple units were to be mass produced and installed close to big cities on coasts close to deep water (*e.g.* in Japan), such systems could store hundreds of GWh, and be cost competitive.

Other ways to use potential energy include *compressed air* (or nitrogen), *liquefied gas* (air, nitrogen), and *heat*.

Compressed air energy storage (CAES) was first used in the 1870s. This has progressed to utility scale installations, *e.g.* 290 MW (1970) [4]. Standard storage pressure is 70 bar, and the compressed gas is used to drive a turbine coupled to a generator. One must take into account that the air heats up (to about 150 °C) during compression, and indeed use this energy appropriately. Such heat of compression can be recovered and used to reheat compressed air during turbine expansion, to give a cycle efficiency of up to 70% —but it does require storing the heat during the intervening time. As storage density is low this requires very large volumes, pointing to the use of salt caverns. In isothermal CAES, heat is stored at up to 600 K in oil [5]. Another idea is to use wind turbines to compress air with a mechanical link for storage in inflatable underwater bags, for decoupling the generation of power from is production. RWE (D) is planning to use this storage method [6].

Pumped heat (thermal energy) storage (TES) is another method [7]. This uses two large steel tanks filled with mineral particulates (gravel). The system is charged by compressing argon at ambient temperature to 12 bar, heats to 500 °C, storing the energy used to power the motor generator (*A monatomic gas (e.g. argon) is used as the working gas: it heat/cools more than air for the same pressure change, reducing storage cost*). Argon is injected at the top of tank 1, slowly descends, exits at the bottom at room temperature (RT) and $\sim$ 12 bar, whence it passes to the expander, the temperature is reduced to -160 °C and it is injected into the bottom of tank 2 to exit at the top at RT. To recover the energy, the process is reversed. The argon is cooled to -160 °C in tank 2, it is then compressed to 12 bar (at RT) and sent to tank 1 where it is heated and the hot gas expanded gas via a turbine to generate power. The round trip efficiency of such a process is about 75%.

An alternative way to store thermal energy is *liquid air energy storage* (LAES) (also referred to as *cryogenic energy storage* (CES)). The charging system is an air (or N_2) liquefier, using electrical energy (700 litres of ambient air become 1 litre of liquid air). The liquid air is stored in an insulated tank at low pressure, which functions as the energy store. Such equipment is used globally for bulk storage of liquid N_2, O_2 and LNG and can hold GWh of energy. When power is required, liquid air is drawn from the tank(s) and pumped to high pressure. The air is evaporated, is superheated to ambient temperature and the warm high-pressure gas drives a turbine. A proprietary high-grade cold store, to be used later during the liquefaction process to enhance efficiency, captures the cold from the expansion. An example of a pilot system of this type is a 2.5 MWh installation in the UK [8].

3. – Kinetic energy

Flywheels are a confirmed method to store significant quantities of energy for a limited time, *e.g.* energy is stored during deceleration of a vehicle (instead of dissipating it in the brakes), for re-utilization afterwards to provide additional acceleration, or for providing pulsed power for particle accelerators. Flywheels were already years ago used to power buses (using mechanical gearing rather than electricity). Modern composite materials are ideal for flywheels that spin at very high revolution frequencies (tens of thousands of rpm) in a partial vacuum enclosure (to reduce drag), and supported via magnetic bearings. These devices are sometimes used to avoid transmitting pulsed energy to the electrical grid, and for optimizing the power factor. Such *Kinetic Energy Storage Systems* (KESS) are based on an electrical machine joined to a flywheel. To store energy, the machine works as a motor and the flywheel is accelerated until it stores the nominal energy. When the system provides energy, the electrical machine works as a generator and the flywheel decelerates. The aim is to store electrical energy when it the load is light and to provide devices with additional electrical power when they need it, without having to draw excessive power from the network. These systems can also be used to provide ride-through power and voltage stabilization for power quality and power recycling applications. A typical commercial product is a 10 MJ on-board flywheel system for use in trains that can deliver a burst of power of up to 300 kW [9]. The 375 mm diameter flywheel for such a system weighs 350 kg and rotates at 5000 to 10000 rpm: the mass of a complete system is 1.5 t.

At another scale are the large motor generators used in conjunction with powering the pulsed magnets of particle accelerators and tokamaks. The magnetic field in the 101 bending magnets of the CERN proton synchrotron (CPS) is cycled between 0.11 T at injection energy of the protons (1.4 GeV) to 1.25 T at top energy (26 GeV). The repetition time is 2.4 s. About 6 to 8 million cycles are executed each year. The impedance or the magnets string is 0.32 Ω and 0.9 H and the current at top energy is 5.5 kA. The active power for operating this string peaks at 40 MW at the end of the acceleration, at which time the increase in stored magnetic energy also reaches a maximum. In order to avoid perturbing the network, from the start, in 1959, a *flywheel motor-generator system* was used (fig. 2). The mass of the rotor was 90 t, and the maximum stored energy 233 MJ (@ 1000 rpm). The speed of the rotor decreases by 5% during ramp-up while the generator absorbs 6 MW; during ramp-down, the stored energy flowing from the magnets, peaking at 12 MJ, reaccelerates the rotor to nominal speed [10]. The operation of the accelerator relied on this device working: it obviously contained moving parts, so preventive maintenance was crucial. This was outsourced, but over the years it became increasingly difficult to find a competent supplier, and the device was recently replaced by a capacitor bank (see below). The world's largest such device supplied the toroidal magnet of the JT-60 tokamak. The 650 t flywheel, built from 6 carbon steel disks, is installed in a 12 m diameter and 12 m deep vertical shaft, and features a water-cooled heat exchanger [11]. Once every 10 minutes the device can generate a current of 6.9 kA at 18 kV, with a discharge energy of 4 GJ. As for the case of the CPS machine, the main purpose of such a system was to avoid periodic disruptive overload of the network, rather than to store energy *per se*. However, with the advent of more efficient batteries and super-capacitors, interest in the use of flywheels is on the wane.

4. – Energy storage in magnets

There is no doubt that magnetic fields can be used to store significant energy that is directly available in electrical form without passing through a generator. For example, the 3.8 T solenoid used with the CMS experiment at the CERN LHC stores 2.6 GJ, but it is very large and cost over $100M. To put things into perspective, a field of 2 T in a volume of 1 m^3 (of air or vacuum) stores an energy of $(40/8\pi) \cdot 10^6$ J $\approx$ 1.6 MJ, *i.e.* about the quantity of energy stored in a car battery. This is therefore an application requiring higher field (*i.e.* using superconducting magnets). If in the above example the field is pushed from 2 T to 10 T the stored energy is increased to 40 MJ, and one can readily envisage larger volume magnets (toroidal or solenoidal) of a relatively simple design. For superconducting magnets we need cryogenics —an additional headache (and cost). Demonstrator helium-free HTS coils, cooled using cryocoolers, have been made for storing a few MJ. Cost is problem: for 10 MJ at least 1 M$. However, there could be niche applications. For example, the Japanese have been quite active in studying the

Fig. 2. – The CERN PS motor generator (photo CERN).

use of *superconducting magnetic energy storage* (SMES), and the corresponding design of suitable magnets, with regard to

– use in conjunction with medical accelerators (10 MJ scale),

– use in conjunction with large accelerators (100 MJ scale).

The idea is to reduce the power footprint of medical synchrotrons, and the cost/performance of a 4 MJ SMES system to compensate load fluctuation of synchrotrons for medical use has been studied [12]. The study was based on a small scale SMES that had already been demonstrated technically as being suitable for this application. Power is saved by compensating input and output using SMES. The depreciation time of the SMES system (cost \$3M) was found to be about 20 years, assuming an annual maintenance cost of \$50k. It is nevertheless interesting, as the power saving would be about 1.2 GWh per year per facility. As it is planned to construct a medical facility in every prefecture, *i.e.* 50 facilities in Japan, there is the potential for a saving of 60 GWh/year nation-wide. General use of this technique would have to pass through the stage of a practical demonstration. A major difficulty is the lack of off-the-shelf industrial components, which means that the cost is high. And there is the eternal constraint that potential users are of a conservative nature. What we take away from this study is that if the cost of the SMES could be halved, it would clearly be the way to go! There are no moving parts, so maintenance is easy, and it is environmentally friendly. However, from both the economic and environmental standpoints it is important to plan to install the

SMES system at the beginning of the planning of construction. Magnet designers should pay attention and come up with a cut-price solution.

Another study [13], also in Japan, addressed the opportunity of using SMES on a larger scale (as an alternative to the flywheel solution for the energy upgrade of the J-PARC accelerator). For this accelerator, the energy to be delivered by SMES is 21 MJ. If it is assumed that 30% of the capacity of the SMES system is used for this purpose, the system must be designed to store 70 MJ. The study proposed using a modular solution —as there are six power converter modules, there would be six SMES modules, affording some economy of scale. An economic magnet design based on the use of Nb-Ti superconductor in the form of a flat cable (as used for the LHC dipoles) and suitable for series production was also proposed.

5. – Energy storage in capacitors

This is a dynamic sector, due to the development of self-healing components for moderately high voltage applications. The public is becoming aware of the possibilities thanks to the incorporation in "stop-start" in cars, where it reduces the peak demand on the battery when starting the engine. Its use is attractive due to its passive nature (no moving parts, no cryogenics), and the mass production of components. But the energy density is low, even compared with that of batteries, so one needs plenty of space, and there is some concern that the components may be delicate. The nature of this form of energy storage is such that it is often assimilated into the category of batteries.

If it fails, a capacitor will fail in one of three ways:

- *Capacitor short-circuit.*

  The capacitor gets short-circuited with very high loss factor. It occurs when the dielectric breaks down completely and film melts. This is severe case.

- *Capacitor open-circuit.*

  The capacitance value drops by more than 30%. There is no visible damage to the film metallization. This failure is due to poor metal spraying, or to defective welding of terminals.

- *Degradation of capacitance.*

  There can be two reasons for this: excessive concentration of self-healing at the weak spots of the dielectric, leaving other areas unaffected; metal oxidation.

Clearly the essence is very strict quality control. But with no moving parts, no cryogenics, and directly available electrical power, such systems are confidently projected to last for hundreds of thousands of cycles. This was the reason for the choice of this method for the replacement of the motor rotor system for the CERN PS accelerator. Studies of alternative solutions were started in 2003 [14]. The use of batteries was discarded due to the limited lifetime; SMES was discarded due to lack of standard industrial products; energy storage in capacitors was preferred, basically for two reasons:

Fig. 3. – Capacitor energy storage for powering the CERN PS (photo CERN).

– capacitors support practically *unlimited discharge cycles*,

– a modular solution could be based on *industrial components*.

The capacitors are part of a new system, in operation since 2011, which integrates three functions: it converts AC current from the network to DC current as required; it charges the capacitors with energy for pulsing the magnets; when the energy is not needed it is stored in the capacitor banks. The six capacitor banks are connected to the magnet string via six DC/DC converters, which precisely control current and voltage in the magnet circuit, independent of the voltage of the capacitors. The capacitor voltage decreases from 5 kV to 2 kV during the ramping to top energy and increases again to 5 kV during ramp-down. The energy density is low and the banks of capacitors for this system fill six standard 40-foot shipping containers as shown in fig. 3. This exhibits the principal disadvantage of capacitors for storage of large quantities of energy: the very large footprint. In addition, while the capacitors are expected to be reliable and long-lived, one has to consider the reliability of the necessary ancillary equipment. However, as all the components are of an industrial nature, with proper attention to quality control, these issues should be soluble.

6. – Chemical energy with reversible reaction. Batteries

There is an immense effort put into improving rechargeable batteries for large scale energy storage [15], in particular for powering cars and for domestic use (in combination with photovoltaic (PV) power production, this can serve to reduce reliance on networks). As was the case for PV, mass production and competition are driving down costs, opening the market to more users and enhancing the effect. Power utilities in certain cases are even getting interested (note the deal between TESLA and an Australian producer). Batteries are in the news —especially for application to cars, stimulated by the realization that it could be envisaged to replace the internal combustion engine reducing pollution in urban centres and possibly reducing the associated production of CO_2, with its suspected impact on climate change. The vigorous development, increasing demand, and competition, is leading to improved performance and reductions in cost that when extrapolated support the interest in the field. Reported problems due to volatility of material cost should prove to be temporary. The major remaining issues are related to lifetime and reliability.

What is a battery? A device that converts chemical energy into electrical energy. It features a positive terminal (*anode*), a negative terminal (*cathode*), separated by an *electrolyte* that allows ions to move between the electrodes, giving rise to the flow of current. Just as for the other energy storage systems described here, this is not a new idea: Alessandro Volta developed the first battery in 1800. However, advances in technology have led to improved output and reliability of rechargeable battery systems, and economies of scale have led to cost reduction and ever more applications.

- *Lead-acid battery.*

  This has been (and still is) a "workhorse" rechargeable battery. It is based on mature technology, is mass-produced, and its primary component (lead) can be recycled. However, it is heavy, has a relatively short life ($\sim$ 5 years), and rather poor storage density. During the charging cycle hydrogen is produced, which can also be a problem. Most of the other batteries discussed here feature solid-state electrolytes.

- *Lithium-ion batteries* (R&D in 1980s, commercialized in 1991 by Sony and Asahi Kasei).

  The transfer of Li ions between electrodes during charge/discharge [16] characterizes this type of battery. The anode is an oxide (*e.g.* $LiCoO_2$ or $LiMnO_2$), the cathode typically graphite, and the electrolyte an organic carbonate (*e.g.* ethylene carbonate). The battery is sealed to keep out water, which reacts with Li. Advantages with respect to lead-acid batteries are superior energy density, low weight, no acid, and no H_2 during charging, but charge/discharge parameters require closer attention. First developed for use in mobile phones and laptop computers, the emergence of electrically powered vehicles (EV) has served to encourage scale-up of the technology, to improve performance and to drive down the cost. The *Li-polymer* type uses a polymer electrolyte and can be slim and lightweight, making it good for mobile phones, but the energy density is less good and cost is higher than the regular Li-ion type. Thanks to increased demand and vigorous R&D the technology is advancing rapidly, and this type of battery is on the way to dominate the market. Domestic owners of PV installations, eager to reduce reliance on the network, are becoming interested, but the present lifetime (defined as when retention falls to 70%) of about seven years and/or 500 charge/discharge cycles is a drawback. That aspect will need to be substantially improved if one hopes to persuade the networks to invest in this technology.

- *Nickel-cadmium* (Ni-Cd) batteries (in production since 1910).

  The advantage of this type with respect to lead-acid batteries are longer life, near maintenance-free operation, low but acceptable energy density. Historically such batteries have been less costly than lithium (due to long manufacturing experience), but this advantage is now declining. This type of battery has long been used in energy storage systems.

- *Nickel metal hydride* (NiMH).

  This is the type of battery that Toyota uses for its hybrid cars [17]. The traction battery is a sealed 28 module pack storing 1.31 kWh (201.6 V, 5.6 Ah) made by Panasonic. Energy density is 46 Wh/kg, maximum output power 1.3 kW/kg. Over 2 million Priuses with such batteries have been sold, but Toyota now equips higher end vehicles with 15 kg lighter Li-ion packs. NiMH batteries are said to last about 10 years, and this type of battery is also proposed for domestic use, being cheaper than Li-ion alternatives (but heavier and takes more space).

- *Sodium-sulphur (Na/S) batteries* (1960, Ford; now NGK, Japan).

  These batteries operate at 300 to 350 °C, so are not likely to appear in homes. The anode is molten sulphur, the cathode is molten sodium and the electrolyte is sodium alumina [18]. The battery exhibits a remarkable round-trip efficiency (up to 90%), and is widely used for energy storage in Japan (190 sites). The largest installation can store 245 MWh and deliver 34 MW.

Another type of battery is the *flow battery* [6]. This is a rechargeable battery where energy is provided by two chemical components, dissolved in liquids, separated by a membrane, making it similar to a fuel cell. The main discerning feature is that immediate recharge is possible by replacing electrolyte. It is nice to know that the technology exists, and there are certainly uses for this kind of battery.

Energy can also be stored as hydrogen, produced by steam reforming from hydrocarbons (95%) or electrolysis (5%), to be used to either generate electrical power on demand (via a fuel cell or a turbine), or directly for heating and cooking. At about 15% the round-trip efficiency for producing electricity is lower than many other methods, and most of the hydrogen is presently sourced from non-renewables (but these could be replaced by biomass). It may be convenient in some circumstances to produce H_2 by electrolysis where there is abundant solar energy, and as a subsidiary product in conjunction with water desalination plants. Hydrogen gas is highly inflammable, requiring special care in handling and appropriately designed double walled containers [19].

TABLE I. – *Comparison of the approximate energy density of common materials and devices.*

Material/device	Stored energy Volume (kWh/L)	Stored energy Specific (kWh/kg)	Notes
Gasoline[(a)]	9	13	Diesel and kerosene similar
LNG[(a)]	6	14	Propane, butane similar
Coal[(a)]	8	10	Especially dirty
Ethanol[(a)]	6	7	As a fuel
Wood[(a)]	3.5	4.5	Methanol and biomass similar
Liquid H_2[(b)]	2.5	39	Needs a cryostat
H_2 @ 700 bar[(b)]	1.5	39	Needs a pressure vessel
H_2 @ 1 bar[(b)]	0.003	39	Voluminous, highly combustible
Flywheel	0.2	0.15	Mass is tripled by containment gear
Liquid air	0.07	0.06	Needs a cryostat. Voluminous
Lithium-ion battery	0.5	0.3	Better versions in development
NiMH battery	0.25	0.1	As used in the Toyota Prius
Ni-Cd battery	0.15	0.07	Mature technology, robust
Lead-acid battery	0.15	0.05	Mature technology, recyclable
superconducting magnet	0.03	0.02	5 T. Needs a cryostat. Long life
Super-capacitor	0.015	0.01	Self-healing propylene. Long life

[(a)] Efficiency of producing electricity via turbines powered by burning these materials: 30%.
[(b)] Efficiency about 30% if used for heat, and 40–70% if used in a fuel cell, not counting the energy needed to produce the hydrogen (unless it is derived from non-renewable sources).

7. – Energy density

None of the methods to store energy described above can come anywhere near to the density, volume or mass, of energy that is stored in gasoline and similar substances. It is important to have an idea of the orders of magnitude of the energy density to guide rational decisions with regard to the choice of methods for producing energy from renewable sources, taking into account the storage factor, and eventually a judicious mix with a reduced quantity of cleaner types of fossil fuel, *e.g.* liquid natural gas. There is an environmental impact of building the equipment to produce renewable energy: *e.g.* one may ask whether it is preferable to produce storable methane from biomass or to use photo-voltaic (PV) panels in conjunction with a battery (or perhaps a well-chosen combination of the two, gas for cooking and PV for light). Hydrogen is a nice clean fuel —but either a lot of energy is required to produce it, or it is not from a renewable source. Some examples of approximate energy density are given in table I.

This table is very instructive: when looking at the whole picture the convenience and cost efficiency of natural gas-powered production of electrical, and of vehicles using the internal combustion engine cannot be denied. Moving towards a system where a significant fraction of energy come from renewable sources will require an immense effort on the part of society. It is perhaps out-of-fashion, but from the point of view of reducing pollution and production of greenhouse gases, it would be probably be preferable to replace the unavoidable fraction of power delivered from non-renewable sources by nuclear power, preferably based on the intrinsically safe thorium technology [20].

In the case of electric vehicles (EV), it is not obvious that one should aim at storing sufficient electrical energy in batteries for driving 500 km, when for 50 weeks of the year the average daily drive is 65 km. A plug-in hybrid with a range of 100 km fully electric would be lighter and cheaper, and require less infrastructure. Moreover, if in any case we admit that some power will come from fossil fuel this could as well be burnt in an internal combustion engine as in a power station. Alternatively, one could consider using an on-board generator to charge the battery, and which could be installed temporally for long trips.

8. – Are there alternatives to energy storage (ES)?

The reasons for needing to store energy, readily available in the form of electricity, are:

- *To integrate renewable energy into existing networks and energy independence.*

  ES supports the integration of renewable power. By reducing the load on fossil-fuel generation ES helps cut emissions —but more capacity to meet peak demand is required.

- *To reduce risk associated with power outages.*

  Today's electricity grid is vulnerable to threats from nature, terrorists, and accidents.

- *To save consumers money.*

  Energy storage (ES) lets customers avoid premium pricing during times of peak demand.

- *For the economy in general.*

  In addition to reducing economic losses from major and minor annual outages, ES will be a critical technology in the electricity grids of the future, and will thereby create economic activity. In this context it is sometimes said that a next-generation smart grid without energy storage would be like a computer without a hard drive.

However, there are alternatives to storage (or which could be used to complement it to make it more efficient). The problem of requiring very high peak power generating capacity to satisfy all demands at all times can be addressed in other ways. The hard-nosed market-based approach would be to accept some short periods (hours, days?) of power outage: this also provides an incentive to adopt local energy storage, at the personal or communal level. A pure market-based approach would also lead to massive increase of tariffs during bad peaks and serve to "train" consumers to adjust consumption. Or one could use technology to adopt a socially more acceptable approach, *e.g.* to install long distance power transmission lines to flatten demand by distributing consumption, addressing daily peaks by delivering surplus power to locations in other time zones. This can be done using either *high voltage DC power transmission* (800 kV overhead lines are in operation in Korea and China, but problematic in Europe and the USA, more prone to "not-in-my-backyard" (NIMBY) problems), or *high current DC power transmission via superconducting links* which can provide virtually loss-free transmission of GW proportions. Such links can be buried like pipelines. As an example, CERN, needing to power its high current magnet systems from distant radiation-free locations of power converters, has developed a magnesium diboride (MgB_2) superconductor and 20 kA cable for its own use [21]. Further to the CERN initiative, MgB_2 technology was proposed by Prof. Carlo Rubbia, erstwhile scientific director of the Institute for Advanced Sustainability Studies (IASS) in Potsdam, for an innovative transmission line for long-distance transport of power generated from renewable sources, and which has since been taken up by the BEST PATHS European project [22]. The relatively new MgB_2 superconductor, which could be cheaper than other superconducting materials if it were to be manufactured on a large scale, has a critical temperature of 39 K. This means that it can be used cooled to 20 to 25 K, presenting a lower cryogenic load than regular Nb-Ti superconductor that requires 4.5 K, and liquid hydrogen could be used as the coolant rather than the more expensive liquid helium. One could even envisage combining the transmission of liquid hydrogen in the pipeline to increase the transfer of energy. As well as being used to connect renewable energy sources (wind, PV) in flat areas with existing energy storage facilities such as dams in mountainous areas, electrical utilities can call upon superconductivity to minimize costs in other ways, *e.g.* to improve the efficiency of power generation from wind by using superconducting generators.

9. – Impact of markets and regulation on the development of energy storage

The enthusiasm with which the world is embracing the idea of moving to generating an ever-increasing quantity of electrical power from renewable sources is certainly commendable. However, the combination of the intermittent production of a significant fraction of the demand for power, due to the nature of wind and solar components is having a strong impact on the market, based on regulation and contracts drawn up in a previous era. This is in particular, and paradoxically, true for energy storage. Wind and solar need storage capacity, yet the present situation with times where the energy production exceeds the demand, driving down prices, interferes with the business models of generation of hydroelectric power and pumped water energy storage to such an extent that maintenance is being neglected and plans to increase capacity are being put

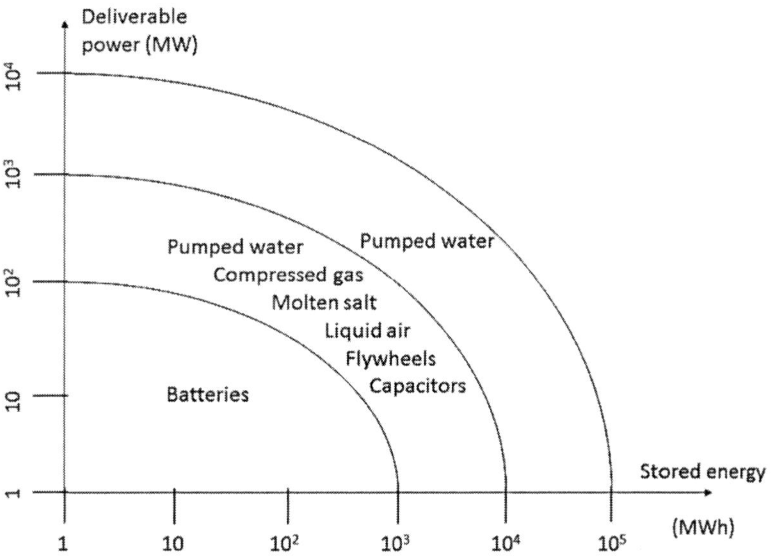

Fig. 4. – Comparative overview of the potential of energy storage methods.

on hold. This is in contradiction with the most pressing societal needs, *i.e.* to increase storage capacity. By far the most efficient way to store large quantities of electrical energy, at the level of the network, is by the pumped water method. It is crucial that this distortion be recognized and the market regulated to counteract the effect, by coupling the production of peaky electrical power with a requirement to invest in energy storage, to render the power less peaky (*e.g.* by taxing surplus power and using the proceeds to invest in pumped water storage). In addition, individual households and communities should be encouraged to invest increasingly in battery storage, as the price of such equipment comes down, and one could probable use efficiently the storage capacity of batteries in electric vehicles when they are not being used (which is most of the time...). These batteries could be charged during the daytime when power from PV parks is (should be) cheap, to be used in the evening or early morning. However, this will require that the distributors of power, which presently profit inordinately by charging customers historical daytime rates, pass on the low prices. As the distributors usually benefit from a monopoly position, this will require some regulation.

10. – Comparative overview of the field of energy storage

As seen today, it would appear that the two methods that will dominate the approach to storing electrical energy generated from renewable resources in the near future will be pumped water and batteries (see fig. 4). The vigorous R&D devoted to lithium batteries is such that these will rapidly dominate the market for low to medium power application, *e.g.* for electric vehicles and local electrical storage. The sodium-sulphur, lead acid, and

NiMH will continue to be used because of the long experience with production and application, but the market for these types is unlikely to expand. For storage of large quantities of electrical energy, the traditional method of pumped water, either between reservoirs or in and out of pods located in trenches under the sea, will continue to be the preferred technology. It is to be hoped that ways will be found to couple this resource to energy derived from wind and PV in such a way as to enhance its overall quality. The other methods referred to in this report will continue to have niche applications, but it seems unlikely that in the near future they will rise to challenge the new generation of batteries and the established dominance of pumped water storage.

11. – Conclusion

From the very beginning, the electrical grid system designers have been aware of the importance of finding ways to store energy. Most of the methods have been in use for quite a long time, but (apart from pumped water) on a relatively small scale. This was also because fossil fuel is cheap. But the realization that the side effects of massive use of fossil fuel (pollution, possible effect on climate change, and eventually limited supplies), the world has woken up to the fact that we should move to using renewable sources, *e.g.* wind, PV and biomass. Electrical power from wind and PV fluctuates widely, focusing on the importance of storage, but electrical power and electronic engineering has advanced to make DC-DC, AC-DC and DC-AC conversion highly efficient, and with components readily available many new possibilities are opening up. Economies of scale have made generation of power from renewable sources competitive. This is already a highly dynamic field of activity. According to market research, the energy storage market is set to rise to an annual installation size of 6 GW in 2017 and over 40 GW by 2022, from a base of 0.34 GW installed in 2012 and 2013. Over a thousand companies serve the energy storage industry. Pumped water, heat, flywheel, battery and capacitor systems are operating today in the competitive ancillary services power market with fast and accurate response to distribution signals. The market for solar panels —which was less than $200 million in 2012— will be about $19 billion in 2017, and much of the energy produced will have to be stored- at least for a few hours. Energy storage is already big business, and it is set to become much bigger still. One should nevertheless pay attention to other technologies that could affect storage requirements, *e.g.* by increasing long distance power transmission capacity and smartening the grid.

REFERENCES

[1] Letcher T. M., *Storing Energy* (Elsevier) 2016.
[2] Jones G. M. (Editor), *Pumping Station Design* (Butterworth-Heinemann, Elsevier) 2011.
[3] `https://www.energiesystemtechnik.iwes.fraunhofer.de/`.
[4] Lockerby R. W., *Compressed Air Energy Storage* (Vance Bibliographies) 1983.
[5] Breeze P., *Solar Power Generation* (Elsevier) 2016, p. 39.
[6] Crawley G. M. (Editor), *Energy Storage* (World Scientific) 2017, p. 92.

[7] DINÇER I. and ROSEN M. A., *Thermal Energy Storage: Systems and Applications* (Wiley) 2011.
[8] http://www.highview-power.com/technology/.
[9] http://elytt.com/#!modules/pages/home.
[10] BURNET J.-P., Capacitive energy storage replaces flywheel, p. 68 in http://www.worldscientific.com/worldscibooks/10.1142/9921#t=toc.
[11] MATSUKAWA T. *et al.*, "A 215 MVA flywheel motor-generator with 4 GJ discharge energy for JT-60 toroidal field coil power supply system", *IEEE Trans. Energy Conversion*, **EC-2**, No. 2 (1987) 262.
[12] SATO H. *et al.*, "Application of energy storage system for the accelerator magnet power supply", *Proc. Int. Particle Accelerator Conference (IPAC-10)* (2010) p. 3236.
[13] SATO H., SHINTOMI T. *et al.*, "Electric power compensation of the large scale accelerator using SMES", *IEEE Proc. Particle Accelerator Conference (PAC'07)* (2007) p. 239.
[14] FAHRNI C., BURNET J. P. *et al.*, "A multilevel power converter with integrated storage for particle accelerators", in *Proc. Power Converter Conference (PCC'07)* (2007) p. 1480.
[15] MENICTAS C., SKYLLAS-KAZACOS M. and LIM T. M. (Editors), *Advances in Batteries for Medium- and Large-scale Energy Storage* (Woodhead Publishing, Elsevier) 2015.
[16] http://batteryuniversity.com/learn/article/types_of_lithium_ion.
[17] http://www.hybridcars.com/hybrid-car-battery/.
[18] SUDWORTH J. L. and TILLEY A. R., *The Sodium Sulphur Battery* (Chapman and Hall, London) 1985.
[19] HIRSCHER M., *Handbook of Hydrogen Storage* (Wiley-VCH) 2010.
[20] REVOL J.-P., BOURQUIN M. *et al.* (Editors), "Thorium Energy for the World", *Proc. ThEC Conference, Geneva, Switzerland, October 2013* (Springer, Berlin) 2013.
[21] BALLARINO A., Final design report, *CERN-ACC-2015-0134* (2015); https://cds.cern.ch/record/2063726/files/CERN-ACC-2015-0134.pdf.
[22] www.bestpaths-project.eu/.

EPJ Web of Conferences **189**, 00010 (2018) https://doi.org/10.1051/epjconf/201818900010
LNES 2017

CO_2 neutral fuels

Adelbert P. H. Goede(*)

Dutch Institute for Fundamental Energy Research DIFFER
PO Box 6336, 5600 HH Eindhoven, The Netherlands

Summary. — CO_2 is a valuable resource, life on Earth depends on it. Rather than wasting it to the atmosphere, or burying it underground, CO_2 can be combined with water and turned into valuable chemicals and fuels, the process being powered by renewable electricity. Renewable electricity generated by wind and photovoltaics (PV) is making big strides, but is limited by ill-matched supply and demand. In addition, electricity only makes up 20% to 30% of total energy demand. Domestic heating, high temperature/pressure Industrial processes and mobility/transportation gobble up the rest. Mobility and transportation prove particularly difficult to decarbonise. Aviation is a case in point. Battery-powered aircraft are unlikely to become feasible by 2050. Hydrogen has too low an energy density and is haunted by safety issues. Current policy, therefore, is directed at bio fuels. One problem, there is not enough of it. The Fuel *vs.* Food *vs.* Flora trilemma of bio-based fuel is unlikely to gain public acceptance. By converting renewable electricity into fuel, power to molecules (P2M), two birds are killed with one stone: providing fuel for long haul transportation and enabling long-term, large-scale energy storage to cover the seasonal mismatch between supply and demand of renewable electricity. Feedstock consists of air-captured carbon or nitrogen and water. Chemically combined, it creates a liquid fuel with greatly enhanced energy density, such as kerosene or ammonia, or gaseous fuel like methane which can replace natural gas in the existing gas network. Direct air capture of CO_2 is currently being commercialised. The conversion technology of water and CO_2 by electrolysis has recently been extended to novel plasma technology, the subject of this paper. For CO_2 splitting by plasmolysis, the reduced

(*) E-mail: a.p.h.goede@differ.nl

electric field has been identified as the key parameter explaining and improving the energy efficiency. Energy efficiency by plasmolysis is similar to that of electrolysis, but offers advantages in energy density, upscaling and switching in response to intermittent power with no use of scarce material. A simple model explains the inverse relation between energy efficiency and particle conversion and relates input microwave power to CO_2 gas density, plasma dimension and ionisation degree, allowing design parameters for a 100 kW pilot reactor to be specified. Recycling CO_2 in combination with P2M is a game-changing technology to meet overall CO_2 emission reduction targets. It takes advantage of existing, inexpensive infrastructure for energy storage, transport and distribution. Existing internal combustion engine technology can be maintained where necessary. Close coupled to a remote solar array or an off-shore wind farm it offers a solution to decentralised renewable fuel production at the renewable electricity source.

1. – The energy transition

The next 20 years will mark a fundamental shift in the way energy is generated, stored, delivered, valued and purchased. Unlike nuclear, this energy transition will be market-driven, for governments can regulate and facilitate, industries can produce the goods, but consumer demand will ultimately decide the fate of this energy transition. This view is shared by the European Commission. Germany, pioneer in the energy transition, has added further requirements: the energy must be clean, secure and affordable. Targets for 2030 and 2050 have been set by International Agreement (UNFCCC), by the EU (RED) and by National policy [1, 2]. The road to get there, however, is largely unchartered.

The driver of the energy transition is climate change. Mitigation policy is based on reduction of greenhouse gas emissions, notably CO_2 emissions from the use of fossil fuels. Currently, the rise in CO_2 emissions show signs of levelling off, see fig. 1a). The cause is attributed, not to an increased share of renewables, but to a slowing of the economy and an improved energy efficiency. What counts, however, is not CO_2 emission, it is the CO_2 concentration in the atmosphere that determines greenhouse radiative forcing. This CO_2 concentration shows no sign of levelling off, see fig. 1b). Governed by the coupling between atmosphere and ocean, it may take decades, if not centuries, before the effects of CO_2 emission reduction become tangible, ocean inertia being the rate determining factor in the climate system.

Mitigation of climate change and sustainability of System Earth call for renewable energy. Renewable energy is energy obtained real-time from our Solar System, either directly in the form of solar photons, or indirectly through wind, rain, ocean currents and waves, or driven by the moon through tides. Photosynthesis has energised the biosphere throughout geological times by converting solar photons into a finite amount of biofuel, which has allowed humankind to develop over the past million years. Bio energy derived from, for example, crops, algae, wood and bio-waste is thus renewable.

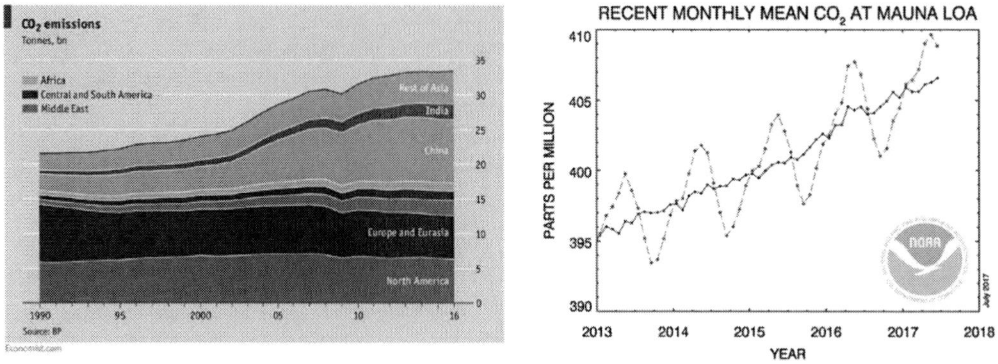

Fig. 1. – a) Global CO_2 emissions from energy use roughly flat over 2014–2016. Economic downturn and improved energy efficiency are partly responsible. Source: BP and The Economist, 17 June 2017. b) Atmospheric CO_2 concentration measured at Mauna Loa ground station. Red is monthly mean, black is seasonally corrected moving average. Source: NOAA.

Faced with a growing world population and an increasing level of consumption, bio energy is in competition with food and flora and, in any case, will not suffice to meet future world energy demand [3]. Fossilised biomass, created over several 100 million of years is resource limited and not renewable over human life span.

Alternative renewable energy sources are thus needed. Renewable energy schemes pursued to date are based on either direct conversion of solar photons into fuel, or indirectly, through the intermediate of renewable electricity. The direct conversion route, or artificial photosynthesis, mimics nature by combining solar energy conversion and fuel synthesis into one device, whereas the indirect route converts to electricity first, followed by fuel synthesis in a separate step powered by renewable electricity. Feedstock, in both cases consist of air and water. Most renewable energy sources deployed to date generate electricity, hence tailored to the indirect solar fuel production concept.

Electricity covers only about 20% to 30% of primary energy consumption, the remaining 70–80% going into low-temperature domestic heat (usually natural gas), high-temperature industrial heat (usually coal) and transportation (mainly oil). Typical CO_2 emissions produced over the various sectors are listed in table I, presented for the case of The Netherlands. It shows that CO_2 emissions are roughly equally attributed over all sectors at around 20%. Hence, in order to meet CO_2 emission reduction targets, it is paramount to shift current focus away from the electricity system towards the energy system as a whole.

The challenge is *not* how to achieve a 100% renewable electricity system, the challenge is how to transform all sectors of the economy into sustainability. Civil aviation is a case in point. The UNFCCC agreed on an overall 80–95% CO_2 emission reduction target, as compared with 1990 values, to be reached by 2050. For transportation this target has been relaxed to 60% emission reduction [4]. Aviation is exempted by a further reduced 40% sustainable fuel target in 2050, whilst ICAO, the UN civil aviation organisation, has

TABLE I. – CO_2 *emissions attributed to the various sectors of the economy for The Netherlands. The Netherlands accounts for* 0.5% *of global* CO_2 *emissions, the EU for* 10.5%*, see also fig.* 1*a) for world* CO_2 *emission distribution.*

CO_2 emission budget	The Netherlands	0.5%
Power and lightning	electricity	26%
Low-temperature domestic heat	gas	18%
High-temperature industrial heat	coal	19%
Transport (cars, ships, airplanes)	oil	17%
Non-energy emissions (agriculture, waste, chemical feedstock)		20%
Typically CO_2 emission distribution Source: Energy Agenda, Min. Econ. Affairs,	20:20:20:20:20 Dec 2016/97015	

agreed on a 50% CO_2 emission reduction by 2050 as compared with 2005 [5]. These down rated CO_2 emission targets illustrate the difficulty faced in transforming the transport sector into sustainability.

2. – Hydrogen economy

One radical option to transform the transport, heat and chemical sectors into sustainability is a hydrogen-based economy. This option has been around for a while and could be instrumental in the coupling of the electricity sector to the other sectors of the energy system. Hydrogen, however, is a highly flammable and volatile gas with a low-energy density of 0.01 MJ/ltr at ambient pressure and temperature. Also, it brittles metal containers. This limits the ability to store and transport hydrogen and to use hydrogen as an energy carrier. Combined with the high cost of producing renewable hydrogen, deployment of hydrogen as an energy carrier has been limited so far.

Several options to overcome the low-energy density problem of hydrogen are being explored, including hydrogen compression at 700 bars for storage in vehicles or at 200 bar for large scale storage in salt caverns, liquefaction at 20 K and storage in metal hydrides, Metal Organic Frameworks (MOF), Liquid Organic Hydrides (LOHC) or carbon nanotubes.

Another, attractive option is to chemically bind hydrogen to another atom, forming molecules in liquid or solid phase at ambient temperature and pressure, thereby increasing the power density by orders of magnitude. This is the case for binding hydrogen to nitrogen or carbon. Coupling hydrogen to carbon vastly increases its energy density. For example, the energy density of kerosene is 37.4 MJ/ltr, more than three orders of magnitude higher than hydrogen at ambient condition. Hydrocarbons offer a well-characterised, reliable and inexpensive solution to long-term and large-scale energy stor-

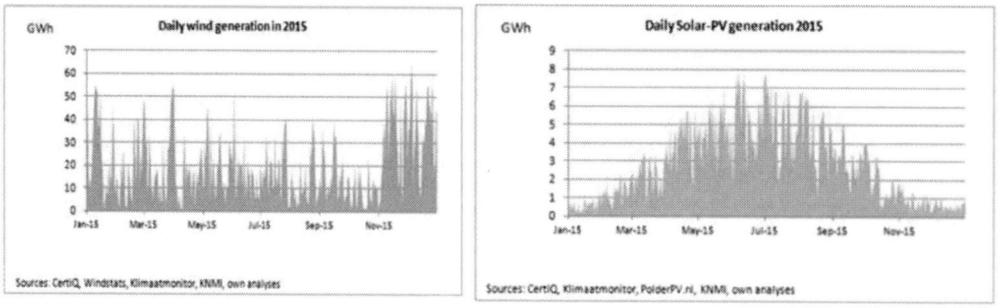

Fig. 2. – a) Wind intermittency over N-W Europe showing a large dynamic range (factor 100) on a day to day basis. Seasonal variations of a factor 2 to 3 is in phase with heat demand but not quantitatively. b) Solar PV electricity is in phase with diurnal demand, but not quantitatively with surplus during the day, shortage at night. Dramatically out of phase with renewable PV supply is seasonal heat demand over Northern EU latitudes, winter insolation being only 10% of summer whilst heat demand is up by a factor four. Courtesy Martijn Visser of Energy Academy Groningen, The Netherlands.

age and transportation, benefiting from existing infrastructure, whilst the well-qualified Internal Combustion Engine (ICE) technology remains to serve long haul transportation.

Hydrocarbons, however, require carbon capture (CC) in order to close the carbon cycle and render the fuel cycle carbon neutral. No carbon is involved in storing hydrogen in ammonia (NH_3). Apart from the pungent smell, ammonia is characterised by low flammability, high hydrogen storage density of 17.6% by mass and 107g H_2/ltr and a high energy density of 11.5 MJ/ltr. Furthermore, ammonia is an essential element in the food chain being at the basis of fertiliser production. The fixation of nitrogen to hydrogen has allowed human population to grow without running out of food, overcoming the 19th century Malthusian dilemma of population growth outpacing food supply.

Modern ammonia plants employ steam methane reforming (SMR) to produce the hydrogen needed to react with atmospheric nitrogen under high pressure and high temperature, the Haber-Bosch process. This is an energy intensive process, and in addition, relies on fossil feedstock. Current ammonia production accounts for 1.4% of global CO_2 emission. For ammonia to become sustainable as an energy carrier both the formation process and the feedstock must become CO_2 free. Powering the process by renewable electricity and deriving hydrogen from electrolysis of water would be an option within reach of existing technology. However, it would be highly desirable to develop a less energy intensive ammonia synthesis process, mimicking nature by departing from air and water. Electro-chemistry and non-equilibrium plasma chemistry may hold the key to such nitrogen fixation pathway.

3. – Energy storage

Conversion of renewable electricity (RE) into gas or liquid fuel not only opens up a route to sustainability for these sectors, it also creates a buffer in the energy system that can handle the mismatch between renewable electricity supply and demand. This buffer

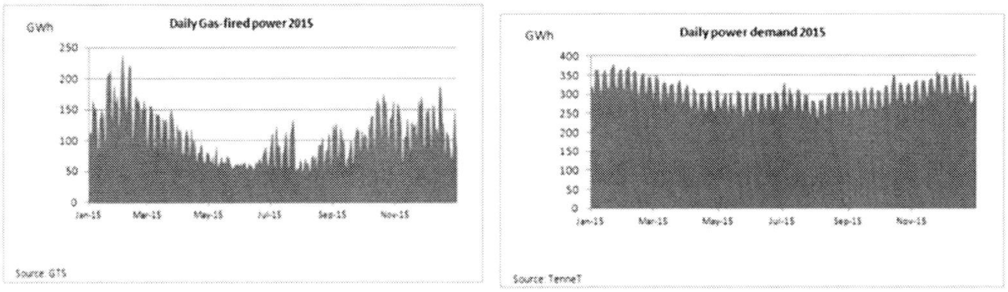

Fig. 3. – a) Gas demand [GWh/day], mainly heat, shows seasonal variation of factor 4. b) Electricity demand [GWh/day] almost flat at ±10% throughout the year. Courtesy Martijn Visser, Energy academy Groningen, The Netherlands.

is needed for several reasons, illustrated by figs. 2 and 3. On the supply side wind power creates large peaks and troughs in electricity generation, a dynamic range of a factor 100 between lulls and gusts in a matter of minutes is not unusual, wind power scaling with wind velocity cubed. PV displays a more predictable diurnal cycle but, obviously not matched to demand at night. Electricity overproduction during the day must be tidied over to cover night-time demand. An energy buffer acting on a short time scale of msec to days is necessary to smooth out the fast variations in the generation of renewable electricity.

Power to Heat (P2H), employing heat pumps and storage heaters to siphon off surplus electricity, may well provide a solution to cover the diurnal mismatch between RE supply and demand in domestic heating. But P2H would offer no solace in covering seasonal RE mismatch. Seasonal heat demand over Northern Europe is shown in fig. 3a), gas serving as a proxy to heat demand, again for the Dutch situation. Demand for heat in winter exceeds that of summer by a factor of four on a weekly basis. On shorter times demand is even more volatile. Electric power demand, on the other hand, is almost flat, see fig. 3b) showing a ±10% variation over the year. A doubling in electric power transmission to cope with seasonal heat demand variation will be costly and such grid will run idle for most of the time being massively oversized during summer. The current practise of curtailment of subsidised wind power during summer will be difficult to justify towards tax payers.

The energy supply - demand mismatch requires a buffer covering seasonal to inter-annual time scales. Existing technologies, including batteries, super capacitors, flywheels, pumped hydro and CAES (Compressed Air Energy Storage) suit the requirements of short-term energy storage, but offer no solution to large-scale seasonal and inter-annual energy storage requirements.

A third reason for an energy buffer to be included in any renewable electricity (RE) system, is that energy scenarios project a substantial surplus in electricity accumulating with increased penetration of renewable electricity in the electricity grid. The German RE scenario expects 34.5 TWh electricity surplus by 2030, increasing to 110–148 TWh

TABLE II. – *Comparison of capital investment cost per kW power transported for electricity and gas transport lines in comparable situation. Gas is a factor* 20 *cheaper. Source: Gas Unie, The Netherlands.*

	Power: BritNed	Gas: BBL
Length	260 km	230 km
Investment	600 M€	500 M€
Capacity	1 GW	20 GW
Specific investment costs	€ 230/kW/100 km	€ 11/kW/100 km

Nord Stream gas pipeline: € 9/kW/100 km (Russia to Europe via Baltic).

in 2050 [6]. For the Netherlands, RE surplus is expected to amount to 1.5 TWh in 2025 ($\sim$ 1% total energy production) increasing to 30–55 TWh in 2050 [7]. For France, these numbers are 15 TWh for 2030 and between 44 and 91 TWh for 2050, respectively [8].

4. – Power to M

Converting renewable electricity into energy dense gaseous or liquid fuel presents a solution to long-term, large-scale energy storage. This is the Power-to-Molecules (P2M) scheme, where molecules stand for synthesised gas or liquid fuel or chemicals. P2M has the powerful ability to couple the electricity sector to the gas, the oil and the chemical sector, leading the way to CO_2 reduction of the energy system at large. Existing inexpensive infrastructure for energy storage and transport is readily available. The Dutch gas net alone unlocks a massive 550 TWh energy storage capacity. Transportation of energy by gas is more than a factor 20 cheaper than energy transport by electricity, see table II.

Currently, Europe runs about 50 pilot and demo P2G projects, about half of them in Germany, with efforts accelerating over the past years. These projects anticipate a substantial RE surplus generated in all RES scenarios by 2050 and recognise the difficulty to store and transport this surplus electricity on a large scale. The existing gas grid provides both storage and transport infrastructure that meets seasonal requirements for heat and overcomes the inherent limitations in storage of renewable electricity.

For transportation similar dilemmas are encountered. Batteries are free from carbon emission during operation. However, they suffer from low-power density compared with hydrocarbon fuel, which limits their range. Taking again the extreme example of long haul aviation: to replace the energy content of a 260 tonnes of kerosene load of an Airbus 380 at take-off, the most advanced Li-ion battery would weigh 14.000 tonnes. The plane will never take off. Long haul aviation powered by batteries is unlikely to become feasible by 2050.

Electric cars may well offer a solution to urban mobility. Both air pollution and noise will be diminished and battery charging infrastructure is relatively easy to install. Added to self-driving and ride sharing it may just cause the disruptive change the automotive

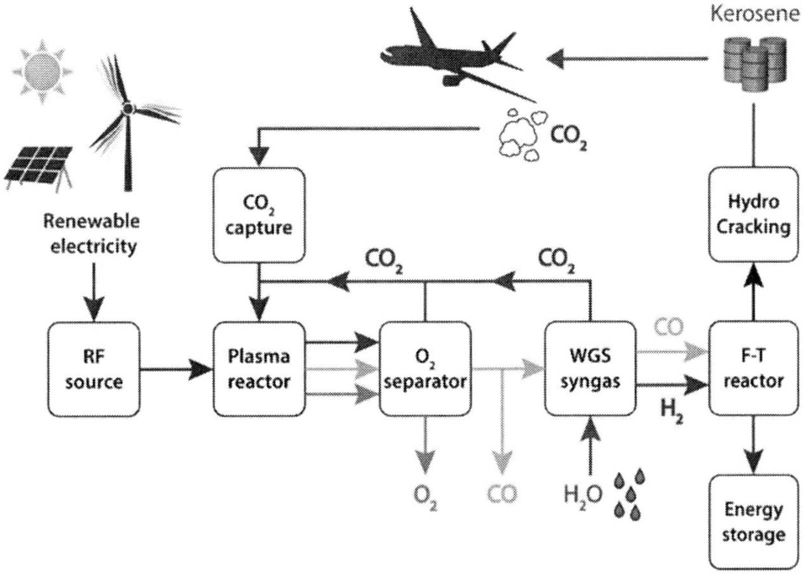

Fig. 4. – The CO_2 neutral fuel cycle of an internal combustion engine driven by hydrocarbon fuel, illustrated for the case of aviation. Sustainable aircraft grade kerosene is produced from water and air powered by renewable electricity through the splitting of CO_2, formation of syngas and Fischer-Tropsch synthesis. By recapturing the CO_2 directly from air, the CO_2 cycle has come full circle. Source: EU H2020 funded KEROGREEN project [9].

industry anticipates. However, batteries will not meet long-distance road transportation requirements, in particular not for heavy-duty lorries. The hydrocarbon powered internal combustion engine is unlikely to lose its grip on the motorways by 2050.

5. – Conversion of renewable electricity into fuel

The conversion step of electricity into fuel and chemicals is the critical step, both technically and economically. Basic research and development of cost-competitive electro-chemical solutions are subject of current R&I effort. Electro-chemical conversion is based on the electrolysis of water and CO_2 into syngas. Syngas is the key mixture around which the synthesis of hydrocarbons revolves. The process is known as Fischer-Tropsch (FT) synthesis. Both electrolysis and FT technology have been around for about 100 years. Deployment so far has been limited to strategic use in South Africa and Nazi-Germany when oil embargos threatened fuel supply. At this point in time, however, high cost and upscaling is the limiting factor to large-scale implementation.

A typical P2M-CC fuel cycle is illustrated for the case of jet powered aviation in fig. 4, borrowed from the EU H2020 funded KEROGREEN project [9]. The CO_2 neutral fuel cycle starts with renewable electricity generation. This electricity powers a microwave plasma reactor in which the CO_2 molecule is split into CO and O_2. Subsequently, the oxygen is separated out and the CO partly converted into hydrogen by the water gas shift reaction creating syngas, a mixture of CO and H_2. Out of syngas hydrocarbons are formed by Fischer-Tropsch (FT) synthesis. The wax produced is further processed by hydrocracking to create kerosene. FT synthesised fuel at 50% blending ratio is ATSM certified as jet grade fuel. Upon combustion, CO_2 and H_2O are emitted. By recapturing the CO_2 directly from air, the fuel cycle is closed, a CO_2 neutral fuel cycle is created which does not add to the atmospheric CO_2 concentration.

Synthetic fuels, based on C-H-O compounds, usually display absorption bands in the infrared and are thus greenhouse gases (GHG). Compared with CO_2 their Global Warming Potential (GWP) maybe higher, but their relatively short atmospheric lifetime will diminish their long-term impact on global warming. For example, methane is a more potent GHG than CO_2 on a molecule by molecule basis. However, the atmospheric lifetime of methane is about 10 year, far less than the several hundred years that CO_2 stays in the atmosphere. Manmade methane emissions may result from leaking gas pipelines, estimated to amount up to 10% of total gas flow for Siberian gas transport. Methane release from the vast Siberian permafrost region may be instrumental in kick-starting climate change before CO_2 greenhouse warming takes over, a situation encountered in the Quaternary periods when ice ages and warm spells succeeded in quick succession.

On air pollution synthetic fuels meet the most stringent regulation; there is no soot because the synthesised fuels contain no aromatics, there is no sulphur because synthesis is based on H_2O and CO_2 only, and NO_x formation is lowered by tuning the ICE combustion temperature.

Carbon neutral fuels offer energy security and reduce the dependency and cost of imported fossil fuels. Feedstock consists of abundant CO_2 and H_2O, and power derives from indigenous renewable electricity. The fuel can be stored at large scale for an extended period of time, transported and distributed to customers by existing, inexpensive infrastructure, with no need to expand the electricity grid.

6. – Electro chemistry

Electro-chemical conversion of H_2O and recently CO_2 is traditionally based on alkaline electrolysers. The technology is commercially available. Power density is low ($0.1\,\mathrm{W/cm^2}$) which means sizable plant lay-out. Proton Exchange Membrane (PEM) electrolysers are at early commercial stage and show higher energy efficiency and power density ($1\,\mathrm{W/cm^2}$), allowing more compact sized equipment and up-scaling to MW level within reach [10]. However, PEM employs an expensive membrane and noble metal electro-catalysts, *i.e.* platinum and iridium, resulting in cost and sustainability issues. Degradation causes limitations to membrane operational life time.

Solid Oxide Cells (SOC) are at the R&D stage and offer higher energy efficiency and power density ($\sim 5\,\mathrm{W/cm^2}$) at high operating temperature (700–800 °C) and pressure (50 bar) to produce hydrogen at over 80% energy efficiency. Recently, syngas (H_2-CO) was produced in a SOC in one step co-electrolysis of CO_2 and H_2O [11, 12]. Upscaling of SOC is hampered by cell degradation and stack lifetime. The physics and chemistry at the triple phase boundary layer between gas reactant, electrode and catalyst is poorly understood [13]. Both alkaline and SOC do not respond well to intermittent renewable electricity.

7. – Plasma chemistry

Plasma-chemical conversion or plasmolysis increases the power density and consequent productivity by more than an order of magnitude compared with SOC. Electro-chemical conversion is a surface process taking place at the gas-electrode interface. By contrast, plasma chemical conversion is a volume process and hence scales with dimension cubed rather than dimension squared of a surface process. Combined with high gas flow rates, upscaling to high CO production rates at 100 kW power level is feasible. Plasma chemical conversion does not employ scarce materials and a plasma responds instantly to intermittent input power.

The plasma eases conditions for splitting CO_2 through vibrational excitation of the molecules, not through direct electron impact dissociation, electron dissociative attachment or through heating of the gas [14, 15]. As a consequence, vibrational temperature exceeds the translational and the rotational temperatures, the CO_2 gas is in a thermodynamic non-equilibrium state. The CO_2 gas is weakly ionised —only one in 10^5 molecules is ionised— and is relatively cold, similar to a fluorescent lamp. The energy to produce the plasma is a relatively small fraction in the energy balance.

By tuning the electron energy distribution function EEDF, energy is directed mainly at the CO_2 vibrational degree of freedom, not to translational or rotational energy. A non-equilibrium condition is created whereby the vibrating molecules impart vibrational energy on each other, ratcheting up vibrational energy of one molecule at the expense of others, ultimately breaking the CO_2 molecular bond. Most effective in vibrational dissociation is the CO_2 asymmetric stretch mode, stretching the O atom away from the CO part of the linear CO_2 molecule. This mode is excited most effectively by relatively low energy electrons in the range of 0.4–2 eV [16].

Dissociation to CO proceeds in two steps: first one molecule CO and one atomic oxygen O are formed at 5.5 eV dissociation energy. Subsequently, the atomic oxygen reacts with another exited CO_2 molecule to form a second CO at 0.3 eV net energy, yielding a net enthalpy H for the production of one CO molecule $H = 5.8/2 = 2.9\,\text{eV}$. This is much lower than direct electron dissociation and lower than the $> 7\,\text{eV}$ required for an electron induced Franck-Condon transition [15]. The loss of vibration energy by vibrational-translational relaxation is reduced by cooling the gas, notably for the asymmetric stretch mode [17]. Low gas temperature ($T_0 < 1000\,^{\circ}\text{C}$) also prevents the reverse CO_2 forming reaction from taking place. Low temperatures are obtained by supersonic expansion of the gas through a nozzle.

7˙1. *Plasma experiments.* – CO_2 plasmolysis experiments are conducted in an microwave powered reactor equipped with a cylindrical quartz flow tube of length 124 mm and 26 mm diameter, placed on axis of the RF cavity [18, 19]. The CO_2 gas is injected tangentially at the entrance of the reaction chamber to create a vortex gas flow aimed at stabilising the plasma by creating a radial pressure and temperature gradient. Gas composition measurements are based on quadrupole mass spectrometry (QMS) and optical emission spectroscopy (OES). Gas pressure is measured at the entrance and at the exit of the reaction chamber. A typical plasma discharge is shown in fig. 5b) [20-22].

Experimental results show production rates of 7.5 slm CO at 200 mbar pressure at a CO_2 gas flow rate of 75 slm. The energy efficiency η, defined in sect. **11**, reaches 60% at 3 kW microwave power (915 MHz) absorbed. The CO_2 to CO conversion ratio is relatively low at 10% for safety reason. However, conversion ratio CO_2 to CO is shown to scale linearly with RF power and inversely with gas pressure to reach 70%, see fig. 5c) and fig. 12. A typical CO spectrum is shown in fig. 5d). Linear scaling of CO concentration with power is confirmed by CO line strength measurement.

7·2. *Plasmolysis vs. electrolysis.* – The effectiveness of plasma against electro chemistry is illustrated by comparing the physical size of the plasma reactor itself with an SOC electrolyser producing the same amount of O_2 flux. Plasma O_2 production rate per kW input power is 1.25 slm O_2, which corresponds to $0.93 \cdot 10^{-3}$ mole O_2/s. At typical SOC transport of 1 A/cm^2 this requires about 358 cm^2 SOC surface area (Faraday constant $1/4\,F$ to account for transport of 2 double negatively charged O atoms). By contrast, the surface area surrounding the plasma is about 30 cm^2. Hence, plasma productivity is up by a factor 10 by volume compared with SOC electro-chemical conversion for the same surface envelope.

The relevant parameter in achieving high vibrational excitation and subsequent high energy efficiency and conversion, turns out to be the reduced electric field E/n, see sect. **10** and subsect. **14**·2. Although known from theoretical studies [23], these experiments show direct experimental evidence for the first time [21]. CO_2 molecules are vibrational excited by low energy electrons in the range of 1 eV where the reaction rate peaks at about 10^{-8} cm^3/s. This is achieved by adjusting the mean free path of plasma electrons accelerated in the externally applied RF electric field E and tuning the neutral particle density n to obtain $E/n \sim 10^{-16}$ V cm^2. Current CO_2 plasmolysis development is directed at optimisation of the reduced electric field E/n and reduction of the CO_2 gas temperature, both aiming at an increased energy efficiency.

7·3. *Plasma modelling.* – Computational studies support the experimental effort to improve energy efficiency and CO yield. This modelling requires detail knowledge of the plasma physics and the plasma chemistry, referring to a large number of chemical reactions with vibrational excited cross-sections largely unknown. Models of the reaction kinetics developed so far contain only crude approximations. At DIFFER a zero-dimensional computational model for CO_2 plasma dissociation is being developed based on a Monte Carlo solution of the Boltzmann equation for electrons and ions and a hybrid discrete-continuum model for vibrational up pumping [24]. Employing a fast computational solver, the vibrational and non-equilibrium electron energy distributions are computed in the presence of molecular and atomic carbon and oxygen species.

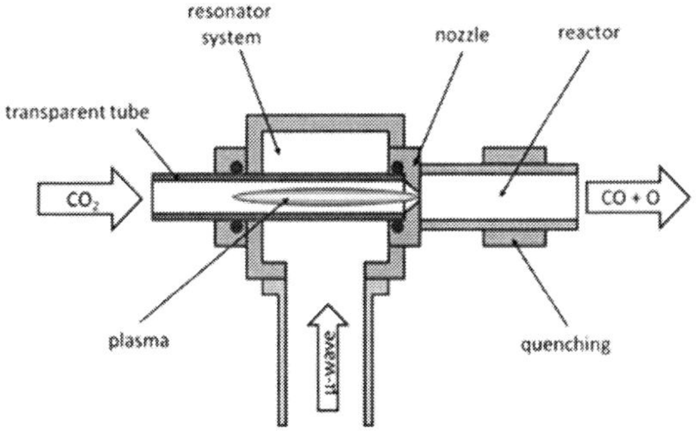

(a)

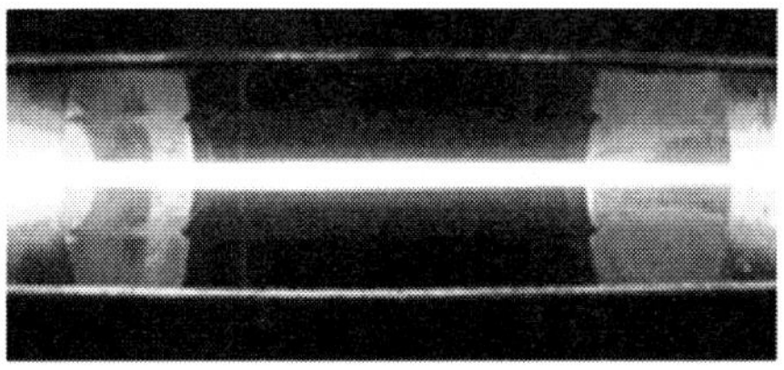

(b)

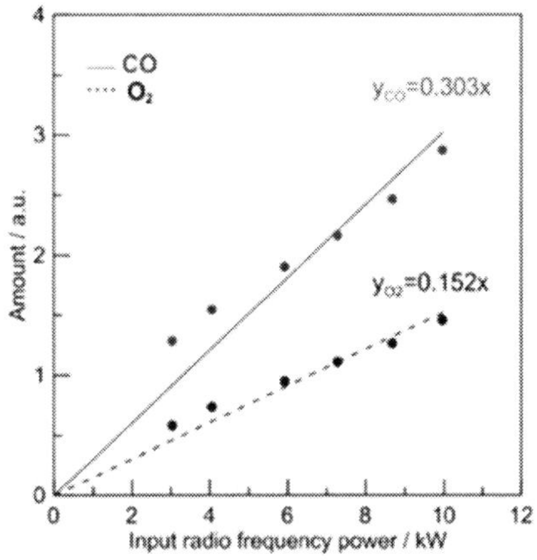

(c)

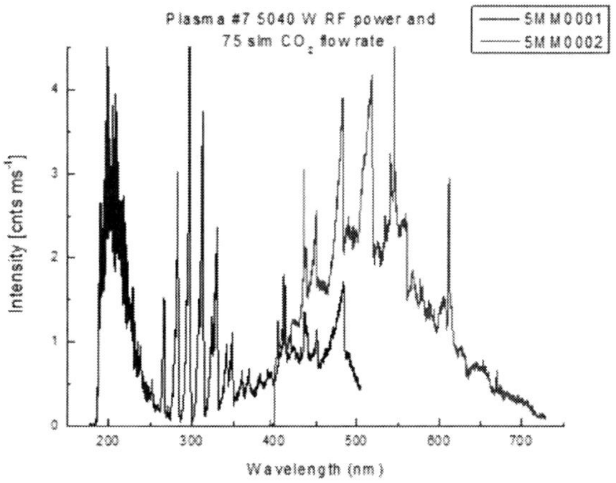

(d)

Fig. 5. – a) Schematic of the 30 kW 915 MHz plasmolysis reactor for CO_2 splitting. Microwave power absorbed creates a stable plasma column at a pressure of 200–500 mbar, contained within a cylindrical quartz tube of length 12.4 cm and 26 mm diameter. CO_2 gas injected tangentially at the entrance of the tube creates a vortex flow to stabilise the plasma and is expanded at the exit by a nozzle to cool the gas. Diagnostics include Optical Emission Spectroscopy and Quadrupole Mass Spectrometry. b) High-density contracted CO_2 plasma driven by microwave power at 3 kW absorbed and a CO_2 gas flow rate of 75 slm (Type II discharge). The blue light emitted is from the CO 3rd positive bands, see panel d) Combined with the C_2 Swan band it produces an intense white light. Enthalpic energy efficiency reached is 50–60% while conversion ratio ranges from 10% to 70% scaling with power. c) Linear scaling of CO_2 conversion with input microwave power. The conversion ratio of CO_2 to CO is the determining factor in system productivity and cost. Drawn lines are stoichiometric fits to CO and O_2 ratio. d) A typical CO spectrum at high density (Type II) operation mode. Left are the 4th positive singlet and 3rd positive and $5B$ triplet CO bands. Between 400 and 650 nm the CO Ångström and triplet system is observed together with an upcoming C_2 Swan band emission at high gas density. All CO lines observed result from direct excitation by plasma electrons.

8. – Gas separation

A major challenge is the separation of O_2 from the CO_2, CO, O_2 gas mixture produced by plasmolysis of CO_2. Notably the separation of the oxygen molecule from the nearly equally sized CO molecule, with kinetic diameters of 0.346 nm and 0.376 nm, respectively, is not trivial. Various schemes have been considered, based on paramagnetic properties of oxygen, cryogenic distillation and membranes. The membrane scheme includes polymeric membranes, MOFs, ceramic microporous membranes, mixed ion electron conducting membranes and electrochemical oxygen pumps.

The latter, considered here, is a solid oxide cell (SOC), similar to that used in high-temperature fuel cells and electrolysers. It consists of a thin electrolyte sandwiched between two oppositely biased electrodes, one of these usually providing mechanical support. An SOC operates in the temperature range 700 to 800 °C where the material becomes conductive to negatively charged oxygen ions. If the SOC is operated as an electrolyser, the plasma electrode displays high electro-catalytic activity in reducing CO_2. If the SOC is operated as an oxygen selective pump it supresses the back reaction of CO by the electrochemical O_2 reduction reaction and transport of ionic oxygen away from the electrode surface into the electrolyte.

SOC composed of perovskite electrode material and an Yttria stabilized Zirconia (YSZ) electrolyte have shown remarkable properties that can be fully exploited in oxygen gas separation. They act as an oxygen extractor and a barrier to other species in gas mixtures when subjected to electrical potential gradients. Issues include the oxidation of CO by gaseous or lattice oxygen at the hot plasma electrode surface in competition with desired catalytic reduction of CO_2 at that surface. Furthermore, chemo-mechanical compatibility of electrode-electrolyte materials and negative oxygen ion conductivity are key parameters. The reaction kinetics of oxygen reduction $O_2 + 4e^- \rightarrow 2O^{2-}$ at the plasma electrode is influenced by both the plasma sheath electric field and the externally applied voltage across the SOC membrane. Both compete in pulling electrons out of the plasma electrode interface.

8·1. *Perovskite membranes.* – Lanthanum-based perovskite membranes yield superior oxygen flux values for oxygen separation. State-of-the-art materials for solid oxide fuel cells are based on an electrolyte of Yttria stabilized Zirconia (YSZ) and a Ni/YSZ cermet fuel electrode. The oxygen electrode is a $La_{0.85}Sr_{0.15}MnO_{3\pm\delta}$ (LSM) or $La_{0.6}Sr_{0.4}Co_{0.2}Fe_{0.8}O_{3-d}$ (LSCF) perovskite that forms a good combination with YSZ or GDC (gadolinium-doped ceria). The Ni/YSZ fuel electrode displays excellent catalytic properties for fuel oxidation and good current collection, however, carbon deposition and poor redox cycling degrades performance. Furthermore, electrolysis of H_2O and/or CO_2 oxidises the Ni leading to NiO/YSZ which is a poor conductor and therefore not suitable for the important functionality of current collection, hence Ni/YSZ cannot be employed in our case. Current research concentrates on alternative redox stable perovskites.

One candidate material is $La_{0.75}Sr_{0.25}Cr_{0.5}Mn_{0.5}O_3$ since it has comparable electrochemical performance as the Ni/YSZ cermet and is redox stable [25-27]. This material

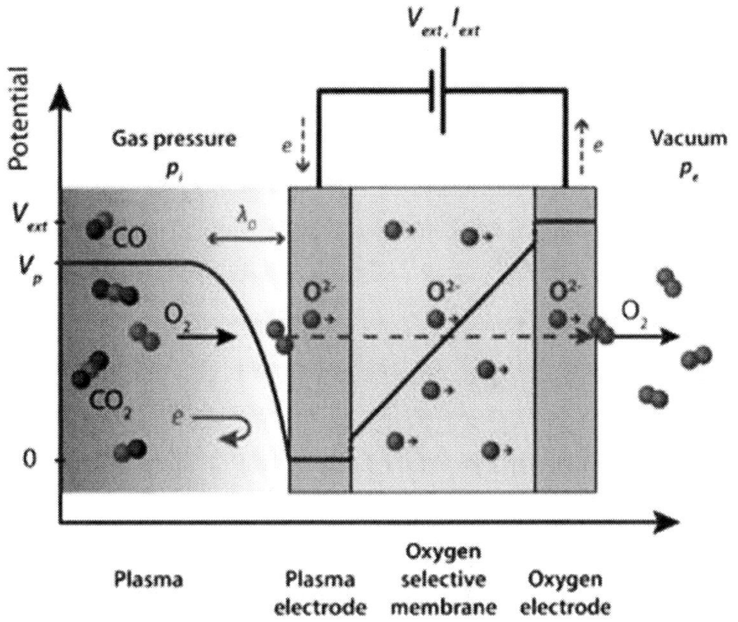

Fig. 6. – Voltage distribution of the plasma across the plasma sheath and across the solid oxygen membrane. An externally applied electric field across the membrane transports the negative oxide ions to the oxygen electrode. Plasma-activated CO_2, CO, O_2 molecules, radicals, atoms and plasma electrons incident on the plasma electrode and subjected to the plasma sheath electric field influence the electro-chemistry at the plasma perovskite interface, synergistically boosting CO formation and oxygen transport.

will be the starting point for the development of an efficient plasma electrode which will be inactive for CO oxidation and will exhibit oxygen separation fluxes upon mild polarization. On the oxygen side, state of the art material LSCF or LSM in combination with YSZ or GDC will be used, while YSZ will be the electrolyte.

8·2. *Plasma electrochemical interaction.* – Novel is the presence of plasma adjacent to the SOC cathode, see fig. 6. The plasma leads to activated species including vibrational and electronically excited molecules and atoms, radicals, ions and electrons. This will activate the chemistry at the plasma electrode interface. In addition, the positively charged plasma and the polarisation of the SOC will enhance reactivity. Preliminary hybrid plasma-SOC experiments conducted at the Tokyo Institute for Technology indicate that synergism occurs between a SOC membrane and a plasma-activated CO_2 gas, lowering the SOC operating temperature to 600 °C and improving conversion ratio as compared with pure electrolysis [28, 29].

8·3. *Two SOC design concepts.* – Two SOC design concepts are being explored, one based on flat membranes employing heritage from Fuel Cell technology, the other based

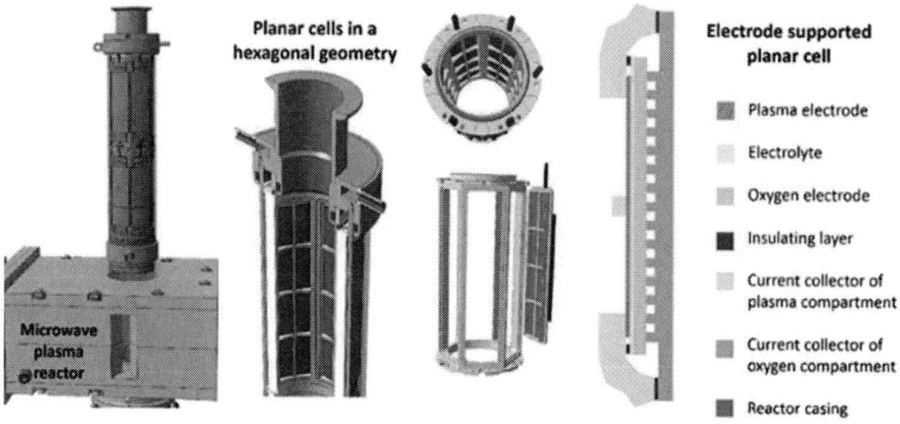

Fig. 7. – Schematic of the oxygen separator module based on planar membranes arranged at the plasma exit in hexagonal geometry. The oxygen outlet is indicated in blue. Detail on the right hand side shows the membrane assembled into a cell. Indicated are the perovskite plasma electrode (red), electrolyte (yellow) and the oxygen electrode (green). This membrane sandwich is mechanically supported by the castellated metal surface also serving as oxygen gas transport conduit. The SOC Cell is equipped with current feed through connected to the plasma and oxygen electrodes, respectively.

on capped tubular membranes which offer intimate plasma to SOC coupling [30, 9]. Requirements include gas tight separation of oxygen to plasma compartment and Galvanic isolation of the electrodes. Tubes are thermo-mechanically more stable and benefit from proven metal jointing technology. If failed, they can be individually replaced. The main issue is the fabrication of tubular membranes and their current collection. Here, the planar membrane concept offers advantage. The deposition of thin layers is simpler than for tubular membranes. Commercial Fuel Cell technology is predominantly based on the fabrication of planar membrane cells and their integration into stacks, offering a well proven technology. A conceptual design is shown in fig. 7.

At the typical plasma conditions discussed, 1.25 slm O_2 needs to be transported out of the plasma per kW power input, corresponding to 0.93×10^{-3} mole/sec. The membrane surface area required is $358\,\text{cm}^2$ at a typical SOC current density of $1\,\text{A/cm}^2$, taking account a 4 times Faraday constant to account for $2O^{2-}$ ion transport. This figure forms the basis of membrane surface area specification of the O_2 separator.

Computational studies carried out at DIFFER assist in development of new perovskite material for the efficient separation of O_2 from the plasma CO_2, CO, O_2 gas mixture and associated atoms, radicals, ions and electrons [31]. The strategy is to use high-throughput computation based on *ab initio* quantum chemical computation to study the performance of variant perovskite materials. Density Function Theory (DFT) can effectively approximate the electron distribution that governs fundamental physical material properties. DFT calculations will be carried out using the *Vienna Ab Initio Simula-*

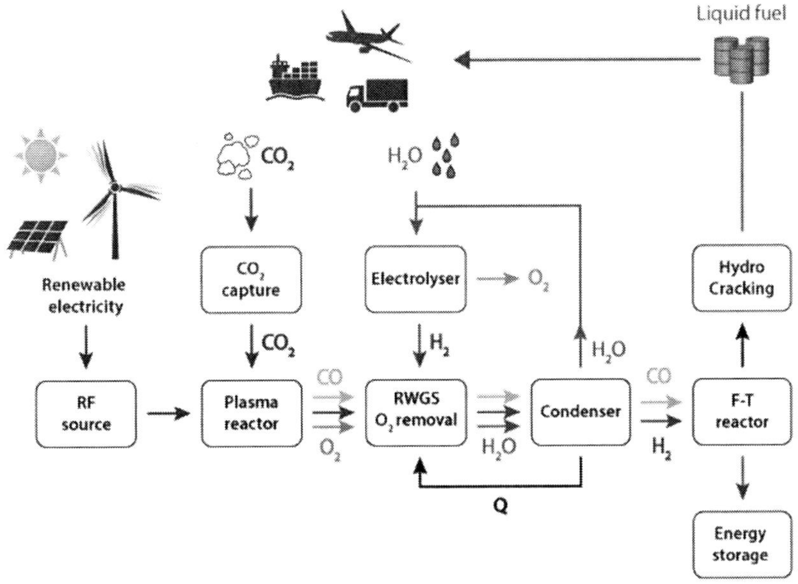

Fig. 8. – Alternative scheme to oxygen separation: replace by proton feed. This option does away with the water gas shift and CO purification units, however, adds a water electrolyser unit.

tion Package (VASP) [32]. Computation is directed at newly doped quinary perovskite electrodes, that are expected to perform better than the existing electrode materials such as $La_{0.75}Sr_{0.25}Cr_{0.5}Mn_{0.5}O_3$ (LSCM) [33, 34].

8'4. *Alternative to oxygen separation.* – A proton conducting cell [30] is employed as an alternative to oxygen separation, see fig. 8. In this approach water electrolysis occurs at the oxygen electrode and is followed by injection of protons into the plasma electrode. This is expected to lead to selective water formation at the plasma electrode interface through oxygen reduction. The intermediate operation temperature of proton conducting SOC (500–600 °C) is expected to reduce the kinetics of the undesired back reaction of CO oxidation to CO_2 at the plasma electrode. If CO oxidation kinetics dictate a further lowering of the operation temperature, the scenario of employing commercial PEM electrolysers becomes an option. These electrochemical cells operate in the temperature range of 20–200 °C, however, at the expense of costly and less-sustainable noble metals needed for the electrode materials.

9. – Basic parameters

In order to understand the plasma chemistry of CO_2 dissociation it is useful to gain insight into the characteristic time and length scales of the weakly ionised CO_2 plasma and its interaction with the driving electro-magnetic field. This includes collision times and mean free paths for the CO_2 gas and for the plasma constituents. For the plasma, the electric shielding distance (Debye length), plasma frequency and electro-magnetic wave dispersion add to pertinent time and length scales. Some simplified considerations will provide an order of magnitude ordering of the relevant processes.

9·1. *The* CO_2 *molecule*. – Carbon dioxide is a linear molecule exhibiting 4 vibration and 2 rotation oscillation modes: the symmetric stretch mode $(\nu, 0, 0)$, the asymmetric stretch mode $(0, 0, \nu)$ and two bend modes $(0, \nu, 0)$ in two perpendicular directions, hence 9 degrees of freedom in total including translational motion. The radius of a CO_2 molecule is $a_0 = 0.116\,\text{nm}$ (the equilibrium C-O distance). The electron dissociation energy of CO_2 is $D = 5.451\,\text{eV}$, the cross-section reaches a max $10^{-16}\,\text{cm}^2$ above 30 eV. Similarly, the cross-section for dissociative ionisation and formation CO^+ starts at 25 eV reaching a max $0.39 \times 10^{-16}\,\text{cm}^2$ at $\sim 90\,\text{eV}$ electron energy. The ionisation energy of CO_2 is 13.777 eV, reaching a maximum at an electron energy $\sim 100\,\text{eV}$ [17].

9·2. *Spectrum of* CO_2. – The absorption/emission spectrum of CO_2 vibration-rotation modes is situated in the IR, see the HITRAN database [35]. The fundamental of the asymmetric mode is found at $\lambda = 4.24$–$4.28\,\mu\text{m}$ (750 THz, $\tau = 14\,\text{fs}$, $h\nu = 0.29\,\text{eV}$) with an absorption cross-section of $1.4 \times 10^{-17}\,\text{cm}^2/\text{molecule}$. Overtones are found at $\lambda = 2.68$–$2.78\,\mu\text{m}$ and $\lambda = 2.0$–$2.02\,\mu\text{m}$. The fundamental of the bend mode is at $\lambda = 14.9$–$15\,\mu\text{m}$ with cross-section $5 \times 10^{-18}\,\text{cm}^2/\text{molecule}$. The symmetric mode has no IR transition (no dipole).

Electronic excitations in the eV range produce optical emission of the molecules involved [36]. The CO_2 term scheme, representing the electronic energy levels, is given by [37]. As it turns out, there is no emission of CO_2 in the UV visible. Only in the far UV 140–160 nm some weak absorption bands exist. Emission of CO_2^+ exist for the transitions $A^2\Pi_u > X^2\Pi_g$ at 393.3–438.4 nm and $B^2\Sigma_u^+ > X^2\Pi_g$ at 218.9–226.8 nm [36]. These emissions have not been observed in our experiments.

9·3. *Emission spectra of* CO, C_2, O, O_2 *reaction products*. – These spectra, by contrast, exhibit emission spectra in the UV-visible. The CO molecule energy diagram is given in [38]. In our experiments, the UV 4th positive singlet system $A^1\Pi > X^1\Sigma^+$ (200–250 nm) and 3rd positive and $5B$ triplet system $b^3X^+ > a^3\Pi$ (260–390 nm) have been identified. In the visible, the Ångström singlet $B^1\Sigma^+ > A^1\Pi$ and the Triplet system $d^3\Pi > a^3\Pi$ (400–650 nm) are also observed. A shift from 3rd and 4th positives to the Ångström and Triplet system is observed at high gas pressure. The plasma turns from pale blue (Type I) to intense white (Type II).

The C_2 Swan band has been observed in the same spectral interval (430–620 nm) as the CO Ångström and Triplet bands, showing further decomposition of CO_2 to carbon taking place.

Molecular oxygen O_2 has a number of absorption and emission bands in the vis-NIR range, notably the oxygen absorption bands at 680–690 nm and 760–764 nm and the emission band at $1.27\,\mu\text{m}$ [35]. These lines are outside the spectral range ($< 725\,\text{nm}$) of the OES spectrometer, hence not observed in our experiments. Unfortunately, also the 680 nm oxygen line is not observed.

Atomic oxygen lines $^3D^0$ and $^5D^0 > {}^5P$ around 615 nm are observed in the measured spectrum. These small peaks are only measurable in the high-reduced E-field condition. At low reduced E-field the Swan band outshines them. Atomic oxygen spectral lines could elucidate their vibrational origin, deriving either from the O(^1D) state or the lower energy O(^3P) dissociation branch of CO_2. The O(^1D) $2p^4$ transition to ground state O(^3P) has an emission at 636 nm, whilst O(^3P)3s to O(^3S) has an emission at 844.7 nm [39]. Unfortunately, these lines could not be identified from our measurements.

In conclusion, the CO molecule produced in the CO_2 dissociation process is in the $X^1\Sigma^+$ singlet ground state. Hence, no electronic emission representing this state is available. Atomic oxygen produced is either in the ground state O(^3P) or in the excited state O(^1D). A transition from the excited singlet D level to the ground state would reveal its origin. The line corresponding to $^1D \rightarrow {}^3P$ at 630 nm has not been observed. Similarly, transitions $^5S^0$ and $^3S^0$ to the ^3P ground level at wavelength 130 and 135 nm, respectively, would produce tell-tale signs, but are not within spectral range of the spectrometer and anyway are blocked by absorption in air. Hence, all CO and O lines observed result from direct excitation by plasma electrons and not from vibrational excited CO_2 energy levels.

9·4. *The microwave field.* – The electro-magnetic field in the cylindrical cavity is in the TM_{010} mode, hence the electric field maximises on-axis to which it is aligned. A typical electric field strength on axis of the RF cavity calculated by COMSOL is $E = 27\,\mathrm{kV/m}$ @ 915 MHz and $P_{\mathrm{RF}} = 3\,\mathrm{kW}$ [19]. The distance travelled by an electron accelerated in vacuum by this E-field during the half cycle $\tau = 0.546\,\mathrm{ns}$, is $s = 1/2(eE/m)t^2 \sim 720\,\mu\mathrm{m}$. By contrast, CO_2^+ ions move only 9 nm, remaining essentially stationary during RF-field oscillation. The speed acquired by electrons in this half cycle ($v = at$) becomes $v = 2.6 \times 10^6\,\mathrm{m/s}$, equivalent to an energy of $\sim 20\,\mathrm{eV}$.

The de Broglie wavelength $\lambda = h/p$ of a 20 eV electron is 0.28 nm, hence of the size of the CO_2 diameter (0.232 nm). Above this energy, the electrons start seeing individual atoms, rendering them ineffective for excitation of molecular vibration.

9·5. *Electron-CO_2 gas collisions.* – The cross-sections of vibrational excitation of a CO_2 molecule by electrons at around 1 eV electron energy is of order $\sim 10^{-16}\,\mathrm{cm}^2$ for all modes, dominated by the asymmetric stretch mode. The cross-section for the asymmetric mode peaks at 0.4 eV and drops characteristically as $\sigma \approx 1/v$ with increasing energy [17]. This implies a constant collision frequency for the asymmetric stretch mode over the range from 0.5 to over 30 eV electron energy.

Considering the total collision cross-section for electrons in CO_2, $\sigma_{\mathrm{tot}} \sim 10^{-15}\,\mathrm{cm}^2$, the mean free path $\lambda_{\mathrm{tot}} = 1/n\sigma_{\mathrm{tot}}$ at low pressure Type I discharge (1 mbar) is $\lambda_{\mathrm{tot}} \sim 370\,\mu\mathrm{m}$, *i.e.* $\lambda_{\mathrm{tot}} \approx s$ at low pressure Type I discharge and the electron is able to gain relatively high energy between collisions.

TABLE III. – *Characteristic time scales and lengths of the electron neutral CO_2 gas interaction.*

P mbar	$\tau_{\rm exc}$ fs	$\tau_{\rm vib}$ fs	$\tau_{\rm tot}$ ps	$\tau_{\rm ev}$ ps	$\tau_{\rm nn}$ ns	$\lambda_{\rm nn}$ μm	$\lambda_{\rm ev}$ μm	$\lambda_{\rm tot}$ μm	s μm
1	0.39	14@	62	12E3	130	50	74E3	370	720@
100		$\lambda = 4.2\,\mu$m	6	62	1.3	0.5	37	3.7	$E = 27\,$kV/m

At high-pressure Type II discharges ($> 100\,$mbar), the situation is very different since $\lambda_{\rm tot} \approx 4\,\mu$m and therefore $\lambda_{\rm tot} \ll s$, *i.e.* the electron is never able to pick up speed between collisions. At less than 1/100 of the RF cycle, the electron is knocked off-course by collision with a CO_2 molecule. A constant drift velocity establishes with $v_{\rm D} \approx E\lambda \approx E/n$, where the proportionality factor is the electron mobility defined as $\mu = e/(m\nu)$.

The interaction time of an electron with a CO_2 molecule $\tau_{\rm exc} = 2a_0/v_{\rm e} = 0.39\,$fs for a 1 eV electron ($v_{\rm e} = 5.93 \times 10^5\,$m/s), is much faster than the vibration time $\tau_{\rm vib} = 14\,$fs ($\sigma = \pi/4a_0^2 \approx 10^{-16}\,{\rm cm}^2$). Hence, the molecule can be regarded as frozen during the inelastic electron collision; the Frank-Condon principle applies. The electron acts as a clapper ringing the CO_2 molecular bell.

The electron neutral collision time follows from the total cross-section averaged over the electron velocity distribution $\tau_{\rm etot} = 1/n\langle\sigma_{\rm tot}v_{\rm e}\rangle$ yielding $\tau_{\rm tot} = 6\,$ps at $p = 100\,$mbar and 1 eV electron energy. Similarly, for the excitation of the asymmetric vibration mode the electron collision time becomes $\tau_{\rm ev} = 60\,$ps at $p = 100\,$mbar, both much smaller than the RF oscillation time.

The mean free path for inter-molecular CO_2-CO_2 collision at 300 K is $\lambda_{\rm nn}[{\rm m}] = 4.5 \times 10^{-5}/p$ [mbar]. At a pressure of 1 mbar (Type I discharge) this yields a mean free path $\lambda_{\rm nn} = 50\,\mu$m. At a pressure of 100 mbar $\lambda_{\rm nn} = 0.5\,\mu$m. This path length results in a collision time with $T_{\rm gas} = 300\,$K ($v_{\rm th} = 378\,$m/s) of $\tau_{\rm nn} = \lambda_{\rm nn}/v_{\rm th} = 130\,$ns at 1 mbar and $\tau_{\rm nn} = 1.3\,$ns at 100 mbar. These times become a factor 3 shorter at $T_{\rm gas} = 3000\,$K and another factor 5 at a pressure of 500 mbar, more typical for the Type II discharge. Hence, typical n-n collision times of Type II discharges are $\tau_{\rm nn} < 0.1\,$ns, *i.e.* shorter than the oscillation period. The interaction time of two CO_2 molecules is $\tau_{\rm int} = 4a_0/v_{\rm th} = 1.2\,$ps, *i.e.* slow compared with the vibration oscillation time. Interacting molecules are locked together for some 100 vibrations.

Results on characteristic times and mean free path calculations are summarised in table III, leading to the following ordering in time and length scales pertaining to the CO_2 plasma discharge:

$$\tau_{\rm exc} \ll \tau_{\rm vib} \ll \tau_{\rm int} \ll \tau_{\rm etot} \ll \tau_{\rm ev} \ll \tau_{\rm nn} \ll \tau_{\rm osc}.$$

Inter-molecular energy exchange acts on a time scale much longer than the electron-molecule collision time scale. However, electrons are light and not very effective in transferring energy to molecules. When two CO_2 molecules collide the energy exchange is a factor $2\,{\rm M/m} \sim 10^5$ larger and, albeit less frequent (factor 20 to 5000), neutral-neutral collisions dominate the rate of energy transfer between molecules.

Once the CO_2 molecule is vibrational excited in the ground state $^1\Sigma^+$ by high rate electron impact, it needs to climb the vibration energy ladder in order to overcome the dissociation threshold energy of 5.5 eV. The necessary energy derives from the kinetic energy of intermolecular CO_2-CO_2 collision. Energy is transferred in high-energy vibration modes at points where the nuclear coordinates of the different molecular energy levels match according to the Frank-Condon principle. This VV exchange process populates the higher $^1B_2$ vibration energy levels of the CO_2 molecule. It is the limiting factor in the rate of dissociation [15]. Thus, CO_2 molecular bells, gently but repeatedly twanged by electrons, chime in unison until occasionally crashing into each other, transferring sufficient energy to ring overtones. Overtones of sufficiently high energy break the bell.

9·6. *Plasma parameters.* – The degree of ionisation is a critical parameter in the dissociation of CO_2 for being the source of electrons that are instrumental in vibrational excitation and subsequent dissociation. The relaxation efficiency η_{VT} is defined as the fraction of vibration energy going into dissociation over the fraction going into translation energy [15]. Thus, relaxation efficiency η_{VT} must be high for effective dissociation. It increases with the degree of ionisation in the discharge at fixed energy input. At an input energy of $E_v = 0.5$ eV per molecule CO_2 and an ionisation degree $n_e/n_0 = 10^{-6.5}$ the relaxation efficiency $\eta_{VT} = 40\%$. This increases to $\eta_{VT} = 80\%$ for ten times higher ionisation degree $n_e/n_0 = 10^{-5.5}$. At a neutral density $n_0 = 5.38\times10^{18}$ cm^{-3} (pressure of 200 mbar) and an electron density $n_e = 1.7\times10^{13}$ cm^{-3}, a relaxation efficiency $\eta_{VT} = 80\%$ is reached. In RF discharges, the energy efficiency peaks at $E_v = 0.8$–1 eV [15]. Optimum energy efficiency for CO_2 dissociation is reached at $E_v = 0.5$ eV and an ionisation degree $n_e/n_0 \sim 10^{-5}$.

Since the degree of ionisation is important for efficient dissociation of CO_2, plasma parameters add to characteristic time and length scales relevant to the activation system. These parameters include Coulomb collisions, Debye length and plasma frequency.

Coulomb collision times vary as $(m^{1/2}T^{3/2})/n$. Hence, at low electron temperature and high plasma density these times become remarkably short, nearing neutral collision times. Times for electrons relate to times for ions as $\tau_e = (m_e/m_i)^{1/2}\tau_i$. Hence, electron-electron collision times are a factor 3.5×10^{-3} shorter than CO_2 ion-ion collision times. On the other hand, collision times between electrons and ions are approximately equal to the electron-electron collision times: $\tau_{ei} = 2\sqrt{2}\tau_{ee}$. Thus, the following ordering in Coulomb times results: $\tau_{ee} \approx \tau_{ei} \ll \tau_{ii}$, showing that electrons relax by mutual collision much faster than ions. As shown before, Coulomb collisions do not play a role in plasma thermalisation, the plasma is electron-neutral collision dominated.

Coulomb collision times are calculated from the Spitzer formula, including a weak dependence on the Coulomb logarithm $\ln\Lambda$, varying from 7 to 15 in the parameter range considered. The Coulomb logarithm represents the screening of a test particle by the plasma at distances greater than the Debye length, yielding small angle scattering as the dominant collision mechanism,

$$\tau_{ei} = 3(2\pi)^{3/2}\left\{\varepsilon_0^2 m_e^{1/2} T_e^{3/2}\right\} \Big/ \left\{n_i (Ze^2)^2 \ln\Lambda\right\}.$$

TABLE IV. – *Electron-ion Coulomb collision times τ_{ei} vs. electron temperature and ion density.*

n_i	$T_e = 0.1\,\text{eV}$	$T_e = 1\,\text{eV}$	$T_e = 10\,\text{eV}$	$T_e = 100\,\text{eV}$
$10^{18}\,\text{m}^{-3}$	$\tau_{ei} = 1.3\,\text{ns}$	$\tau_{ei} = 32\,\text{ns}$	$\tau_{ei} = 840\,\text{ns}$	$\tau_{ei} = 23\,\mu\text{s}$
$10^{19}\,\text{m}^{-3}$	$\tau_{ei} = 0.15\,\text{ns}$	$\tau_{ei} = 3.6\,\text{ns}$	$\tau_{ei} = 92\,\text{ns}$	$\tau_{ei} = 2.5\,\mu\text{s}$

Results of the Coulomb collision time τ_{ei} and τ_{ii} calculation are shown in table IV and table V, respectively.

Hence, electron-ion Coulomb collision times are longer than electron neutral collision times, even for a cold, dense plasma. Also, the Coulomb collision times are longer than the RF oscillation time. Only collisions with neutrals play a role, the plasma is neutral collision-dominated, but essentially collision-free for electron and ion collisions mutually.

The Coulomb collision mean-free path follows from $\lambda = v_{th}\tau$ with $1/2\,mv_{th}^2 = kT$. For the same temperature $\lambda_{ee} = \lambda_{ii}$. Results are shown table VI.

For low T_e and high density the mean free path becomes of order plasma dimension. Otherwise, no Coulomb collision takes place inside the plasma volume.

The Debye length is given by $\lambda_D = \sqrt{(\varepsilon_0 kT)/(ne^2)}$. Results are tabulated in table VII.

For most conditions the Debye screening length is $< 10\,\mu\text{m}$, *i.e.* small compared with the plasma dimension. For a collisional plasma the skin depth becomes the relevant parameter, depending on the plasma resistivity.

The number of particles in a Debye sphere is given by $N_D = 4/3\pi\lambda_D^3 n_e$. Results are tabulated in table VIII. Hence, at low T_e there are very few particles contained in a Debye sphere in this type of plasma.

Plasma frequency is defined as $\omega_{pe}^2 = n_e e^2/\varepsilon_0 m_e$. Results are tabulated in table IX.

The RF frequency (915 MHz) is lower than the plasma frequency. Hence, the wave cannot propagate. The wave can propagate when collisions are included. The plasma permittivity $\varepsilon(\omega) = 1 - \omega_{pe}^2/\omega^2\{1 - i\nu/\omega\}$ contains the factor $i\nu/\omega$, where ν is the electron neutral collision frequency. Since this ratio is generally larger than 1, the electric field in the RF cavity is displaced by the plasma.

9·7. *Plasma chemistry.* – The reaction proceeds in two steps, starting with CO_2 in the ${}^1\sum^+$ ground state, excited by electron impact to low-level vibrational energy [15]:

$$(1) \qquad CO_2{}^*({}^1\Sigma^+) \rightarrow CO_2{}^*({}^1B_2) \rightarrow CO_2{}^*({}^3B_2) \rightarrow CO({}^1\Sigma^+) + O({}^3P),$$
$$\Delta H = 5.451\,\text{eV net enthalpy.}$$

TABLE V. – *Ion-ion Coulomb momentum transfer-time τ_{ii}.*

n_i	$T_i = 0.1\,\text{eV}$	$T_i = 1\,\text{eV}$
$10^{18}\,\text{m}^{-3}$	$\tau_{ii} = 287\,\text{ns}$	$\tau_{ii} = 9.1\,\mu\text{s}$
$10^{19}\,\text{m}^{-3}$	$\tau_{ii} = 32\,\text{ns}$	$\tau_{ii} = 1\,\mu\text{s}$

TABLE VI. – *Mean free path* $\lambda_{ee/ii}$ *Coulomb interaction.*

n_i	$T_e = 0.1\,eV$	$T_e = 1\,eV$	$T_e = 10\,eV$
$10^{18}\,m^{-3}$	$\lambda_{ee/ii} = 238\,\mu m$	$\lambda_{ee/ii} = 13.8\,mm$	$\lambda_{ee/ii} = 1.5\,m$
$10^{19}\,m^{-3}$	$\lambda_{ee/ii} = 27.5\,\mu m$	$\lambda_{ee/ii} = 2.1\,mm$	$\lambda_{ee/ii} = 0.17\,m$

The higher CO_2 vibrational energy level ${}^1B_2$ is reached by molecular V-V relaxation. Activation energy is lowered by a non-adiabatic singlet ${}^1B_2$ to triplet ${}^3B_2$ spin transition. This happens at the point where energy level and molecular distance of the O=CO bond coincide. It allows molecular break-up at minimum 5.5 eV energy, leaving CO and O formed in the ground electronic state.

The atomic oxygen radical produced follows up by a second reaction with abundantly available CO_2 producing one more CO molecule at 0.3 eV net energy, additional energy being freed by recombination of atomic oxygen into molecular oxygen at 5.2 eV:

$$(2) \qquad O + CO_2 \to CO + O_2, \qquad \Delta H = 0.3\,eV \text{ net enthalpy.}$$

The three-body recombination reaction of atomic oxygen to O_2 has a much lower rate coefficient as compared with the reaction with CO_2. The two reactions in succession yield a net enthalpy for the production of one CO molecule $H = 5.8/2 = 2.9\,eV$, the limiting stage being reaction (1).

10. – Reduced electric field

Electrons gain energy by the applied electric field and are scattered by CO_2 molecules that can be considered stationary (Lorentzian gas). A steady-state electron drift sets up with velocity v_D proportional to the applied electric field with proportionality constant μ_e, the electron mobility. The velocity acquired between collisions is $v_D = (eE/m_e)\tau_{en}$, where $\tau_{en} = (n_0\sigma_{en}v_D)^{-1}$ is the collision time. Combination yields the energy U_e gained by electrons between collisions

$$(3) \qquad U_e = 1/2\,m_e {v_D}^2 = 1/2\,eE/(n_0\sigma_{en}) = 1/2\,eE\lambda_{en},$$

TABLE VII. – *Debye length* λ_D.

n_e	$T_e = 0.1\,eV$	$T_e = 1\,eV$	$T_e = 10\,eV$
$10^{18}\,m^{-3}$	$2.35\,\mu m$	$7.45\,\mu m$	$23.5\,\mu m$
$10^{19}\,m^{-3}$	$0.75\,\mu m$	$2.35\,\mu m$	$7.45\,\mu m$

TABLE VIII. – *Number of particles N_D in a Debye sphere.*

N_D	$T_e = 0.1\,eV$	$T_e = 1\,eV$
$10^{18}\,m^{-3}$	54	1720
$10^{19}\,m^{-3}$	17	550

where λ_{en} is the electron neutral mean free path. This expression yields a relation between electron energy and the reduced electric field:

$$U_e = e/(2\sigma_{en})\{E/n_0\}. \tag{4}$$

This expression permits calculation of the reduced electric field for a given electron energy. For example, the cross-section for vibrational excitation of the CO_2 asymmetric stretch mode peaks at an electron energy of 0.4 eV where the cross-section is $\sigma_{en} = 1.8 \times 10^{-16}\,cm^2$ [16]. Substitution yields $E/n_0 = 1.4 \times 10^{-16}\,V\,cm^2$ for the reduced electric field value, where vibrational excitation by electrons is maximum. This value corresponds with Russian theory [15].

Through collisions the electrons randomise and, albeit non-Maxwellian, could be assigned a temperature. Since $U_e = 3/2kT_e$ the expression for the reduced electric field yields

$$kT_e = e/(3\sigma_{en})\{E/n_0\}. \tag{5}$$

Hence, a linear relation exists between electron temperature and reduced electric field, which is useful in relating measured plasma parameters to the reduced electric field. The above expression differs from [15] by a numerical factor, which takes into account inelastic energy loss during electron randomisation.

11. – Energy efficiency for CO_2 splitting

Energy efficiency is defined as $\eta = H/E_{CO}$, where $H = 2.9\,eV$ is the minimum net enthalpy required for the production of one CO molecule from CO_2 input gas flow and E_{CO} is the actual energy expended per CO molecule produced. Measured is E_v, the energy input per incoming CO_2 molecule, which is proportional to input power P_{RF} over gas flow rate Q or $E_v \approx P_{RF}/Q$. The fraction of CO_2 converted into CO is $\alpha = [CO]_{out}/[CO_2]_{in}$. Since the input energy E_v is distributed over CO_2/α molecules to

TABLE IX. – *Plasma frequency as a function of plasma density.*

n_e	$\nu_{pe} = 8.967\sqrt{n_e}$
$10^{18}\,m^{-3}$	$\sim 9\,GHz$
$10^{19}\,m^{-3}$	$\sim 28\,GHz$

produce one CO molecule, it holds that $E_v = \alpha E_{co}$. Since both α and E_v are amenable to measurement, the energy efficiency η becomes expressed in measured quantities:

$$\eta = CQ/P_{RF} = \alpha H/E_v. \tag{6}$$

Here, C is a constant of numerical factors, in this case $C = 1.93 \times 10^2$ at 300 K. The conversion factor α is determined experimentally from mass analysis of the exit gas stream and supported by spectral line strength measurement. Measured values of η *vs.* E_v are plotted in figs. 11, 12 and 13 applying eq. (6).

12. – Plasma particle balance

Plasma is created by electron ionisation of the CO_2 gas and lost by ambipolar diffusion up to the sheath edge where the ions reach the Bohm velocity. The Bohm velocity is a result of the overriding need for plasma neutrality, which requires fluxes of ions and electrons out of the plasma to be equal. Equating production and loss, the steady-state particle balance is obtained. Since production by ionisation is a volume process and loss by diffusion a surface process, the gas density n_0 and the plasma radius R, are coupled parameters:

$$2v_B = n_0\langle\sigma_{ion}v\rangle R \approx \sqrt{(E/n)} \tag{7}$$

here $v_B = \sqrt{2kT_e/m_i}$ is the Bohm velocity and $\langle\sigma_{ion}v\rangle$ the ionisation rate averaged over the electron velocity distribution, whilst use is made of eq. (5) to relate electron temperature to the reduced electric field. The ionisation rate increases proportionally with the reduced electric field. This expression shows the plasma discharge radius to contract with increasing gas pressure. This phenomenon has been observed experimentally in the transition of going from Type I discharge at low pressure (1 mbar), see fig. 9, to Type II discharge at high pressure (100 mbar), see fig. 5b).

Put differently, the loss of the plasma can be expressed in a plasma electron containment time τ_e as

$$\tau_e\sqrt{(E/n)} \approx R. \tag{8}$$

Hence, increasing electron containment —which is good because ionisation is energetically expensive— lowers the reduced electric field for fixed plasma radius and thereby increases energy efficiency.

The energy balance yields an expression for the electron density n_e

$$P_{RF} = n_e\Sigma_i n_0\langle\sigma_i v\rangle\Delta\varepsilon_i V, \tag{9}$$

where σ_i represents vibration, excitation and ionisation cross-sections summed, causing an electron energy loss $\Delta\varepsilon_i$, V is the volume of the discharge. The above equations

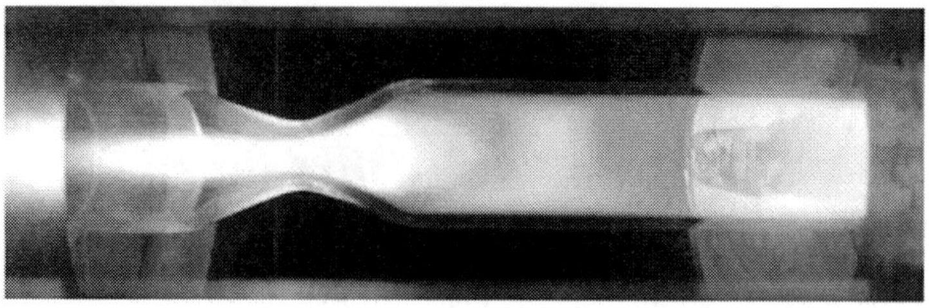

Fig. 9. – Typical Type I discharge with nozzle placed at the entrance of the discharge, characterised by 1 mbar pressure. After expansion striations are visible indicating shock fronts, a tell-tale sign of supersonic flow.

yield basic gas discharge parameters. Typically, the plasma density is in the range 10^{18}–$10^{19}\,\mathrm{m}^{-3}$, ionisation degree 10^{-6} to 10^{-5} and electron temperature in the range 1 to 10 eV.

13. – Vortex flow

Gas is injected tangentially at the entrance of the cylindrical reaction chamber through a small orifice where it accelerates and expands. A strong spiralling gas flow is established along the chamber wall, causing a radial pressure and temperature gradient to build up. This type of flow has been exploited in the Ranque-Hilsch cooling machine [40, 41]. In our case it provides stability of the plasma discharge.

In axial direction, flow is directed through a nozzle after which the gas expands and cools. When the exit pressure is low enough this flow reaches a maximum, called choked flow. Under this condition, in the throat of the nozzle, the gas velocity reaches the speed of sound c_s or Mach $M = 1$. Beyond, the velocity increases and becomes supersonic at the exit of the reaction chamber leading to shock waves to form, see fig. 9. At these sonic gas speeds the vortex rotates at high frequency, 300000 r.p.m. is not unusual. The resulting centrifugal acceleration typically reaches $10^6 g$ at the wall of the reactor vessel.

Assuming $M = 1$ in the throat of the nozzle, *i.e.* chocked flow, the gas velocity equals the speed of sound. The gas flow velocity v_F follows from $nv_\mathrm{F}O = F$, *i.e.* $v_\mathrm{F} = F/nO = (kT/p)(F/O)$, assuming an ideal gas. For $v_\mathrm{F} = c_\mathrm{s}$ to hold, the relation between temperature and gas pressure becomes: $kT = \gamma/m_\mathrm{i}(pO/F)^2$, where $\gamma = (i+2)/i = 11/9 = 1.222$, the number i being the 9 degrees of freedom of the CO_2 molecule. Converting F from particles/sec to slm, using $F = N_\mathrm{A}F_\mathrm{slm}/(60 * 24)\,\mathrm{s}^{-1}$ yields $T = 6.93 * 10^6\{pO/F_\mathrm{slm}\}^2$. As a consequence, at constant flow, the gas pressure scales $p \sim 1/\sqrt{T}$. The pressure is seen to increase linearly with power; hence the gas temperature T increases with power. Gas temperature values ranging between $T = 500\,^\circ\mathrm{C}$ and $3000\,^\circ\mathrm{C}$ have been calculated in this way.

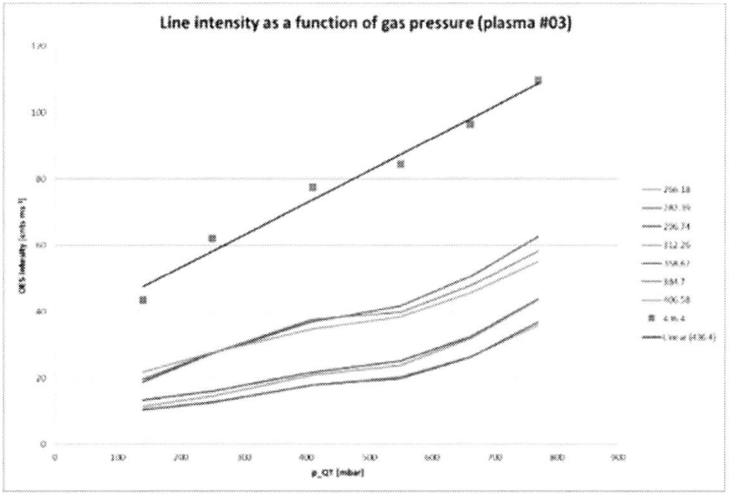

Fig. 10. – The CO 3rd positive triplet band is shown to increase linearly with gas pressure, consistent with CO production rates.

14. – Experimental results

Three types of discharge have been investigated at constant input CO_2 gas flow rate of 11.1 slm:

Type I gas expansion nozzle, 10 mm diameter is placed at the entrance of the cavity expanding the gas into the discharge. Typical gas pressure 1 mbar is characterised by high $E/n \sim 100 \times 10^{-16}\,\mathrm{V\,cm^2}$. Typical energy efficiency $\eta = 15\%$ is found constant with power and particle conversion α increases linearly with power. OES shows dominance of 3rd and 4th positive CO bands to increase linear with power.

Type II gas expansion nozzle is placed at the exit of the cavity freely expanding the converted gas stream. The high-pressure 200 mbar gas discharge is characterised by low $E/n \sim 10^{-16}\,\mathrm{V\,cm^2}$. The discharge diameter contracts with higher pressure in accord with eq. (7), see fig. 5b). Typical energy efficiency $\eta = 35\%$ increases by reducing power. Particle conversion increases with power from 10% upward. The OES is dominated by the CO Ångström and CO Triplet system, see fig. 5d), whilst the Swan system reveals carbon.

Type III exit nozzle diameter has been increased in steps from 5 mm, to 10 mm, to 26 mm diameter. This type of discharge is characterised by an intermediate reduced electric field regime $E/n \sim 10 \times 10^{-16}\,\mathrm{V\,cm^2}$. Intermediate energy efficiencies result.

In order to increase energy efficiency, Type II high-gas-flow experiments are repeated at increased gas flow rate of 75 slm and 3 kW absorbed power, reaching record high energy efficiency $\sim 50\%$ albeit at low conversion fraction 10%, as expected from eq. (6). Power density is typically $45.5\,\mathrm{W/cm^3}$. At 200 mbar the CO output flow becomes $20\,\mathrm{A/cm^2}$ equivalent.

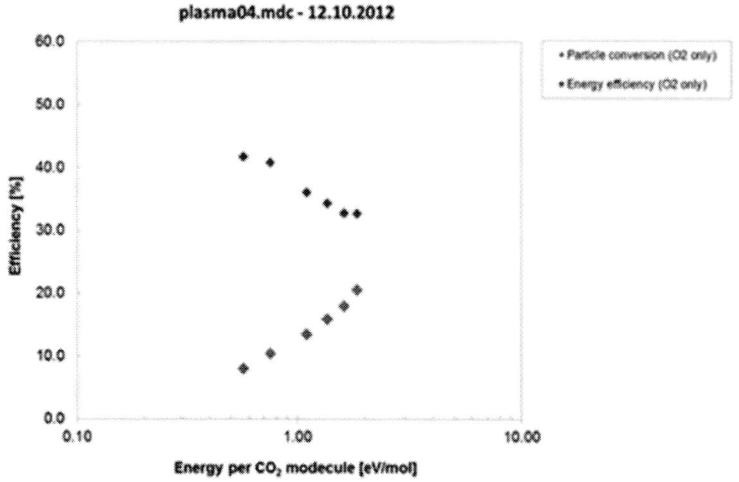

Fig. 11. – Energy efficiency and particle conversion in terms of energy E_v per incoming CO_2 molecule obtained from variation of RF power. Data are constrained by stoichiometric fit, taking the O_2 data as reference for not suffering from cracking products produced inside the QMS.

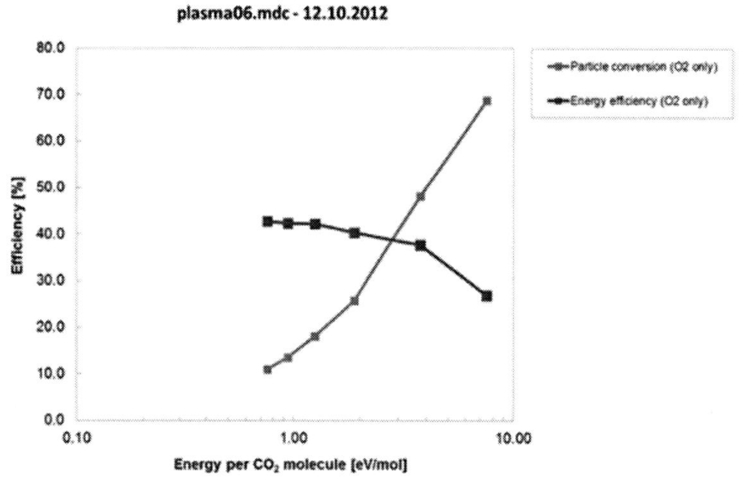

Fig. 12. – shows energy efficiency and particle conversion in terms of energy E_v [eV/mol] per incoming CO_2 molecule, obtained from variation of gas flow rate extending the data range of fig. 11. Conversion ratios of 70% are reached at reduced energy efficiency. A stoichiometric constrained fit is applied.

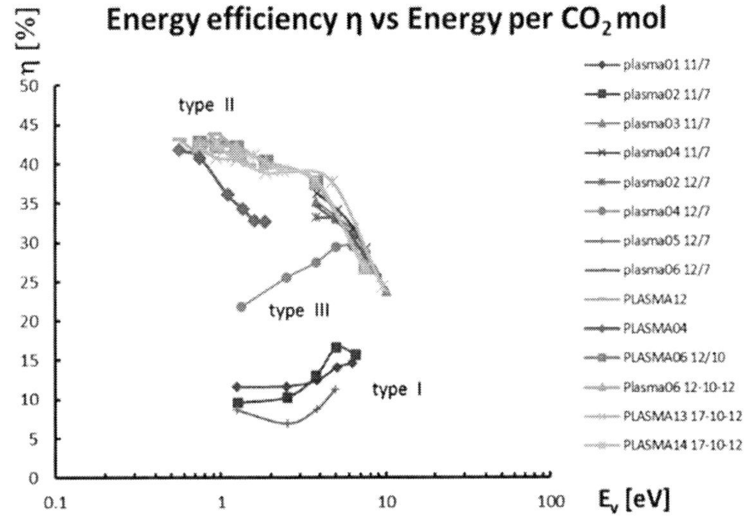

Fig. 13. – Energy efficiency η for CO production increases by lowering the energy E_v per incoming CO_2 molecule.

Increasing the input microwave power P_{RF} at constant flow rate Q shows an approximate linear increase in particle conversion ratio $\alpha = aP_{RF}$ for all types of discharge, see fig. 5c). Since $E_v/H \sim P_{RF}/Q$, where E_v is the energy expended per incoming CO_2 molecule expressed in units of H, the minimum enthalpy, it follows that $\alpha \approx (E_v/H)Q$. Since the energy efficiency $\eta = H/E_{co} = \alpha(H/E_v) \approx P_{RF}(H/E_v) = Q$, it follows that at constant flow rate the energy efficiency is constant with power. In reality there are deviations to linearity of conversion ratio, which explains that calculated energy efficiencies show a slight increase at low power.

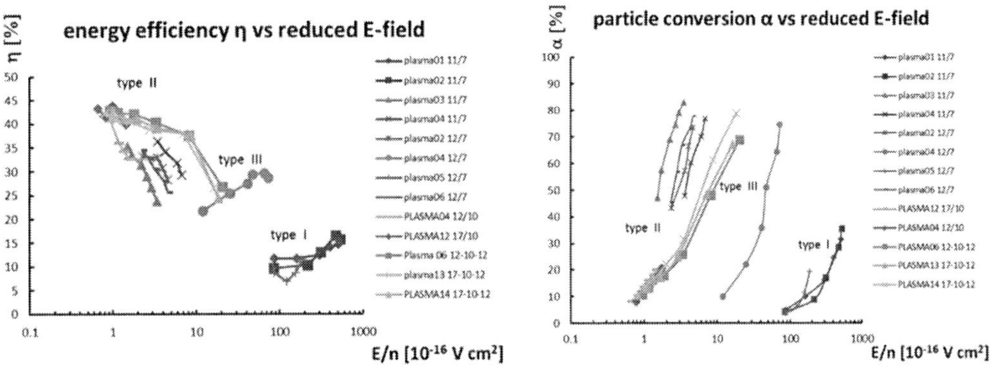

Fig. 14. – Lowering the reduced electric field E/n is key to increased energy efficiency η (left) but goes at the expense of the conversion fraction α, consistent with E_v values below the dissociation energy (right).

The CO emission line strength is shown to increase linearly with RF power, confirming the QMS data, see fig. 10. Similarly, with gas pressure the line strength is seen to increase linearly.

14'1. *Energy efficiency.* – Plotting CO conversion data as a function of input power, fig. 5c) and similar for gas flow variation, in terms of specific energy E_v per CO_2 molecule yields figs. 11 and 12. The typical energy efficiency 50% obtained reduces with increase of power, whilst the conversion to CO increases linearly with power.

14'2. *Reduced electric field.* – The electric field has been calculated by Comsol Multiphysics simulation program [19]. The vacuum field strength on axis of the cavity is $E[\mathrm{V/cm}] = 4.965\sqrt{P_{\mathrm{RF}}}$ [W]. Since the microwave frequency is much lower than the cut-off plasma frequency, power is absorbed by e-n collisions in the plasma. At the plasma density considered the E-field skin depth still penetrates the plasma on axis, albeit at reduced strength. These aspects have been ignored in the E estimate.

Results are collated in figs. 13 and 14. The trend suggests that higher efficiencies may be obtained at lower E/n values. Here, conflicting requirements on sustaining the gas discharge requiring $T_e > 1\,\mathrm{eV}$ and low E/n values requiring $T_e < 1\,\mathrm{eV}$ need to be accommodated, suggesting decoupling of requirements necessary. This may be achieved by a double humped electron energy distribution function (EEDF). Energy efficiency goes up with decreasing energy E_v per incoming CO_2 molecule, whilst particle conversion goes down. This result is consistent with eq. (6) and is tantamount to saying that the available energy per input CO_2 molecule needs to be concentrated in one out of many CO_2 molecules in order to accumulate sufficient energy to attain the net 2.9 eV dissociation energy.

REFERENCES

[1] United Nations Framework Convention Climate Change, `http://unfccc.int/paris_agreement/items/9485.php`.

[2] RED 2009/28/EC, `https://ec.europa.eu/energy/en/consultations/preparation-new-renewable-energy-directive-period-after-2020`.

[3] Mohr A. and Raman S., *Energy Policy*, **63** (2013) 114; Searle S. Y. and Malins C. J. Proc. 20th Eur. Bio. Conf. &.Exh., Milan, Vol. **53**, (2012).

[4] A European Strategy for Low-Emission Mobility SWD(2016) 244 final, Brussels, 20.7.2016 COM(2016) 501 final.

[5] UN International Civil Aviation Organisation, `http://www.icao.int/Meetings/a39/Documents/Resolutions/a39_res_prov_en.pdf`.

[6] BMU Leitstudie 2012 by DLR, IWES, IFNE.

[7] Grond L. J. *et al.*, DNV-KEMA report GCS 13.R.23579 (2013), `www.energeia.nl/docs/P2G_ECN_2014.pdf`.

[8] E&E Consultants, Etude portant sur l'hydrogene et la methanation comme procédé de valorisation de l'electricite excedentaire, Hespul, Solagro (2014).

[9] KEROGREEN, EU H2020 funded project 763909 "*Production of Sustainable aircraft grade Kerosene from Water and Air powered by Renewable Electricity through the splitting of CO_2, formation of Syngas and Fischer-Tropsch synthesis*" by DIFFER, KIT, VITO, HyGear, CerPoTech and INERATEC, see `https://www.differ.nl/network/kerogreen`.

[10] Carmo M., Fritz D. L., Mergel J. and Stolten D., *Int. J. Hydrogen Energy*, **38** (2013) 4901.

[11] Jensen S. H., Sun X., Ebbesen S. D., Knibbe R. and Mogensen M., *Int. J. Hydrogen Energy*, **35** (2010) 9544.

[12] Zheng Y., Wang J., Yu B., Zhang W., Chen J., Qiao J. and Zhang J., *Chem. Soc. Rev.*, **46** (2017) 1427.

[13] Irvine J. T., Neagu D., Verbaeken M.C., Chatzichristodoulou C., Graves C. and Mogensen M. B., *Nature Energy*, **1** (2016) 15014.

[14] Legasov V. A. *et al.*, *Dokl. Akad. Nauk.*, **238** (1978) 66, *Sovj. Phys. Dokl.*, **23** (1978) 66.

[15] Rusanov V. D., Fridman A. A. and Sholin G. V., *Sov. Phys. Usp.*, **24** (1981) 447, *Usp. Fiz. Nauk.*, **134** (1981) 185.

[16] Itikawa Y., *J. Phys. Chem. Ref. Data*, **31** (2002) 749.

[17] Azizov R. I., Vakar A. K., Zhivotov V. K., Krotov M. F., Zinov'ev O. A., Potapkin B. V., Rusanov A. A., Rusanov V. D. and Fridman A. A., *Sov. Phys. Dokl.*, **28** (1983) 567.

[18] Leins M. *et al.*, 31th Int. Conf. Phen. in Ionized Gases, Granada (2013).

[19] Leins M., Thesis, Ch 2.3.2 (2010), Universität Stuttgart.

[20] Goede A. P. H. and van de Sanden M. C. M., *Europhys. News*, **47**, issue No. 3 (2016) 22, `http://dx.doi.org/10.1051/epn/2016304`.

[21] Goede A. P. H. *et al.*, *EPJ Web of Conferences*, **79** (2014) 01005, DOI:10.1051/epjconf/20147901005.

[22] Goede A. P. H., *Joint EPS-SIF International School on Energy*, edited by Cifarelli L. and Wagner F. (SIF, Bologna) 2015, pp. 357-380, ISBN 978-88-7438-094-7.

[23] Nighan W. L., *Phys. Rev. A*, **2** (1970) 1989.

[24] Diomede Paola, DIFFER communication.

[25] Tsampas M. N. *et al.*, *Cat. Sci. Techn.*, **5** (2015) 4884, `http://dx.doi.org/10.1039/C5CY00739A`.

[26] Sapountzi F. M. *et al.*, *Solid State Ionics*, **277** (2015) 65, `http://dx.doi.org/10.1016/j.ssi.2015.05.003`.

[27] Vernoux P. *et al.*, *Chem. Rev.*, **113** (2013) 8192, `http://dx.doi.org/10.1021/cr4000336`.

[28] Tagawa Y. *et al.*, *Kagaku Kogak Ronbunshu*, **37** (2011) 114.

[29] Tun L. L., Matsuura N. and Mori S., *22nd Int. Symp. Plasma Chemistry* (2015) p. O-15-3.

[30] Tsampas Mihalis, DIFFER Internal Communication.

[31] Suleyman E. r., DIFFER Internal Communication.

[32] Kresse G. and Furthmuller J., *Phys. Rev. B*, **54** (1996) 11169, DOI:10.1103/PhysRevB.54.11169.

[33] Tao S. and Irvine J. T. S., *Nat. Mater.*, **2** (2003) 320, DOI:10.1038/nmat871.

[34] Yue X., Arenillas A. and Irvine J. T. S., *Faraday Discuss*, **190** (2016) 269, DOI:10.1039/C6FD00001K.

[35] HITRAN (2004) Harvard Smithsonian, Cambridge, `http://vpl.astro.washington.edu/`.

[36] Pearse R. W. B. and Gaydon A. G., *The Identification of Molecular Spectra*, 4th edition (Chapman and Hall) 1976, pp. 109–112.

[37] Herzberg G., *Spectra of Polyatomic Molecules*, Vol. **III** p. 503, CO^{2+} energy diagram.

[38] Herzberg G., *Molecular Spectra and Molecular Structure*, Vol. **1** *Spectra of Diatomic Molecules*, 2nd edition, p. 452.
[39] Herzberg G., *Atomic Spectra and Atomic Structure* (Dover books) 2010, p. 163.
[40] Liew R., Zeegers J. C. H., Kuerten J. G. M. and Michalek W. R., *Phys. Rev. L*, **109** (2012) 054503.
[41] Liew R., Zeegers J. C. H., Kuerten J. G. M. and Michalek W. R., *Exp Fluids*, **54** (2013) 1416.

EPJ Web of Conferences **189**, 00011 (2018) https://doi.org/10.1051/epjconf/201818900011
LNES 2017

Fuel cells for electrochemical energy conversion

RYAN P. O'HAYRE
Metallurgical and Materials Engineering, Colorado School of Mines
1500 Illinois St, Golden, CO 80401, USA

These lecture notes were originally published in *EPJ Web of Conferences*, **148** (2017) DOI: 10.1051/epjconf/201714800013 as a contribution to the Proceedings of the 5th course of the MRS-EMRS "Materials for Energy and Sustainability" and 3rd course of the Joint EPS-SIF International School on Energy held in 2016. They are here reproduced as a good summary of the material presented in July 2017 during the course "Advances in Basic Energy Issues".

EPJ Web of Conferences **189**, 00012 (2018) https://doi.org/10.1051/epjconf/201818900012
LNES 2017

Smart grids: enabler for the energy transition

Diana Moneta

Ricerca sul Sistema Energetico—RSE SpA - Via Rubattino 54, 20134 Milano, Italy

Summary. — This paper deals with the concept of "smart grids". Starting from drivers behind the evolution of the power system, an overview of elements and technologies is given: smart grids have to intelligently integrate the actions of all users connected to it — generators, consumers and those that do both — in order to efficiently deliver sustainable, economic and secure electricity supplies. Examples of solutions from demonstration projects represent the base for discussing methodologies to estimate benefits deriving from smart grid solutions, showing how conflicting goals cannot always be fulfilled simultaneously. Finally, future perspectives are discussed: the main thinking regards the awaited pivot role for customers, in the future scenario with an increased utilization of electricity combined with a high penetration of non-programmable resources.

1. – Smart grid: drivers

The European Union in 2007 has started a comprehensive plan to increase the environmental sustainability, the security of supply and the competitiveness of its energy system. This strategy was translated into binding targets for greenhouse gases reduction, renewable energy sources (RESs) and energy efficiency (the so called "20-20-20" package).

In most EU countries, these goals have required a major contribution from electrical final uses in order to compensate lower efforts achievable in transports and thermal usages. In Italy, for example, to reach a 17% of gross final energy uses in 2020 covered by renewable resources, a 26.39% share of electric uses was expected to balance their smaller penetration in other sectors [1]. In fact, the nature of the electricity vector allows to integrate more easily RESs, even if electricity counts for less than one third of the total energy consumption in Europe. Analogous strategies are followed by extra EU countries, depending on the availability of resources and on the market.

In the past decade, the exploitation of renewable sources has been realized by means of small-size generators especially, with a leading role of photovoltaic plants and wind farms and a more limited growth of biomass and run-of-river (without basin) hydro plants. Medium and small generators connected to the distribution grid are referred to as "distributed generation"; some definitions include a power threshold but the typology of the hosting network represents the focus for the scope of this discussion.

The presence of a large number of plants connected to the distribution networks has caused the entire system to undergo a deep evolution, towards the so-called "smart grid" concept. According to the EU definition, a smart grid is a "grid that can intelligently integrate the actions of all users connected to it — generators, consumers and those that do both — in order to efficiently deliver sustainable, economic and secure electricity supplies" ([1]).

2. – Elements and technologies

The electric system is probably the largest machine built by humankind. It binds in a single, interconnected network the European countries with the neighbouring continent, supplying citizens and industries [2].

Electricity is an energy *vector* and it cannot be stored without converting it into another form (chemical, mechanical, etc.). Then, in every time instant, the balance between the energy absorption from loads and the energy injection by generators has to remain very accurate, and only small variations around the (dynamic) equilibrium point are admissible. Since the power system is based on the alternating current (AC) concept [2], the frequency is maintained around 50 Hz (60 Hz in USA and other countries). Analogously, the voltage must be kept within its allowed range [3] to have appliances and devices operating properly.

This system was designed decades ago in an "unidirectional" framework, where large *bulk* power plants, usually based on coal, hydro, nuclear or gas, inject the electricity into the high voltage (HV) transmission lines, then this energy is delivered to the medium and to the low voltage distribution grid to reach the final users. In fig. 1 a simplified scheme of the power system is given: Primary and secondary substations convert the energy to/from the different voltage levels: in transmission lines the voltage V is higher in order to reduce the losses (inversely proportional to V for a given power transfer) while loads in customers' premise operate at low voltage.

([1]) Firstly proposed by the Electricity Networks Strategy Group (ENSG), UK.

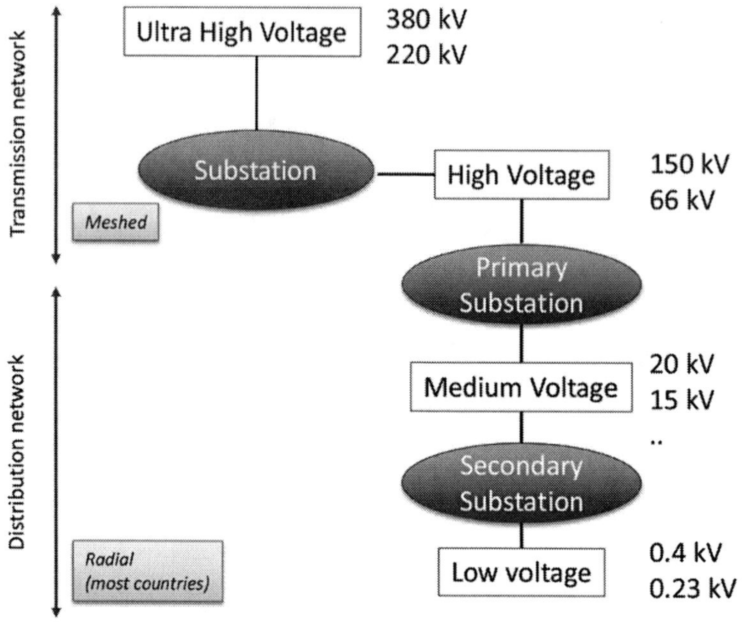

Fig. 1. – Simplified scheme of the power system (voltage levels may vary in the different countries).

The operation of the power system shows differences among the countries, the basic one is that the transmission networks are operated in a meshed topology while distribution grids are usually operated radially even if re-configurations are applied to face faults. Figure 2 schematizes the boundary between transmission and distribution grid, that is the *primary substation* (PS) where the power is converted from the High Voltage (HV) to the Medium Voltage (MV) level. Usually there are two transformers, working each at less than half of the nominal power to have a margin for outages or maintenance of the other one. Some transformers are equipped with the on load tap changer (OLTC) which allows varying the voltage of the MV busbar through the transformation ratio ("taps" are the connection points along the winding).

As mentioned before, most Distribution System Operators (DSO) manage their MV network *radially* even if the grid topology is meshed (fig. 3): in case of outages, automatic procedures exclude the fault section of the network and counterfeed the remaining parts through the working lines, reducing the impact on the customers. Besides, thanks to planning tools, the topology can be optimized to balance the load among the different feeders, for efficiency purposes also. With this architecture, only remote rural loads served by single lines may experience longer outages.

The secondary substations feed the MV customers and, through a transformer, the LV customers (fig. 4). Each secondary substation serves from few users (rural areas) to hundreds of premises (urban centres). The transformer characteristics are selected as a function of the total power of customers, considering the probability of simultaneous

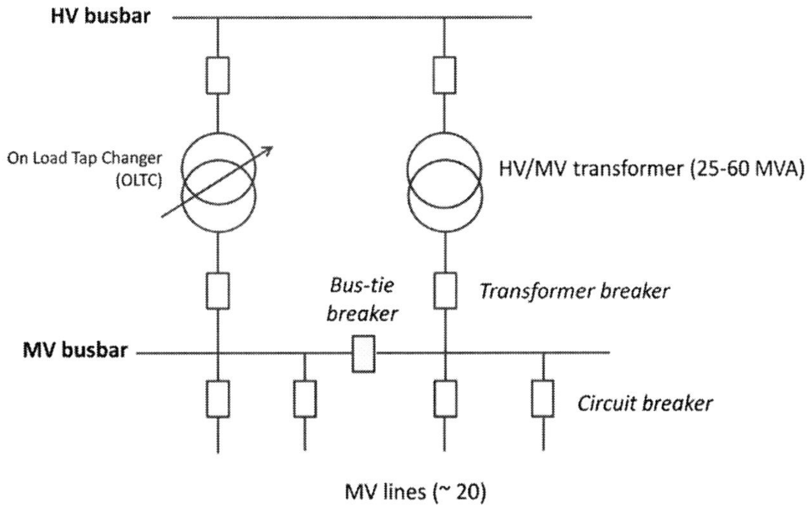

Fig. 2. – Simplified scheme of a HV/MV Primary Substation (PS). PSs usually have 1 to 3 transformers, and in some PSs transformers are equipped with OLTC (on load tap changer) to vary the voltage of the MV busbar independently of the HV busbar. The bus-tie breaker is usually open, it allows to connect all the MV lines under a single transformer in case of outages or maintenance of the second equipment.

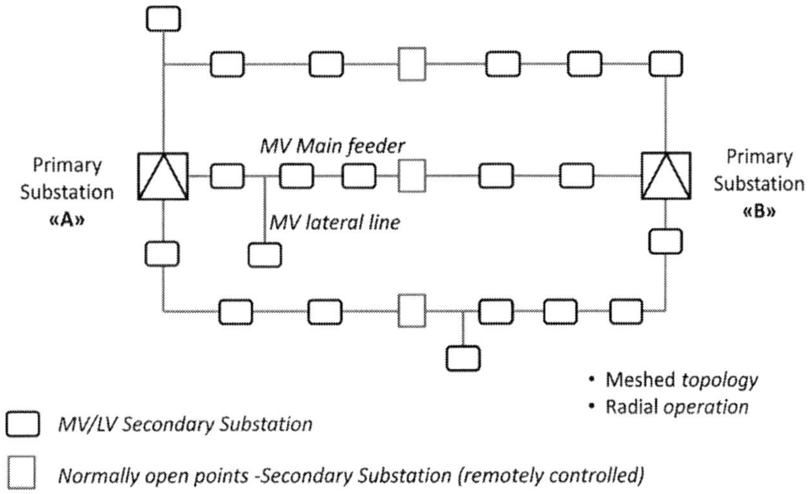

Fig. 3. – Simplified scheme of medium-voltage networks. In case of a fault, the normally open points are closed to re-configure the topology, counterfeeding MV/LV substations from the other Primary substation.

utilization of the loads. The architecture based on the two levels of substations is "scalable" to cover properly the different areas; on the average there is a 200:1 ratio between the number of secondary and primary substations.

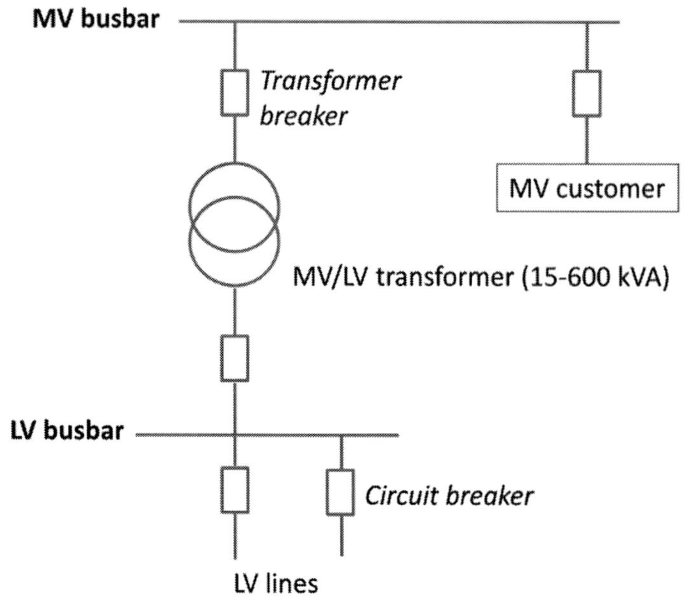

Fig. 4. – Simplified scheme of a MV/LV Secondary Substation (SS).

In very few words, the Transmission System Operator (TSO) is in charge of the security of supply and balancing, so it keeps the *frequency* around 50 Hz (in Europe, 60 Hz in USA and other countries) acting on *balancing* resources, acquired on the ancillary service market. Traditionally, this flexibility has been provided by *bulk generators* able to modulate their active power production within few seconds, in order to counterbalance the frequency deviations. In addition, thermal power plants have large rotating masses that pose a significant *inertia* to the variations of the power frequency.

To avoid duplication of the infrastructure, for a given geographical area there is only one DSO in charge of the supply of electricity. It has the obligation, deriving from the concession, of connecting all the customers, and it has to guarantee an adequate quality of supply in terms of continuity and stability of the voltage, and to keep currents in lines below their rated value. Besides, in most countries DSOs are in charge of metering, that is the collection of consumption data for billing purposes: these data are delivered to the *vendor*, or retailer, *e.g.* the commercial counterpart of customers for the electricity supply. According to EU rules, DSO and the vendor are separate entities so customers can choose the preferred contract on the market.

Automation has strongly contributed to improve the continuity and quality of supply in the past years, having a significant reduction of the number and the duration of outages. In fig. 5 the trend in the Italian case is reported: it can be seen how, thanks to the premium/penalty scheme set by the National Regulation Authority, the overall average duration of interruptions attributed to DSOs has been reduced; conversely, the interruptions deriving from other causes (weather, fraud) have increased in the recent years. Similar patterns are observable in other countries.

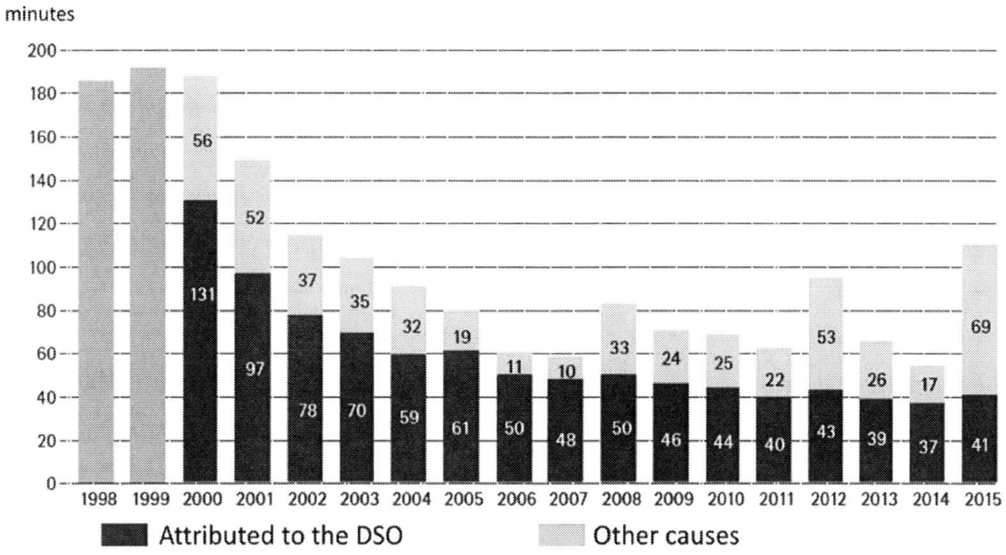

Fig. 5. – Long interruptions without notice — average duration per Low-Voltage Italian customer. Dark bars represent interruptions attributed to DSO and included into the penalty/premium scheme, other causes include weather events and frauds (source: Italian National Regulation Authority for Energy [4]).

As mentioned before, the support to RES-based generation, implemented through implicit and explicit incentives([2]), has led to a high penetration of Distributed Generation (GD) in MV and LV networks.

Besides the environmental aspects, the presence of DG has other tangible effects: firstly, generation is installed close to loads, with less involvement of the transmission lines([3]). In particular, PV production occurs in daylight hours (when the load is greater) and reaches a peak in the summer season (when the load is higher, as is the case in Italy). During the middle of the day the energy prices are lowered, due to the PV energy production: for the year 2014, the estimated price reductions on the Italian day ahead market was between 5.8 and 24€/MWh, with savings for consumers accounting to 896 M€. These effects on the prices is evident around 2 PM in fig. 6, which plots the change in the day ahead market electricity prices over several years.

The impact of DG on the distribution system has been deeply discussed in the literature. Up to a few years ago, DGs were connected under the so-called "fit & forget" approach, where no functional contribution from generators was expected towards the system. With large generation units, the transmission system operator is able to monitor

([2]) See the contribution by Clerici to this School.
([3]) It is worth mentioning that the power system was born in a decentralized architecture: at the beginning generators were built close to buildings and factories, and only when alternating current transmission technologies became available, power plants were located further from towns.

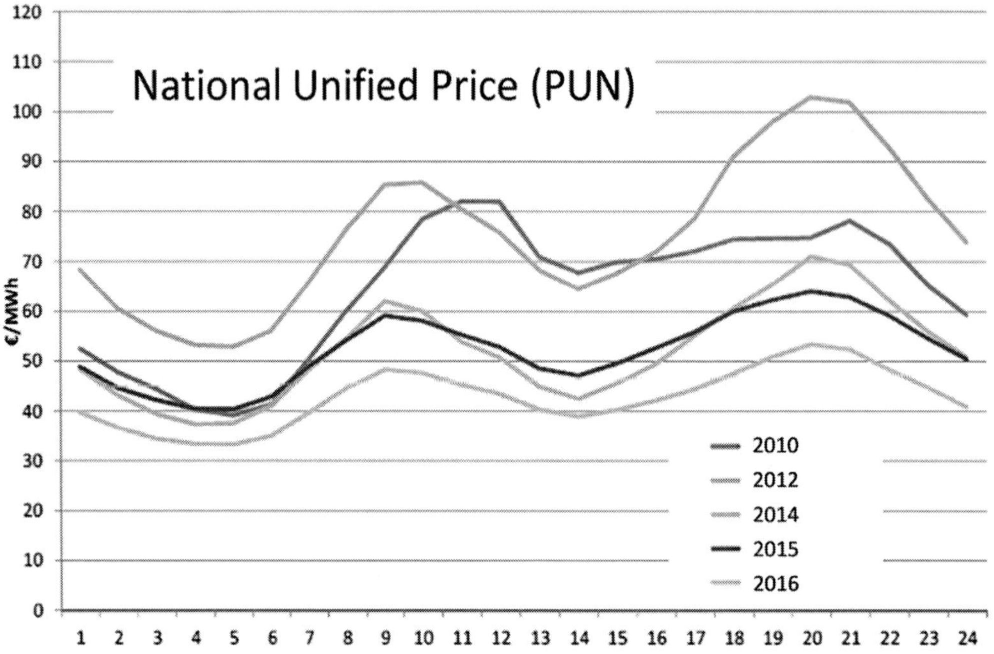

Fig. 6. – Effect of PV production on the day ahead market price in Italy, working days. In 2010 the total PV capacity was 3.47 GW while in 2013 it had already exceed 18 GW (elaboration from GME data).

and plan the production (ordering a curtailment if needed, producers still get paid as usual), while distributed generation, according to the traditional approach, cannot be dispatched in this way: network operators are legally forced to accept their production whatever it is. This may lead to limit the diffusion of generators, given the most valuable effects they could entail at both local and global levels:

- Voltage variations: injection of active power increases the voltage profile in the neighbouring nodes. Voltage rise represents actually the main limitation to the penetration of DG in rural areas [5].
- Harmonic distortion (noise at a multiple of the fundamental frequency 50 Hz); this can interfere with power line communication used for of smart meters [6].
- Grid overload (thermal limits of transformers and lines).
- Impact on protection devices (interference with the automatic procedure of identification and exclusion of faulty areas).
- Poor observability of the system: generators are connected to networks not monitored by the TSO. This combines with the lack of programmability of most RES generation, entailing the availability of additional flexibility resources ("reserve")

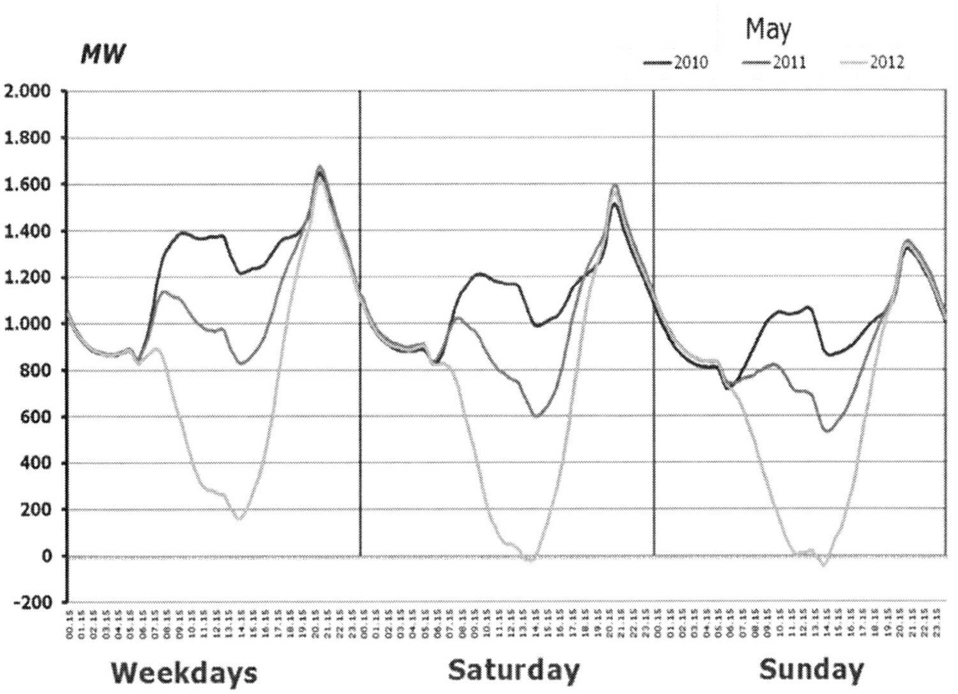

Fig. 7. – "Summer sunny day syndrome": in several HV/MV substations located in areas with high PV penetration, the power exchange from the transmission level to the distribution network has reduced and even reversed in the central hours. This graph refers to the Puglia region, in Southern Italy (source: ENEL). In 2016, more than 25% of primary substations had a reverse flow for at least 1% of hours.

to compensate real time fluctuations of primary sources. Bulk generators, the traditional provider of such flexibility, are progressively displaced from the market because of low prices([4]): some gas fired plants are kept online only for providing this balancing service.

- Reverse flow from the MV to the transmission networks (fig. 7). With high presence of PV production, the net load during the middle of the day is lowered, while in the afternoon it follows a steep rise because the reduction in solar radiation combines with the evening load peak (*Summer syndrome*, fig. 7).

- Reduced inertia. RES based generators, connected to the system with power electronic interfaces, displace traditional thermal plants so variations of the frequency are not effectively dumped. Besides, before updating the connection rules, DGs could be disconnected during over — and under — frequency events, worsening the unbalancing and contributing to large blackouts in Europe [7].

([4]) See the contribution by Erdmann to this School.

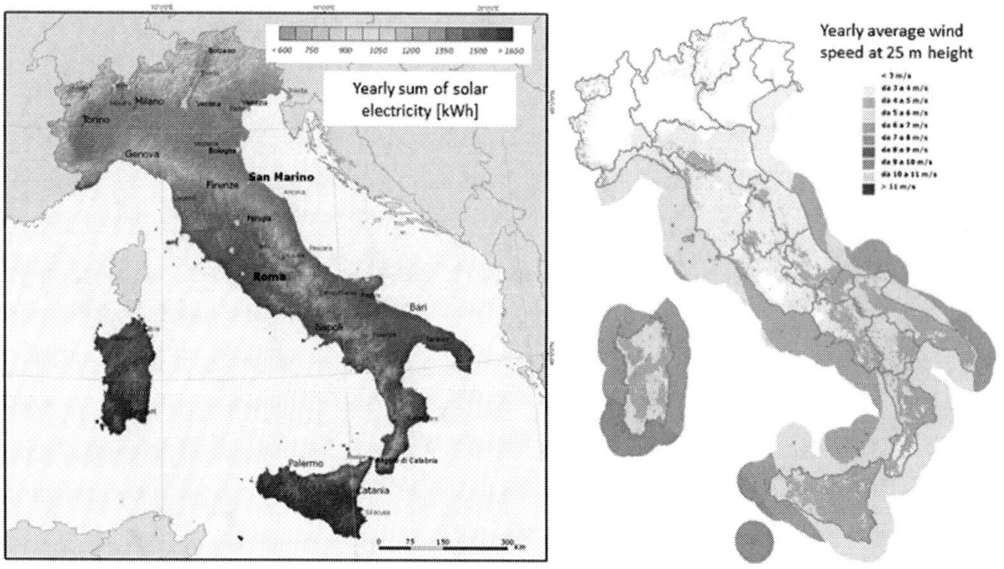

Fig. 8. – Renewable resources usually present a pronounced variability on a geographical basis (source: left PV GIS – JRC [8], right: RSE wind atlas [9]).

Another aspect worth mentioning is the *location* of renewable resources: in some cases, wind and solar radiation can be effectively exploited in areas far from loads. As example, one can cite the wind farms in Northern Germany and industrial areas in the South, and conversely the wind and solar availability in Southern Italy *vs.* the Northern loads (fig. 8). Dense urban areas can be more efficient than sparse centres, however they can hardly rely on local wind and solar resources only since (apart from the visual impact) the occupied area is not sufficient to cover the electricity demand. Therefore adequate transmission lines are needed to convey the renewable energy from the generation to the consumption sites. Moreover transmission lines could improve the integration of energy markets for neighbouring countries, allowing to exchange the excess of RES production over a larger geographic area.

3. – Smart grids: definition and technologies

The widespread diffusion of small-scale generation requires a gradual evolution of the distribution networks from a *passive* structure to an *active* one (bidirectional), similar to the transmission network. As mentioned before, this evolution is labelled as "smart grids".

Several elements are involved in this transition, as summarized in fig. 9; in few words a threefold strategy can be implemented:

- smartening the grid ("software" approach, complementary to "hardware" investments on copper and steel);

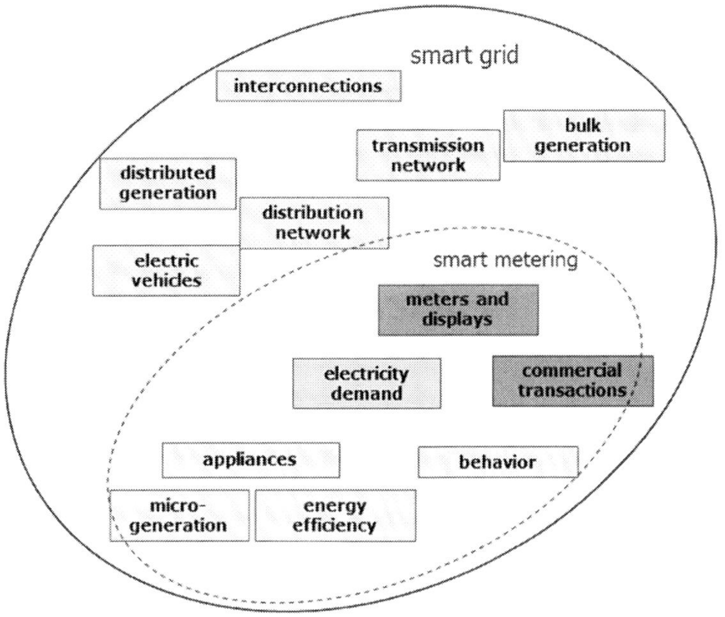

Fig. 9. – Elements of a smart grid (from ERGEG-CEER *Position paper on Smart Grids*).

- smartening the generators (from simple connection to *integration* into the system);
- smartening the loads (awareness and participation to the markets).

Regarding the *distribution grid*, advance monitoring and control techniques include the adoption of protection, measurement and control devices, in order to get a reliable picture of the system's functioning, and to keep currents and voltages within their admissible ranges. In addition, DG's presence requires a stronger coordination between the TSO (in charge of balancing the entire power system, keeping the frequency around 50 Hz) and the DSO (in charge of the quality of supply), for guaranteeing the observability of the power system, especially if distributed resources can participate to the ancillary service market. These advanced monitoring and control approaches rely on DSO's assets, such as OLTC, but prospectively on the contribution from distributed resources as well. In liberalized markets, the utilization of storage devices owned and managed directly by network operators is under discussion, since the market rules separated the role of grid operator from the role of generator.

Secondly, *generators* can be effectively integrated into the system by means of:

- Advanced protection device (to avoid undesired trips).
- Voltage control (exchange reactive power to counterbalance voltage rise caused by injection of active power); this could happen on a *local* basis, *i.e.* only on the measurement of parameters at the point of coupling with the network, or following

set points sent by the DSO. This latter *centralized* approach could include also modulation of active power for solving congestions, if rules allow DSO to do this.

- Frequency control: automatic reduction of generated power in case of over-frequency (not rewarded if required for system security purposes, usually), or modulation of active power following a signal sent by the TSO (participation on a market basis).

Moving from a system with large bulk generation following the loads, to a framework with the presence of non-programmable generation, entails the need for flexibility from the demand side as well. With *Demand Demand Side Integration* (DSI)([5]), customers change their absorption profile after receiving a *price* signal or a *system* signal (security of the system). This allows avoiding over-sizing on the generation side, increasing the efficiency of the generation, reducing the balancing reserve provided by generators, and planning distribution networks on the average load instead of critical peaks. Increased awareness is the first step towards this involvement: a simple in-home-displays can reduce consumption from 3% to 11% [10].

"Prosumers" (producer+customers) can get benefits from increased self-consumption, first of all in terms of reduction of the energy bill. Distributed Energy Resources (DER) — including loads, generators and storage devices — can further participate to both local and global flexibility services, also by aggregation, as summarized in table I.

In fig. 10 an example of centralized voltage control for active MV, or LV, networks is given [11]. The control is based on an optimization algorithm that calculates in each time period the set points for the controllable resources with the constraints of maintaining voltages and current within the admissible range. This controller can operate on DSO's owned resources (OLTC, etc.) or third-party equipment (controllable generators/storage units/loads). The algorithm searches for the minimum of the objective function, composed by the sum of "costs" of dispatching the controllable resources. In case there is a regulatory framework allowing DSOs to ask for (remunerated) support, these represent real cost, otherwise the parameters set the priority in the utilization of resources. A typical case where the centralized control is particularly effective is exemplified by the condition of unbalanced MV feeders: one is "active", *i.e.* the generation is higher than the load, and one "passive", therefore the OLTC cannot set a proper voltage of the MV bus bar to prevent under/over-voltages at the nodes of both feeders. The controller calculates the reactive power modulation from generators and the coordinated operation of OLTC (voltage at MV busbar) in order to solve the issue. It should be observed that voltage controls based on reactive power from generators can increase the losses [12]; the optimization algorithm used in this controller has the additional goal of increasing the efficiency, while satisfying the voltage and current constraints. As a further resource of flexibility for active and reactive power, an electric storage system

([5]) DSI refers to all activities focused on advancing the efficiency of electricity utilization, including demand response and energy efficiency.

TABLE I. – *Traditional solutions and "active" solutions for solving various technical issues.*

Technical Issue	*BAU Distribution Network*	*"Smart Grids"* *(Active distribution networks)*
Voltage rise/drop	• Limits/bands for demand and generation connection/operation • Generation tripping • Capacitor banks	• Coordinated volt-var control • Static var compensators • Coordinated dispatch of DER • On-line reconfiguration
Hosting Capacity	• Network reinforcement (*e.g.*, lines/transformers)	• Coordinated dispatch of DER • On-line reconfiguration
Reactive Power Support	• Dependency on transmission network • Capacitor banks • Limits/bands for demand and generation connection/operation	• Coordinated volt-var control • Static var compensators • Coordinated reactive power dispatch of DER
Protection	• Adjustment of protection settings • New protection elements • Limits for generation connection • Fault ride through specifications for generation	• On-line reconfiguration • Dynamic protection settings
Ageing	• Strict network designs specifications based on technical and economic analyses	• Asset condition monitoring

(ESS, *e.g.* a battery) was included; as mentioned earlier, it is under discussion whether DSOs can own ESSs. Testing of this algorithm in real MV networks has demonstrated its effectiveness: depending on the available regulation resources, *i.e.* OLTC only or also the reactive modulation by dispersed generators, it could increase the Hosting Capacity of more than 20% with respect to the baseline condition without control [13].

For a comprehensive view of all European projects on smart grid, a survey is published periodically by the Joint Research Centre-JRC [14]: the outlook offers a snapshot of the state of play and of the latest developments in the field of smart grids in Europe. The analysis in the last edition was based on a database of 950 R&D and demonstration projects; according to this report, in 2015 a cumulative effort of 5 billion Euros was reached (fig. 11). The document states that "Most of the projects in the database are national projects (66%), *i.e.* projects carried out in one country with the exclusive participation of organisations from that country. Multinational projects, *i.e.* projects that see the participation of organisations coming from different countries, are less numerous but larger in average investment size (EUR 7.5 million *vs.* EUR 4.7 million of national projects) [...]. The correlation between a high total number of projects and a

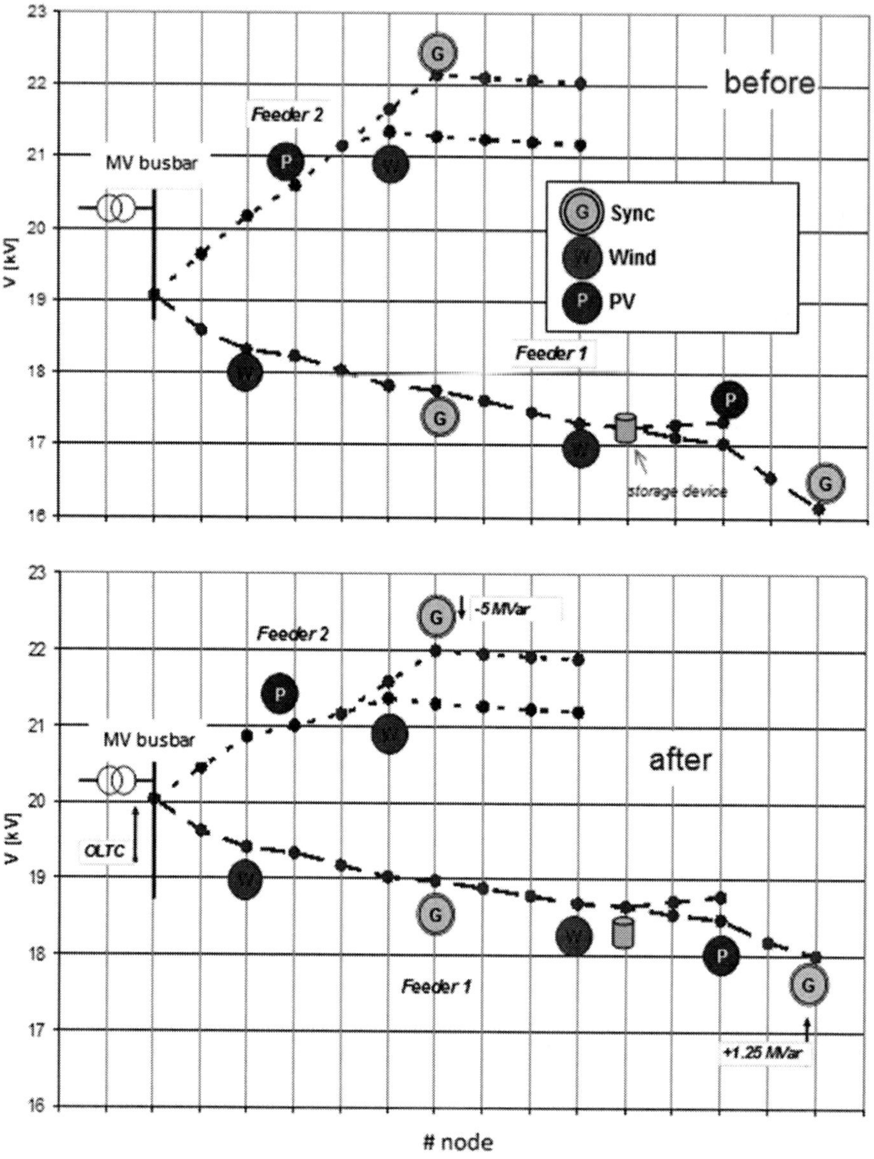

Fig. 10. – Example of centralized voltage control on 20 kV feeders under a primary substation (elaborated from [13]). In case of unbalanced MV feeders (one is "active" and one "passive"), the OLTC cannot solve under/over-voltages (top). The controller requires reactive power modulation from generators and coordinates the operation of OLTC (bottom) (source: RSE).

high number of national projects might be a sign of a mature smart grid environment, where different smart grid stakeholders have had the opportunity to test new technologies and business models thanks to previous R&D and demonstration experiences and to the existence of a supporting national and regulatory framework".

https://doi.org/10.1051/epjconf/201818900012

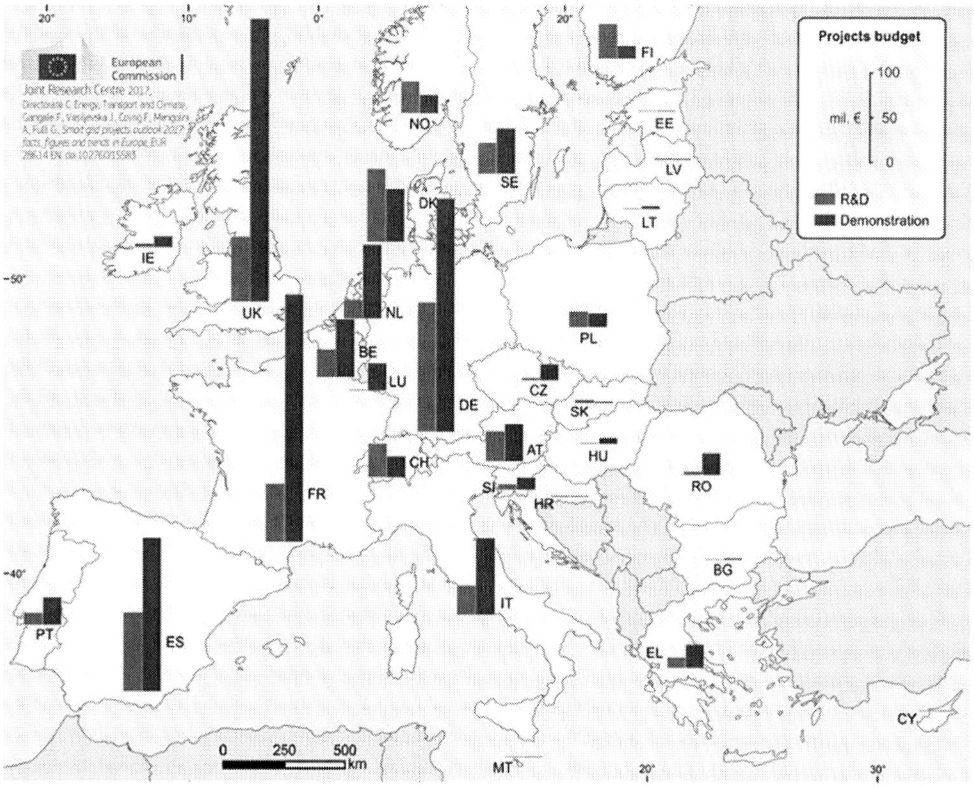

Fig. 11. – R&D and demonstration investments in the EU (source: JRC [14]).

A worldwide initiative on cataloguing smart grid activities was activated by the International Energy Agency Implementing Agreement for a Co-operative Programme on Smart Grids – ISGAN [15].

4. – Estimation of benefits from smart grids

In a complex scenario, the need of understanding emerges, which technologies among a large number of options are "the best" for solving the given problems. In case of distribution networks, whose investment costs are translated directly to customers' bill, this need is even more acute. In this lecture the focus is limited on the technical benefits, the first step for carrying out the cost-benefit analyses (CBA).

Smart grid solutions are generally evaluated by means of an estimation of the impacts on the performance of the electrical grids. A widely used indicator assesses the quantity of distributed generators that can be connected to a given network without affecting the power quality ("hosting capacity"). Other key performance indicators (KPI) consider the energy efficiency, the voltage stability, and so on; KPIs, and relative calculation methodologies, are now used also by Authorities for rewarding smart grid investments.

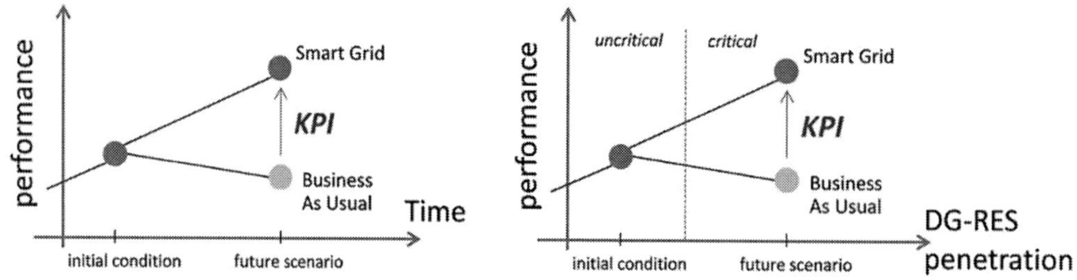

Fig. 12. – Key Performance Indicator calculation schemes: with reference to the time (left) or to the level of RES penetration (right) [16].

With the *output based* approach, investments are rewarded with reference to the benefit they can bring to the system, not simply on the costs incurred by operators.

KPIs are applied to compare the Business As Usual (BAU) condition with the Smart Grid solution in a given scenario, differing in time or in the level of RES penetration [16]. In this latter approach, the "critical scenario" is characterized by a very high DRES penetration that can reduce the performance of the grid (fig. 12).

The performance evaluation based on data from real networks is likely to be inaccurate due to the fact that there are operation constraints (the presence of real users limits the number and type of tests that can be carried out on the field), as well as the duration and extent of the experimental phase (which would lead to inaccurate statistical indices for the small size of the observed sample). Even in case the real field testing does not pose limitations on the validation of the different parameters, simulations are needed for estimating the alternative conditions not observed in the field test (with *vs.* without controller).

Given the uncertainty associated with the future scenarios, since the connection of DG is beyond the scope of DSO's assignment and follows other rules for development (availability of primary sources, incentives), a Monte Carlo approach returns the proba-

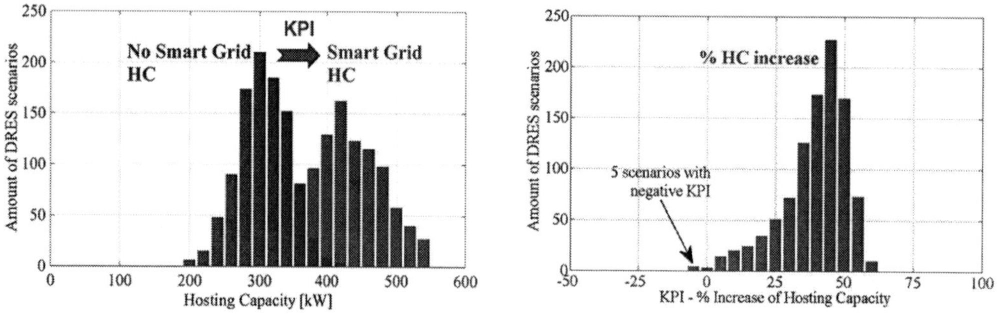

Fig. 13. – Comparison of distribution of Distributed RES scenario as a function of the Hosting Capacity (HC) in the BAU and Smart grid conditions (left); distribution of scenarios as a function of HC increase with the adoption of the "smart grid" solution (right). Some scenarios show a decrease of HC, *i.e.* a negative KPI [18].

bility distribution of the indicator [17]. For example, the Hosting Capacity is calculated in several scenarios with distinctive size and site of generators' allocation, using realistic profiles for loads and generation. Thanks to this method, it is also possible to conduct sensitivity analysis about allocation of DGs in order to identify combinations of networks and control systems that may, under certain conditions, lead to a decrease of the KPI in the smart grid condition if compared to the BAU situation (fig. 13).

The joint analysis of the different KPIs also allows to extract more information on the different aspects of running active networks such as the inability — in some cases — to increase performance both in terms of hosting capacity and loss containment. This demonstrates how smart grids are *enablers* for achieving certain goals but they do not represent a "panacea" to solve all issues simultaneously.

5. – Perspectives

Transitions in the power system last about 50 years, the time necessary to replace generators and upgrade infrastructures. The shift towards renewable resources (RES) observed in these years shows a quicker pace of change than past experiences: smart grids are one of the pillars to enable this transition. Energy systems supplied 100% RES are probably not reasonable because of several reasons [19], not to mention the availability of sustainable technologies([6])([7]), but till 2050 a significant increase is expected both in developing and in OECD countries.

Exploitation of renewable energy resources has led to the diffusion of a large number of small-size generators, connected to distribution grids. Not only farms and factories, but even citizens can now install their own generator, and have "smart" controllers to combine consumption with generation. These small generators could be aggregated to bid their flexibility on the market, competing with conventional large plants for providing ancillary services. This participation, currently limited to few activities, will turn into a necessity within the next decades, when RES penetration will reach higher percentage and fewer thermal plants may remain into service. Integrating variable generation entails additional costs for monitoring and control of distribution networks, and also for balancing the power system, with an increased coordination between the TSO and the DSO.

The progressive reduction of generators' size entails more "lay people" entering into complex interactions with multiple actors (DSO, vendor, aggregator), so shortened and future proof solutions should be defined, also for the administrative procedures. In most countries simplified schemes are in place for rewarding the energy produced by small plants, and more flexible architectures are under discussion (*e.g.* "energy communities" [20]).

Besides, the electrification of transports and thermal applications may bring a higher RES penetration in the final uses and overall a greater energy efficiency. Liberalization

([6]) See the contribution by Vidal to this School.
([7]) See the contribution by Terrasi to this School.

of retail market allows customers to choose their preferred tariff among a large variety of offers, but this freedom converts into an advantage only if users are aware of their actual consumption habits. Smart meters, delivering consumption data close to real time, support this awareness and allow designing tariff schemes with variable prices, for example depending on the actual RES production in the power system [21].

Buildings are responsible for about 40% of energy consumption in Europe, and automation is now required in new large buildings together with insulation to reduce the energy wastage [22]. Operators from ICT sector are now proposing appealing applications for energy management: those let customers choose "friendly" solutions, well integrated with other owned devices (smartphones primarily). Smart appliances are allowing people to remotely programme their utilization easily, also by means of complex but plain energy management systems. On the other hand, such solutions are often based on proprietary protocols for collecting and managing data in remote *cloud* servers; apart from privacy and security issues, this may lead to a lack of *interoperability* among solutions (customers cannot easily integrate devices from different manufacturers), and in the worst case a loss of functionalities when the internet connection is absent, or the manufacturer discontinues the software support.

Basically, smart grids are composed of the superimposition of the power grid and the ICT layer: development of successful and cost effective solutions needs the close collaboration of experts from both sectors, and even the support of social sciences to actually enable customers' participation apart from pure economic reward. Last but not least, regulation authorities have the duty of proposing shared methodologies to assess benefits and costs of smart grid development, and fairly allocate these costs to the different actors. In any case, "smart technologies" are the way to develop a sustainable energy system, not the goal.

REFERENCES

[1] National Renewable Energy Action Plan (NREP), February 2011 `http://www.eea.europa.eu/highlights/renewable-energy-production-must-grow`.

[2] Ohler C., "The physics of power systems operation" in *Lectures Notes - Joint EPS-SIF International School on Energy*, Course 2, *Energy: Basic Concepts and Forefront Ideas, EPJ Web of Conferences*, **98** (2015) 03002.

[3] CENELEC, technical standard EN 50160, "Voltage Characteristics of electricity supplied by Public Distribution Systems", 2010.

[4] Autorità di Regolazione per Energia Reti Ambiente (ARERA, former AEEGSI), "Relazione annuale 2016", 2017 (in Italian), `www.arera.it`.

[5] Delfanti M., Pasquadibisceglie M. S., Pozzi M., Gallanti M. and Vailati R., "Limits to Distributed generation on Italian MV networks. Electricity Distribution – Part 1", in *Proceedings of the 20th International Conference and Exhibition*; 8–11 June 2009 (CIRED) 2009, pp. 1–4.

[6] Kotsampopoulos P. *et al.*, "EMC Issues in the Interaction Between Smart Meters and Power-Electronic Interfaces", *IEEE Trans. Power Delivery*, **32** (2017) 822.

[7] Union for the Co-ordination of Transmission of Electricity UCTE, "Final Report - System Disturbance on 4 November 2006", `www.entsoe.eu`.

[8] HULD T., ŠÚRI M., DUNLOP E., ALBUISSON M. and WALD L., "Integration of HelioClim-1 database into PVGIS to estimate solar electricity potential in Africa", in *Proceedings of the 20th European Photovoltaic Solar Energy Conference and Exhibition, 6-10 June 2005, Barcelona, Spain*, 2005 `http://re.jrc.ec.europa.eu/pvgis/`.
[9] Ricerca sul Sistema Energetico – RSE SpA, ATLAEOLICO, `http://atlanteeolico.rse-web.it/`.
[10] DARBY S., "Communicating energy demand: measurement, display and the language of things", in *Engaging the Public with Climate Change: Behaviour Change and Communication*, edited by WHITMARSH L., O'NEILL S. and LORENZONI I. (Earthscan) 2010.
[11] BOSISIO A., MONETA D., VESPUCCI M. T. and ZIGRINO S., "A procedure for the optimal management of medium-voltage AC networks with distributed generation and storage devices", *AIRO Proceedings, Procedia Soc. Behav. Sci.*, **108** (2014) 164.
[12] SAUER P. W., "Reactive Power and Voltage Control issues in Electric Power Systems", in *Applied Mathematics for Restructured Electric Power Systems*, edited by CHOW J. H. and WU F. F. (Springer, Boston) 2005, pp. 11–24.
[13] STEIN D., CONSIGLIO L., VIGANÒ G., MICHELANGELI C., CARLINI C. and MONETA D., "Real field testing results of the innovative Medium Voltage control system developed in the Italian Demonstrator of GRID4EU", *Proceedings 46th Session CIGRE*, Paris, 2016, `www.cigre.org`.
[14] GANGALE F., VASILJEVSKA J., COVRIG F., MENGOLINI A. and FULLI G., "Smart grid projects outlook 2017: facts, figures and trends in Europe", (Publication Office of the European Union) 2017.
[15] ISGAN, "Global Smart Grid Inventory" `http://www.iea-isgan.org/bbs/content.php?co_id=sub2_2a`.
[16] IGreenGRID Europeana Project, "KPIs calculation procedure and results", `http://www.igreengrid-fp7.eu/`.
[17] ROSSI M., VIGANÒ and MONETA D., "Hosting capacity of distribution networks: Evaluation of the network congestion risk due to distributed generation", in *Proceedings of the 2015 International Conference on Clean Electrical Power* (IEEE ICCEP) 2015.
[18] JARELA J., HATZIARGYRIOU N., PUGLISI L., ROSSI M., ABART A. and BLETTERIE B., "The IGREENGrid Project: Increasing Hosting Capacity in Distribution Grids", *IEEE Power Energy Mag.*, **15** (2017) 30.
[19] WAGNER F., "Electricity generation by intermittent sources", in *Lectures Notes - Joint EPS-SIF International School on Energy*, Course 2, *Energy: Basic Concepts and Forefront Ideas, EPJ Web of Conferences*, **98** (2015) 04006.
[20] European Commission, Communication to the European Parliament, the Council, the European Economic and Social Committee, the Committee of the Regions and the European Investment Bank, "Clean Energy For All Europeans" (Winter package'), COM(2016) 860 final, Brussels, 2016.
[21] MONETA D. and MAURI G., "Factors Influencing Adoption of Smart Meters", in *Smart Grid Handbook*, edited by LIU C. C., MCARTHUR S. and LEE S. J. (John Wiley and Sons Ltd, Chichester, UK) 2016, pp. 1119–1131.
[22] Directive 2012/27/EU of the European Parliament and of the Council of 25 October 2012 on energy efficiency, amending Directives 2009/125/EC and 2010/30/EU and repealing Directives 2004/8/EC and 2006/32/EC.

EPJ Web of Conferences **189**, 00013 (2018) https://doi.org/10.1051/epjconf/201818900013
LNES 2017

Energy from nuclear fission

M. Ripani

INFN - Genova, Italy

Summary. — The main features of nuclear fission as a physical phenomenon will be revisited, emphasizing its peculiarities with respect to other nuclear reactions. Some basic concepts underlying the operation of nuclear reactors and the main types of reactors will be illustrated, including fast reactors, showing the most important differences among them. The nuclear cycle and radioactive nuclear waste production will be also discussed, along with the perspectives offered by next-generation nuclear assemblies being proposed. The current situation of nuclear power in the world, its role in reducing carbon emission and the available resources will be briefly illustrated.

1. – Physics of fission

What is fission and why can it produce energy? The basic reason is the behaviour of the nuclear mass and nuclear binding energy for different nuclear species. As is known from relativity, the mass of a bound system is given by the sum of the masses of its components, plus the binding energy of the system. The binding energy is by definition a negative quantity, *i.e.*, the mass of the composite system is always less than the sum of the masses of its components. When applied to nuclear masses, this concept can be expressed through the equation

$$M(Z,A) = ZM_p + (A-Z)M_n + B(Z,A), \tag{1}$$

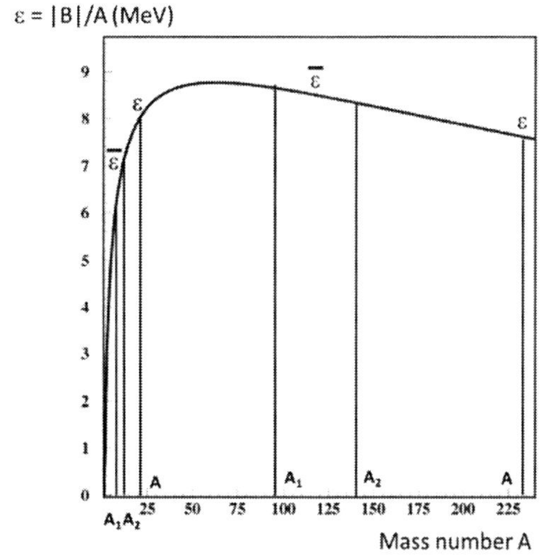

Fig. 1. – Absolute value of binding energy per nucleon, as a function of the mass number A.

where $M(Z, A)$ is the mass of a nucleus with Z protons and A protons and neutrons (*i.e.*, $A - Z$ neutrons), M_p is the proton mass, M_n is the neutron mass and $B(Z, A)$ is the binding energy of the system, with $B(Z, A) < 0$, *i.e.*, $M(Z, A) < ZM_p + (A - Z)M_n$. Z is the atomic number, while A is the mass number. The absolute value of the binding energy is the energy that must be supplied from outside to overcome the binding nuclear force and set all nucleons free. We can also define the binding energy per nucleon, *i.e.*, the average quantity of energy that must be supplied per each single nucleon when separating all nucleons from one another. As one can see from fig. 1, where the absolute value of the binding energy per nucleon is reported, $\epsilon = \frac{|B|}{A}$, such value increases from the lightest nucleus, the deuteron, to the region of iron, then it decreases towards the heavier nuclei. In other words, iron is more bound than the light nuclei, while, *e.g.*, uranium is less bound than iron.

1·1. *Energy balance.* – For a nuclear reaction to give up more energy than it absorbs, it has to move in the direction of stronger binding. Indeed, if a system goes from lower binding energy to higher binding energy, the total mass will decrease in going from the initial to the final state, so that the difference in mass will correspond to energy being released by the system. Again looking at fig. 1, it is clear that two light nuclei undergoing nuclear fusion, *i.e.*, a nuclear system moving from the left-hand side of the graph in fig. 1 to the right, will move towards higher binding, thereby releasing nuclear energy, *i.e.*, giving rise to an *exoenergetic* nuclear reaction, namely a reaction which yields a net amount of energy. This is the mechanism at the basis of nuclear fusion [1], *e.g.*, in the reaction $^2\mathrm{H} + {}^3\mathrm{H} \to {}^4\mathrm{He} + n$, where about 17.5 MeV are released in the process. On the other hand, heavy nuclei above iron can undergo fission, where the nuclear system splits into two nuclei, moving from the right-hand side of the graph in fig. 1 to the left, again

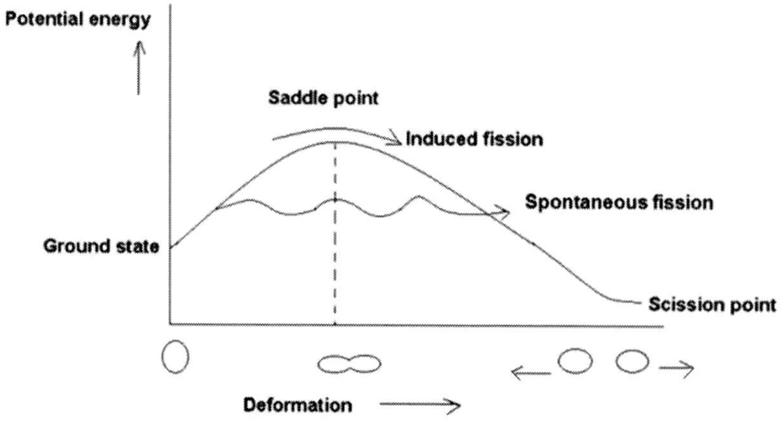

Fig. 2. – Nuclear potential energy as a function of the distance between the two fragments.

moving towards higher binding. This is called the process of nuclear fission and can also be exoenergetic. Let us consider in detail the energy balance of a fission process where a nucleus with Z protons and $(A - Z)$ neutrons splits into two nuclei,

$$(Z, A) \to (Z_1, A_1) + (Z_2, A_2), \tag{2}$$

where obviously $Z = Z_1 + Z_2$ and $A = A_1 + A_2$ and the masses of the nuclei are $M(Z, A)$, $M(Z_1, A_1)$ and $M(Z_2, A_2)$, respectively. The energy released in the reaction is called the Q-value and it is given by definition by

$$Q = M(Z, A) - M(Z_1, A_1) - M(Z_2, A_2). \tag{3}$$

By using eq. (1) we obtain

$$\begin{aligned} Q = B(Z, A) - B(Z_1, A_1) - B(Z_2, A_2) = \\ -\epsilon A + \epsilon_1 A_1 + \epsilon_2 A_2 = -\epsilon A + \bar{\epsilon} A = (\bar{\epsilon} - \epsilon) A, \end{aligned} \tag{4}$$

where

$$\bar{\epsilon} = \frac{\epsilon_1 A_1 + \epsilon_2 A_2}{A_1 + A_2}. \tag{5}$$

Again looking at fig. 1, it is easy to see that if the initial nucleus is a heavy one (as indicated on the right of the picture), its fission into two medium-mass nuclei is such that $Q > 0$. By repeating a similar calculation for two light nuclei undergoing fusion, it is easy to see that also in that case one can get $Q > 0$.

The fact that fission of heavy nuclei can be energetically favored does not mean that fission will happen easily in a spontaneous manner. Indeed, by looking at fig. 2, imagine the two daughter nuclei (called *fission fragments*) being bound together in a potential well: to be separated from each other, they will have to pass through the Coulomb repulsion barrier. This happens in the so-called *spontaneous fission* and the lifetime for the nucleus to decay via spontaneous fission will vary depending on the probability for the two nuclei to pass through the barrier via quantum tunneling. Clearly, if energy is supplied to the initial heavy nucleus from the outside, the system will jump to a more excited quantum state, *i.e.*, a higher energy state of the two fragments in the potential well. Therefore, the probability for the two fragments to penetrate the barrier will increase: this is what we call *induced fission* and can actually happen with appreciable probability for nuclei with $A \gtrsim 230$. Now, suppose that one more neutron is added to such a heavy nucleus (Z, A): in general, it will feature a significant binding energy (a few MeV) in the newly formed $(Z, A+1)$ nucleus. This means that when a neutron is captured by a heavy nucleus, a few MeV of energy will be released and made available for the two fragments inside the nucleus to approach the top of the barrier, thereby increasing the probability of the $(Z, A+1)$ nucleus to undergo fission by orders of magnitude. However, given the specific dependence of the binding energy on Z and A, typically for odd-A nuclei the binding energy of the extra neutron captured is higher than for even-A nuclei. The former nuclei are called *fissile* and they can undergo fission with high probability even when capturing a very slow neutron (even for a neutron in thermal equilibrium with the surrounding medium). The latter nuclei are called *fissionable* and they can undergo fission with significant probability only when capturing a neutron with kinetic energy of the order of 1 MeV or more. Two famous examples of the above are the fissile $^{235}_{92}$U and the fissionable $^{238}_{92}$U (as can be seen in the following).

Fission is not the only exoenergetic nuclear reaction induced by neutrons. Just to make two examples, the reaction $n + {}^3\mathrm{He} \to {}^3\mathrm{H} + p$, when initiated by a very slow neutron (whose kinetic energy we can neglect), can release about 0.76 MeV in form of the kinetic energy of the two final nuclei. The reaction $n + {}^6\mathrm{Li} \to {}^3\mathrm{H} + {}^4\mathrm{He}$ can release about 4.78 MeV, again taken by the motion of the final nuclei. These features are used in ^{3}He tubes and in scintillation detectors with ^{6}Li coating, in particular to detect slow neutrons. Then what is so special about fission of heavy nuclei? First of all, it can happen with very high probability for fissile nuclei when exposed to slow neutrons. Secondly, the fission process is accompanied by the emission of a few *prompt neutrons*, *i.e.*, in addition to the splitting of the initial nucleus into two fragments, a few neutrons are emitted by the fragments in a very short time after the splitting. Such neutrons are fast, *i.e.*, they have high speed and high kinetic energy (about 2 MeV on average) and, as we will see shortly, if properly slowed down, they can be very effective in inducing another fission, and so on and so forth. Therefore, fission, under certain conditions, can give rise to a *chain reaction*, where the neutrons emitted by one fission can produce additional fissions, with a global very significant release of energy.

1·2. *Energy released and reaction products.* – As one could in principle derive from eq. (4), when a uranium nucleus fissions into two nuclear fragments, about 0.1% of uranium mass appears as fission energy of ~ 200 MeV. Such an energy release is bigger than any other exoenergetic nuclear reaction by more than an order of magnitude (see, for instance, the above example of neutron capture on ^{6}Li) and bigger than chemical reactions by several orders of magnitude (just for comparison, in the chemical reaction where a methane molecule is burnt, $CH_4 + 2O_2 \rightarrow CO_2 + 2H_2O$, about 9.2 eV of energy are released, more than 7 orders of magnitude less). Together with the fission fragments, an average of 2.5 prompt neutrons are emitted, with a mean kinetic energy per neutron of about 2 MeV (the total mean kinetic energy of all prompt neutrons being about 4.8 MeV). The average number of neutrons emitted is one of the many important parameters characterizing fission and it is called ν, whose value in general is of the order of 2–3, depending on the nucleus undergoing fission and on the incident neutron energy. Out of the 200 MeV released per fission, about 180 MeV appear at the time of fission as kinetic energy of the emerging two nuclei, plus kinetic energy of the prompt neutrons, plus energy in form of prompt gamma rays (photons) emitted when nuclear fragments are produced in a nuclear excited state that decays electromagnetically in a time so short that is considered negligible (say, typically less than a femtosecond, to give an idea, but sometimes much longer, depending on the nuclear state). Additional energy is released later because most nuclear fragments undergo β-decay (a process caused by the nuclear weak force, where Z of the nucleus increases by one unity, while A stays the same, and concurrently an electron and an antineutrino are emitted by the nucleus; β-decay is a very common process for the decay of unstable light- and medium-mass nuclei). Most of this energy is carried by the electron, the associated antineutrino and by the delayed gammas emitted when the beta-decay transition leads to an excited state of the final nucleus. However, since the antineutrinos will escape from the reactor with negligible interaction with the materials crossed, the amount of energy from β-decay actually contributing to the thermal budget of the reactor is about 13 MeV. When radioactive fragment build-up reaches a steady state, these 13 MeV (6.5% of fission energy) contribute to the instant reactor thermal power. At the exact moment when the reactor is shut down by stopping the chain reaction, the thermal power due to these radioactive decays, called *residual heat*, remains and decays slowly with time (see below).

1·3. *Cross sections and flux.* – The cross-section is a physical quantity and an observable that characterizes a nuclear reaction (elastic, inelastic scattering, etc.). The cross section is connected to the range of the involved forces and represents the effective area of a nuclear target in a certain type of reaction or collision. Here we will only consider the so-called total cross section, defined as follows. Given a flux of particles incident on a single nucleus (target) per unit area $\mathrm{d}S$ per unit time

$$\frac{\mathrm{d}N_{in}}{\mathrm{d}S\,\mathrm{d}t}, \tag{6}$$

and given a reaction rate, which represents the number of interacting particles (scattered or absorbed projectiles) per unit time,

$$R = \frac{\mathrm{d}N_{reac}}{\mathrm{d}t}, \tag{7}$$

then the total reaction cross section σ is

$$\sigma = \frac{\frac{\mathrm{d}N_{reac}}{\mathrm{d}t}}{\frac{\mathrm{d}N_{in}}{\mathrm{d}S\,\mathrm{d}t}}. \tag{8}$$

As is evident from the above equation, σ has the physical dimensions of a surface, which is logical given its interpretation as an effective target area.

It is possible to show that, given a macroscopic target comprising several nuclei with mass density ρ and small thickness x, struck by a particle beam of intensity I (particles/s),

$$R = I\frac{\rho x}{A} N_A \sigma, \tag{9}$$

where A is the target atomic weight and N_A is the Avogadro number. Note that $\frac{\rho x}{A}$ is the number density of nuclei in the target (*i.e.*, number of nuclei per unit volume). For a target of arbitrary thickness, by dividing it in thin slices of thickness $\mathrm{d}x$, one obtains that

$$I(x) = I(0)\exp\left\{-\frac{\rho}{A} N_A \sigma x\right\}. \tag{10}$$

In nuclear physics, the cross section unit is the *barn*, corresponding to 10^{-24} cm^2.

The quantity

$$\Sigma = \frac{\rho}{A} N_A \sigma \tag{11}$$

is called the *macroscopic cross section* and represents the probability of interaction per unit length of material crossed (and, therefore, has dimensions of the inverse of a length). It is easy to show that $1/\Sigma$ is the mean free path of an incident particle in the material and Σv is the frequency at which a projectile with speed v interacts with the crossed material.

1˙4. *Types of nuclear reactions.* – The scattering process between two nuclei A and B can be *elastic*, when the reaction is $A + B \to A + B$ (just a collision without change in the form of the two colliding systems), *inelastic* in the case of reactions like $A + B \to A + B^*$ (where B^* means an excited state of nucleus B), $A + B \to A + C + D$ (where the nucleus B splits into C and D), $A + B \to C + D$ (where A and B, *e.g.*, exchange some protons or neutrons and, therefore, become different nuclei), etc.

The simplest type of nuclear reaction occurring in a nuclear reactor is the so-called *potential scattering*, where neutrons scatter elastically off the nuclear potential without ever penetrating the nucleus itself (similar to billiard ball collisions). By quantum mechanical arguments, it is possible to show that at low energies the cross section for such a reaction is essentially just the geometrical cross section of the nucleus. Such type of collision has a rather flat energy dependence from about 1 eV up to the MeV range.

Another very relevant reaction mechanism is *neutron capture*. Generally speaking, the addition of one more neutron to a nucleus liberates a certain quantity of energy coming from the binding of that neutron. This energy can make the nucleus jump to an excited state, from which it will typically either decay by gamma emission or by fission. The first process, followed by gamma emission, is called *radiative capture* and is clearly in competition with fission. By quantum-mechanical arguments, it is possible to show that at low energies the neutron capture cross-section will follow a $1/v$ law, v being the speed of the neutron. The same is true for fission, but only for fissile nuclei for which fission can occur even following the capture of a very slow neutron.

The capture process can proceed through a so-called resonance reaction, which occurs when the sum of CM energy of the neutron-nucleus system and the binding energy of the captured neutron matches one of the energy levels in the compound nucleus formed after capture. This phenomenon is indicated by sharp peaks in the capture cross section (see fig. 3). Neutrons which go around in a reactor core with an initial kinetic energy above the resonance region, can then "cross" a resonance when they scatter around and lose energy and can therefore be more strongly absorbed also by elements other than fuel.

Since the nuclear radius is roughly 10^{-12} cm, the geometrical cross sectional area of the nucleus is roughly 10^{-24} cm = 1 barn. However, the combination of quantum-mechanical and specific dynamical effects from the nuclear forces can make nuclear cross sections extremely variable over a few orders of magnitude. As an example, in fig. 3 are reported the cross sections for the processes of fission and radiative capture for the two uranium isotopes found in nature and used in nuclear reactors, ^{238}U and ^{235}U, as a function of the neutron energy [2]. It is clearly seen that in ^{235}U, for fast neutrons with kinetic energy around 100 keV (or 0.1 MeV), both fission and radiative capture show a cross section in the order of the barn, but for lower energies the radiative capture for both nuclei and the fission for ^{235}U grow towards much higher values, following the above-mentioned $1/v$ law.

Fission itself can produce fission fragments with very strong radiative capture cross sections: such specific fission products are called *neutron poisons*, *e.g.*, ^{135}Xe ($\sigma \simeq 2\times 10^6$ barns for thermal —see below— neutrons, which is a *very* large cross section in nuclear physics), as they subtract neutrons from the chain reaction.

1·5. *Neutron density and flux*. – We define as neutron density the expected number of neutrons with energy between E and $E + \mathrm{d}E$, in the volume $\mathrm{d}^3\vec{r}$ at the space position $\vec{r}$, at a time t at a certain position and at a certain time in a nuclear assembly. This quantity is indicated by $n(\vec{r}, E, t)$ and can be measured, for instance, in neutrons per cm^3 (or in cm^{-3}). The reaction density $R(\vec{r}, E, t)$ is defined as the number of reactions in the volume $\mathrm{d}^3\vec{r}$ about $\vec{r}$, at a time t, initiated by neutrons with energy between E and $E + \mathrm{d}E$ and is given by the equation

$$R(\vec{r}, E, t) = n(\vec{r}, E, t)\, \Sigma\, v, \tag{12}$$

where again v is the neutron speed. The reaction density is measured in reactions per cm^3 per second. We give a special name to the quantity $n(\vec{r}, E, t)\, v$: it is called the

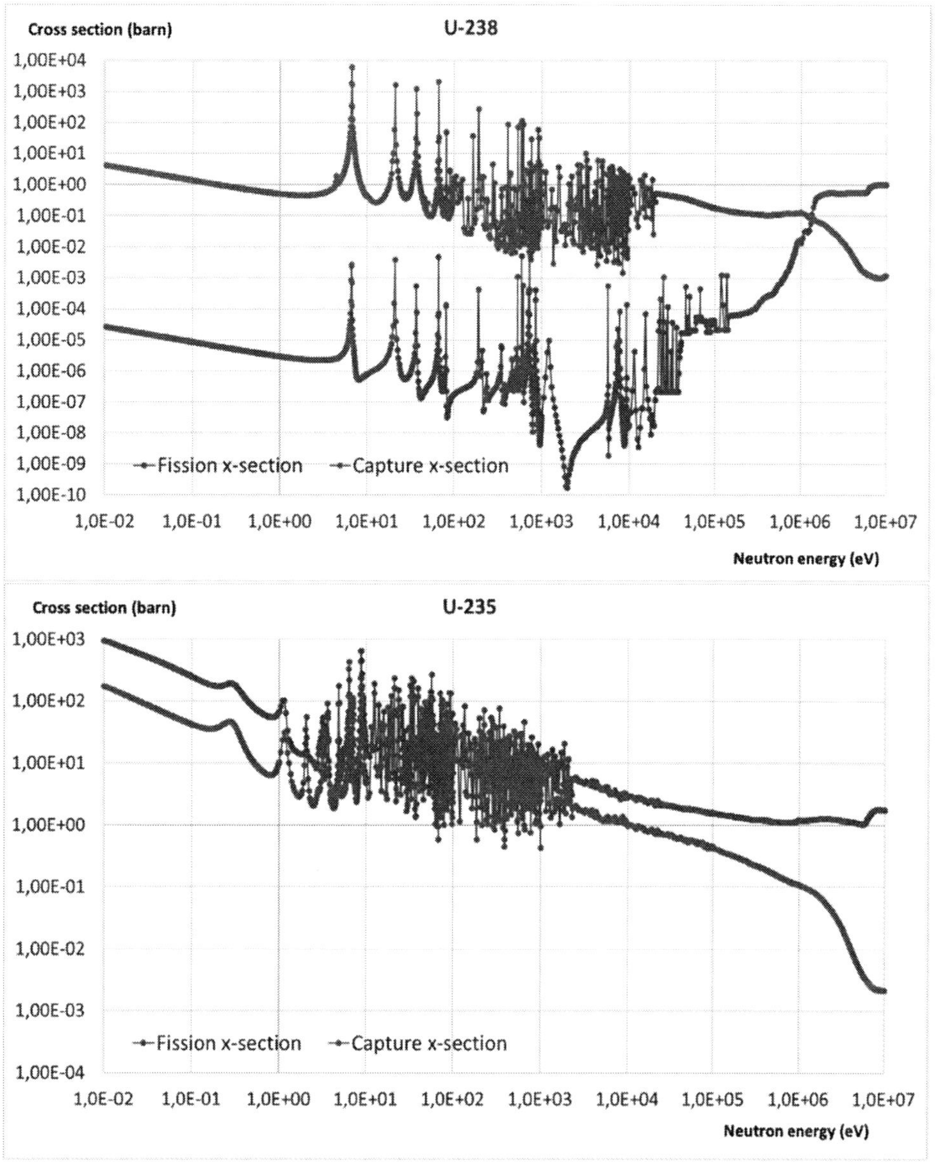

Fig. 3. – Fission (blue) and radiative capture (red) cross sections for the two isotopes of uranium.

neutron "flux" $\phi(\vec{r}, E, t)$ and is measured in neutrons per cm^2 per second. This quantity is essentially the number of neutrons arriving from whatever direction and crossing a unit surface in unit time (it differs from other definitions of flux in physics but we shall not discuss these differences here). Therefore, given a neutron flux ϕ and a macroscopic cross-section Σ, the reaction density, number of reactions per unit volume, R is simply given by $\Sigma\,\phi$. As an immediate example, consider a thermal reactor (see below) with 3 GW thermal power $= 3 \cdot 10^9$ J/s (it is indicated by the symbol 3 GWth and due to the efficiency of conversion to electric power, this corresponds to 1 GW electric power, indicated by 1 GWe). Assume each fission releases an order of 180 MeV energy $= 3 \cdot 10^{-11}$ J at the moment of fission. Then it follows immediately that in the whole reactor the fission rate is about 10^{20} fissions/s, which corresponds to 2–3 $\cdot 10^{20}$ neutrons/s emitted and about $6 \cdot 10^{20}$ neutrinos/s from fragments' β-decay. By using the known value of the fission cross section for thermal (very slow) neutrons on ^{235}U, one can see that the flux inside the reactor must be in the order of $\phi \simeq 10^{13}$–10^{14} neutrons cm^{-2} s^{-1}.

1·6. *Fuel.* – We already encountered the definition of *fissile* (*e.g.*, ^{233}U, ^{235}U, ^{239}Pu) and *fissionable* (*e.g.*, ^{238}U) nuclei. There are also nuclei that can produce a fissile isotope via neutron radiative capture and subsequent β-decay. These nuclei are called *fertile*, meaning that they can be used to produce fuel and essentially comprise ^{238}U, ^{232}Th. As an example, ^{238}U can produce a fissile nucleus via the chain of reactions

$$n + {}^{238}\mathrm{U} \to {}^{239}\mathrm{U} + \gamma \to {}^{239}\mathrm{Np} + e^- + \bar{\nu} \to {}^{239}\mathrm{Pu} + e^- + \bar{\nu}. \tag{13}$$

Natural uranium is composed by 0.7% ^{235}U and 99.3% ^{238}U. Given the cross sections shown in fig. 3, it is easy to understand that in a reactor where most neutrons are in thermal equilibrium with the surrounding materials, *e.g.*, they have energies < 1 eV, ^{235}U is the uranium isotope that actually plays the role of fuel by providing the dominant contribution to the fission rate.

Let us go back to our ideal 1 GWe reactor burning only ^{235}U. We already calculated that about 10^{20} fissions/s occur in the reactor, which means that 10^{20} (^{235}U nuclei)/s disappear (actually somewhat more because of radiative capture). Therefore roughly 40 mg/s of ^{235}U are "burnt" in the reactor. For 1 year of operation at 80% load factor (the actual working time), this means a consumption of about 1 ton of ^{235}U: in volume of pure metallic ^{235}U, this would be a cube of about 36 cm side. For comparison, the same amount of electric power is obtained by burning 1.6 Mtoe, million tons of oil equivalent (considering a slightly higher conversion efficiency of oil-fired power plants), or about 2 billion cubic meters of natural gas, or 2.2 million tons of coal. In a real fission reactor ^{235}U consumption is partly compensated by plutonium (^{239}Pu) breeding according to eq. (13), as ^{239}Pu is also a fissile nucleus. In practice, the initial load of a typical 1 GWe PWR (see below) comprises about 27 tons of enriched uranium, of which about 950 kg are ^{235}U. After 1 year 280 kg of ^{235}U remain: about 560 kg of ^{235}U and 380 kg of other isotopes, mostly ^{239}Pu have been burnt, which is approximately the 1 ton we have been estimating above.

1·7. *Fast and slow neutrons.* – We have seen that the cross sections for nuclear reactions induced by neutrons depend a lot on the energies involved, therefore it is important to introduce some classifications of the neutrons according to their energies. It is customary to adopt the following classification: *slow* neutrons are those with kinetic energy $T_n < 1$ eV, in particular thermal neutrons at room temperature have T_n around 0.025 eV or 25 meV (the value of kT, where k is the Boltzmann constant and T is the temperature of the system); *epithermal* neutrons are those with 1 eV $< T_n <$ 100 keV (0.1 MeV); finally, *fast* neutrons have 0.1 MeV $< T_n <$ 20 MeV. Obviously neutrons in general can have energies above 20 MeV but this is an extreme limit in reactor physics (*e.g.*, neutrons from fusion of a deuteron and a tritium nucleus have 14 MeV fixed energy).

It is easy to show in non-relativistic kinematics that after a scattering off a nucleus with mass number A, *on average* the kinetic energy of the neutron changes according to the ratio

$$\frac{T_n'}{T_n} \simeq \frac{1 + A^2}{(1 + A)^2} \tag{14}$$

(where T_n and T_n' are the kinetic energies before and after the scattering, respectively). For a heavy nucleus $A \gg 1$, one has that $T_n' \simeq T_n$ or, in other words, the neutron has to undergo many collisions in order to significantly lose energy. Consider instead the case $A = 1$ (this is relevant for any medium containing hydrogen, where protons, hydrogen nuclei, will be among the targets). In this case $T_n' = \frac{T_n}{2}$, *i.e.*, on average a neutron will lose half of its energy at each collision and, therefore, few collisions are sufficient to rapidly decrease its energy. For collisions on a light nucleus like carbon ($A = 12$), the situation will be similar to that of hydrogen, although the average loss of energy per collision will be lower. Light materials containing hydrogen (water, paraffin, plastic materials in general, etc.) or containing light nuclei (graphite, etc.) are called *moderators*, for their effectiveness in slowing down neutrons. In a thermal reactor, the fuel is immersed in water, which is used both as coolant and as moderator: the fast neutrons from fission will be slowed down very quickly via collisions with protons and oxygen nuclei in water and, therefore, they will have a very high fission cross section for ^{235}U (see fig. 3).

2. – Chain reaction and simple reactor kinetics

For the reactor to work effectively and safely, the chain reaction must not diverge and must not die away. This means that, amongst the 2–3 neutrons emitted on average in the fission process, precisely one has to induce another fission event, no more and no less. The remaining fission neutrons will then either be absorbed by radiative capture or will leak out from the system. The character of the chain reaction depends essentially on the reactor composition and on its geometry and is defined through the quantity called k, also called multiplication coefficient,

$$k = \frac{\text{number of neutrons in one generation}}{\text{number of neutrons in the preceding generation}}. \tag{15}$$

The condition $k = 1$ corresponds to a *critical* reactor, *i.e.*, a reactor actually working and producing thermal power from fission. For $k > 1$ we have a supercritical reactor (fission reactions tend to run away) and for $k < 1$ we have a subcritical reactor (fission reactions tend to die away). In a "simple minded" reactor time evolution, called *kinetics*, it is relatively easy to show that the number of neutrons in the system evolves as

$$n(t) = n(0) \exp\left(\frac{k-1}{\tau} t\right), \tag{16}$$

where $n(0)$ is the initial number of neutrons and τ represents the average lifetime of a neutron in the system before it disappears due to fission, capture or escape (leakage). Here we clearly see that $k = 1$ corresponds to a steady state, while $k > 1$ implies that the system is diverging and $k < 1$ corresponds to a decreasing population. Based on this simple kinetics, we see that the time constant governing the response of the reactor would be $\frac{\tau}{k-1}$. Since the typical neutron lifetime in a thermal power reactor is of the order of 10^{-4} s, we see that even for $k = 1.001$, the neutron population and, therefore, the reactor power would increase by a factor 2.7 in 0.1 s, which would not be a manageable situation from a practical point of view. But there is one missing piece in the picture of fission discussed above. Actually, there is a small fraction (less than a percent) of neutrons originating from the fission process that are not emitted immediately, but some time later. These neutrons are emitted by highly excited fission fragments on a time scale from milliseconds to seconds and for this reason are called *delayed neutrons*. In simple terms, a nuclear reactor is made critical thanks to such small fraction of delayed neutrons and, therefore, they are the ones that dominate the reactor response time making it much longer, thereby allowing reactor control by neutron-absorbing control rods.

2·1. *A bit more on neutron multiplication and transport.* – By considering in more detail what processes can intervene along the route of the next-generation neutrons emitted after a single fission, one can come up with the following definition of the multiplication coefficient k:

$$k = \eta f P_{NL}, \tag{17}$$

where f is the conditional probability that, if a neutron is absorbed, it will be absorbed in the fuel instead of structural elements; η is the average number of neutrons produced per neutron absorbed in the fuel, in other words the average number of neutrons per fission ν multiplied by the probability that the absorption process leads to fission instead of radiative capture $\eta = \nu \frac{\sigma_f^F}{\sigma_a^F}$, where σ_a^F is the absorption cross section in the fuel (absorption = fission + radiative capture) and σ_f^F is the fission cross-section in the fuel. Finally, P_{NL} is the probability that a neutron does not escape from the reactor (no-leakage). For an infinite reactor clearly $P_{NL} = 1$ and we get $k_\infty = \eta f$. If we consider that fissions can occur both from slow and fast neutrons, there is a correction factor $\epsilon > 1$ which simply tells that fissions can be more than just those produced by slow neutrons (as we saw before, in a thermal reactor the latter will be the vast majority). If we also take into account that neutron moderation to small speeds implies a number of collisions, we see that the neutron may not survive such collisions and be absorbed, so we will add a factor p representing the fraction of fission neutrons that survive moderation without being absorbed. For an infinite reactor where $P_{NL} = 1$, we arrive at the *4 factor formula*:

$$k_\infty = p\epsilon\eta f. \tag{18}$$

If we instead consider a finite reactor and for the sake of precision we consider the probability of no-leakage for thermal (slow) and fast neutrons separately, *i.e.* if we write $P_{NL} = P_{TNL}P_{FNL}$ (T for thermal and F for fast), we arrive at a formula for the *effective* multiplication coefficient, the *6 factor formula*:

$$k_{eff} = p\epsilon\eta f P_{TNL}P_{FNL}. \tag{19}$$

The name "effective" means that we are not considering an infinite, homogeneous medium but a more realistic situation where obviously $k_{eff} < k_\infty$, *i.e.*, in a real reactor one has to consider that the unavoidable leakage due to neutrons escaping from the reactor will make the multiplication in the chain reaction smaller than in an ideal, infinite medium. In practice, to make P_{TNL} and P_{FNL} as close as possible to one and increase k_{eff}, a *reflector* is put around the reactor core containing the fuel, *i.e.*, a thick blanket of material (graphite, Cu, Pb depending on the reactor type) with the purpose of scattering back (reflecting) the neutrons into the core. It is interesting to look at the behavior of the quantity η as a function of neutron energy, shown in fig. 4. It is clear that for thermal neutrons, this number is around 2, which means that effectively there are two neutrons available for further processes, one of which can trigger another fission, while the other can be lost in captures or leak out of the system. In case of capture on a fertile nucleus like ^{238}U, plutonium breeding can occur, but overall this will happen with relatively low probability. On the contrary, for high kinetic energies about 10–100 keV (fast neutrons), η becomes larger than 2 and so more neutrons will be available for breeding new fuel. Therefore, a so-called *fast reactor*, *i.e.*, one in which neutrons are not moderated (see subsection "Nuclear reactor types"), may produce more plutonium than it consumes, in which case we will have a *fast breeder*.

2·2. *Subcritical assemblies.* – From eq. (15), we see that the multiplication coefficient k must be equal to 1 to attain criticality, *i.e.*, to make the reactor work and produce power by a self-sustained chain reaction. What happens if k is less than unity? Clearly, if you start from criticality, then reduce the multiplication by, for instance, inserting control rods significantly, the chain reaction will die away exponentially fast, as indicated by eq. (16). However, the chain reaction can still be kept going if an *external* neutron source is active within the system. It is possible to solve the appropriate equations with such an external source and find a stationary state where the neutron flux and, therefore, the system's power are constant. If we indicate the external source as S_0 (neutrons/cm^3/s supplied by the external source to the reactor core), for an infinite, homogeneous reactor

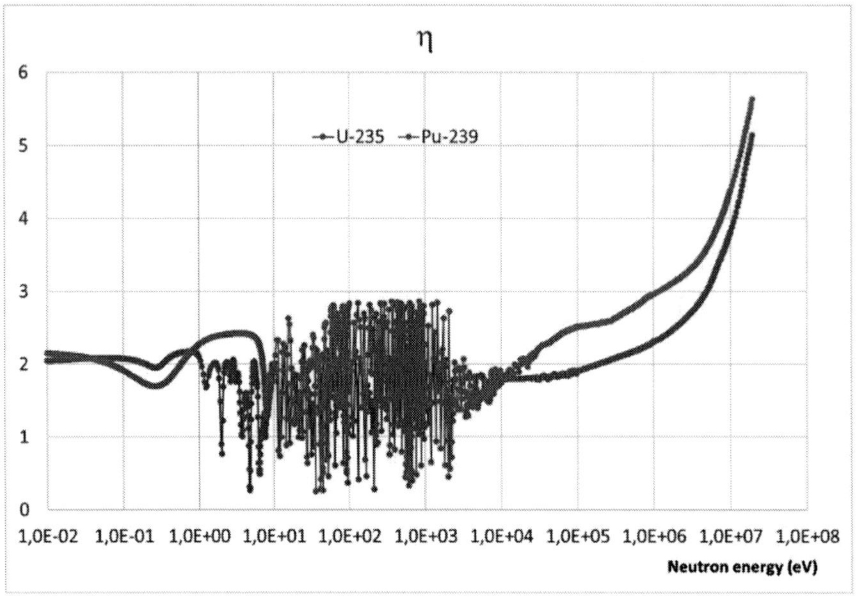

Fig. 4. – The parameter η as a function of neutron kinetic energy for the two main fissile nuclei.

the solution is

$$\phi = \text{const} = \frac{S_0}{k_\infty - 1}\Sigma_a, \tag{20}$$

where Σ_a is the macroscopic absorption cross section. We see that the neutron flux and, therefore, the power are proportional to the intensity of the source and are obviously increasing when the multiplication in the chain reaction k_∞ approaches unity. An external neutron source can be obtained by accelerating electrons or ions (protons, deuterons) in a suitable accelerator, then smashing the accelerated particles onto an appropriate, totally absorbing target. When using high-energy protons with kinetic energies in the order of hundreds of MeV to GeVs on a heavy nuclear target based, for instance, on mercury, tungsten, lead, several neutrons per incoming proton are obtained in the nuclear process called *spallation*. At lower energies, other types of targets can be used but spallation by high-energy ions is the most effective process for producing neutrons. A subcritical reactor core will therefore be comprising a suitable accelerator with a beam penetrating into the reactor core and stopped by a target placed in the middle of the core. This is called an *Accelerator Driven System, ADS* [3]. The interest in this type of reactor will be briefly discussed later.

3. – Main features of a nuclear fission reactor

We saw that neutrons slow down when undergoing collisions with nuclei (in particular with light nuclei), therefore their energies in the core can go from about 10 MeV (usually the maximum energy of fission neutrons), down to as low as 10^{-3} eV. We also saw that neutron cross sections have a strong dependence on neutron energy, in particular absorption cross sections such as capture or fission decrease very strongly with energy for particular species like ^{235}U. When using fuels containing these particular species, it is easier to maintain a fission chain reaction using slow neutrons. Hence most nuclear reactors until now (the so-called Generation I to III+) use low-mass-number materials such as water or graphite to slow down or moderate the fast-fission neutrons, thereby slowing them down to energies comparable to the thermal energies of the nuclei in the reactor core. Therefore, by definition a *thermal reactor* is one where the average neutron energy is comparable to thermal energies (order of several tens of meV). Because of the very-high-fission cross section at thermal energies, this type of reactor requires the minimum amount of fissile material for fueling. As an example, a Light Water Reactor (LWR) can start with 3% ^{235}U + 97% ^{238}U (a mixture enriched with respect to natural uranium). Burn-up (consumption) of ^{235}U is partly compensated by breeding of ^{239}Pu via neutron radiative capture and subsequent β-decays and after 1 year operation, the core may contain something like 1% ^{235}U + 1% ^{239}Pu. On the contrary, in a *fast reactor* the average neutron energies in the core are typically above 100 keV. As previously discussed, this can be accomplished if the core contains mostly high-mass-number materials. This means that for core cooling the choice will be a liquid metal like sodium, lead or a lead-bismuth mixture, or alternatively the coolant should be a low-density material like helium gas. On the other hand, σ_f is smaller for fast neutrons, which means that a much higher percentage of fissile material is needed to sustain the chain reaction. Later on we will consider some possible advantages of fast reactors over thermal reactors. In summary, the typical scheme of a thermal and a fast reactor are illustrated in fig. 5.

3.1. *Decay heat.* – Decay heat is the heat produced in the reactor core as a result of radioactive decay: the energy of the alpha, beta or gamma radiation is converted into atomic motion, but heat production comes mostly from β- and γ-decay of fission products. Decay of the short-lived radioisotopes created by fission continues for some time after shutdown. A practical approximation is given by the formula

$$\frac{P}{P_0} = 6.6 \cdot 10^{-2} \left[\frac{1}{(\tau - \tau_s)^{0.2}} - \frac{1}{\tau^{0.2}}\right], \tag{21}$$

where P is the decay power, P_0 is the reactor power before shutdown, τ is the time since reactor startup and τ_s is the time of reactor shutdown measured from the time of startup (in seconds). At shutdown, the heat power is about 6.5% of the previous core power ($\simeq$ 200 MWth for a 1 GWe reactor), which is sufficient to melt the core. About 1 hour after shutdown, the decay heat will be about 1.5% of the previous core power, after 1 day it will fall to 0.4% and after a week it will be only 0.2%. For this reason, spent fuel rods are kept for a long time in special cooling water pools, before being further processed. Removal of decay heat is therefore a very important aspect in designing a reactor and implementing safety measures, as the Fukushima accident has dramatically stressed.

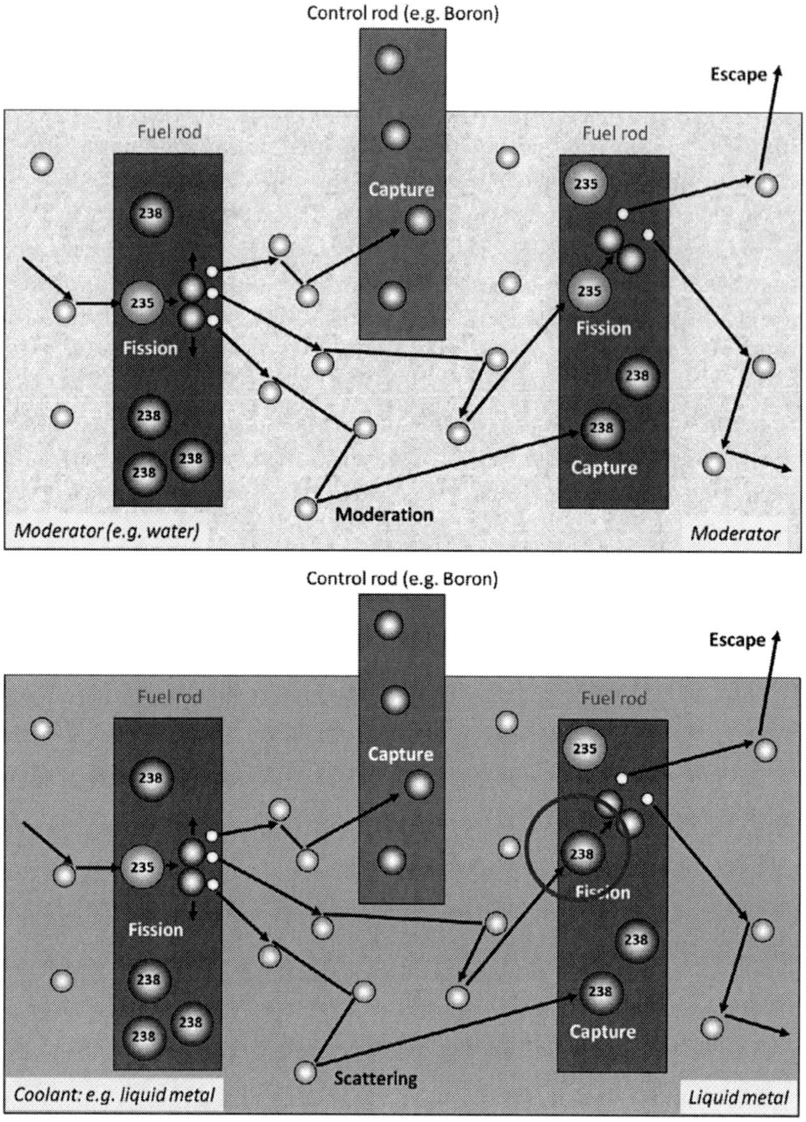

Fig. 5. – Conceptual schemes of thermal (top) and fast (bottom) reactor.

3·2. *Nuclear reactor types.* – Most current reactors are based on fissile fuels, ^{235}U and ^{239}Pu, with fission induced by slow neutrons at thermal energies. Ordinary water serves as both coolant and moderating material in the reactor. There are two major types of Light Water Reactors (LWR): Pressurized Water Reactors (PWR) and Boiling Water Reactors (BWR). In a PWR, the primary coolant is water maintained under very high pressure so that the coolant reaches a high temperature without steam formation within the reactor. Heat transported out of the reactor core by the primary coolant is then transferred to a secondary loop connected to a steam generator. In a BWR, the

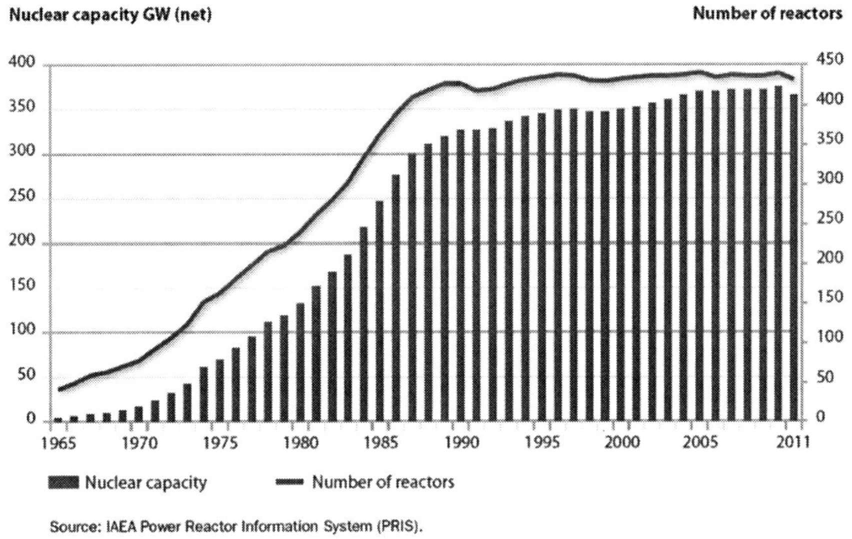

Fig. 6. – Worldwide nuclear generating capacity and number of operating reactors (1965–2011) [4], © OECD 2012.

primary coolant water is maintained at lower pressure, so that boiling and steam formation occur within the reactor core itself, which serves directly as the steam generator, while secondary loop and heat exchanger are not present. In both PWR and BWR, the nuclear reactor itself and the primary coolant are contained in a large steel pressure vessel designed to accomodate the high pressures and temperatures. The heavy-water (D_2O) reactor takes advantage of the lower neutron capture cross section in deuteron with respect to hydrogen. This allows to have more neutrons around in the core and to use low-enrichment uranium fuels (including natural uranium). This concept has been developed for instance in Canada in the CANDU (CANadian Deuterium Uranium) series of power reactors. This type of reactors is indicated by the acronym PHWR (Pressurized Heavy Water Reactor). Yet another design is based on gas as coolant. For instance the early MAGNOX reactors developed in the UK used low-pressure CO_2 as cooling gas. In general this reactor class is indicated by the acronym GCR (Gas Cooled Reactor). A more recent design is the High-Temperature Gas-cooled Reactor (HTGR, USA), that uses high-pressure helium as coolant. Gas-cooled reactors generally use graphite as moderator material to slow down neutrons to thermal energies. Finally, the Fast Breeder Reactor (FBR) does not use a moderator to work with fast neutrons (see sect. **2**) and breeds more fuel than it consumes. The coolant is typically a liquid metal.

4. – Fission energy in perspective

4'1. *Worldwide figures.* – As illustrated in fig. 6, from the 1950s, when the first commercial reactors started to appear, until 1985 the number of operating reactors as

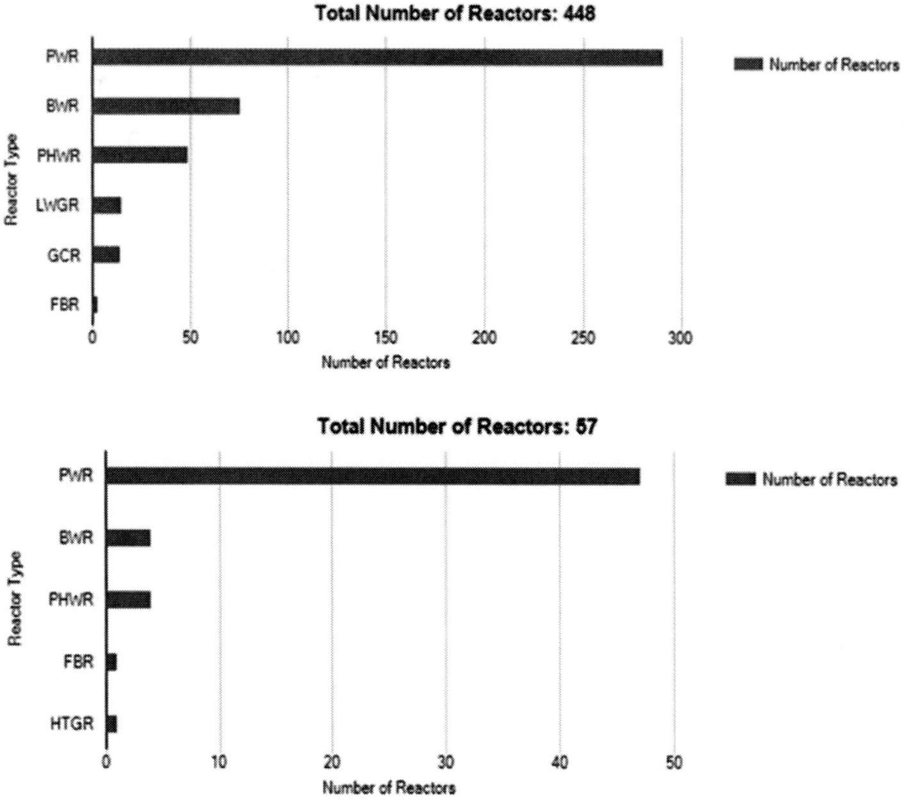

Fig. 7. – Number of operational and long-term shutdown reactors by type (top, as of October 2017). Reactors under construction by type (bottom, as of October 2017) [5], © 2017 International Atomic Energy Agency (IAEA).

well as the nuclear net generating capacity (*i.e.*, the available power) have been increasing about linearly with time, then the number of reactor has remained approximately stable, while the generating capacity has been still increasing somewhat due to the improved efficiency of existing plants and some replacement of older installations [4]. As of October 2017, there are 449 reactors in operation or in long-term shutdown in the world, for a total net installed capacity of about 392 GWe, and 57 under construction, for a total capacity of about 60.5 GWe [5]. In fig. 7, the top and bottom panels show the distribution of reactor types in operation and under construction in the world, respectively. Figure 8 shows the share of nuclear power in electricity production as of October 2017 [5] country by country, while fig. 9 shows the share of nuclear power in the world electricity generation [6].

4·2. *Nuclear fuel cycle, natural resources, cost of electricity and carbon emission scenarios.* – How much uranium is consumed by the nuclear reactor fleet worldwide? And how long will uranium resources last? There are clearly important questions for project-

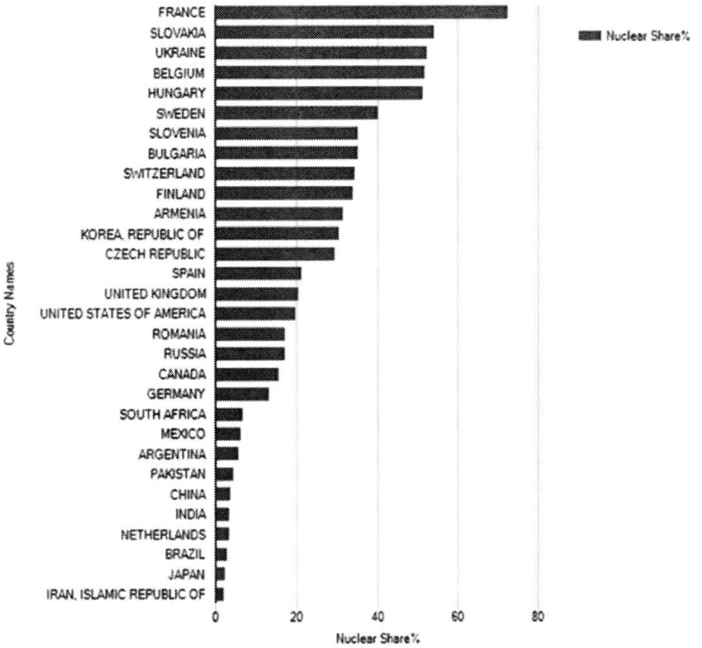

Fig. 8. – Nuclear share of electricity generation in each country (as of October 2017). Note: the share for Taiwan, China was 13.7% of the total [5], ©2017 International Atomic Energy Agency (IAEA).

ing the use of nuclear power from fission to the future. The whole series of activities starting from uranium mining down to disposal of radioactive waste is called the *nuclear fuel cycle*. In the case of the so-called *open fuel cycle* the spent fuel is extracted from the reactor, cooled in specific pools, then transferred to a storage infrastructure. In the case of the so-called *closed fuel cycle*, the spent fuel undergoes a reprocessing procedure, where plutonium is extracted from the fuel rods and used to produce new fuel containing both uranium and plutonium. World annual uranium requirements amounted to 61600 tU (tons uranium) in 2012 [7]. The so-called identified resources appear to be sufficient for over 120 years, based on the 2012 uranium consumption [7]. Consideration of the entire amount of conventional resources would increase the time scale to well over 300 years, but this would require significant efforts in securing all resources for an effective use. Moving to advanced technology reactors and recycling fuel could increase the long-term availability of nuclear energy from hundreds to thousands of years. In addition, thorium, which is more abundant than uranium in the Earth's crust, is another potential source of alternative fuel.

When considering the so-called Levelised Cost Of Electricity, levelised over the estimated plant lifetime, for various world areas, assuming specific discount rates and cost of carbon emissions, especially for a low discount rate scenario, the cost of electricity from nuclear power appears to be still competitive with other sources [8].

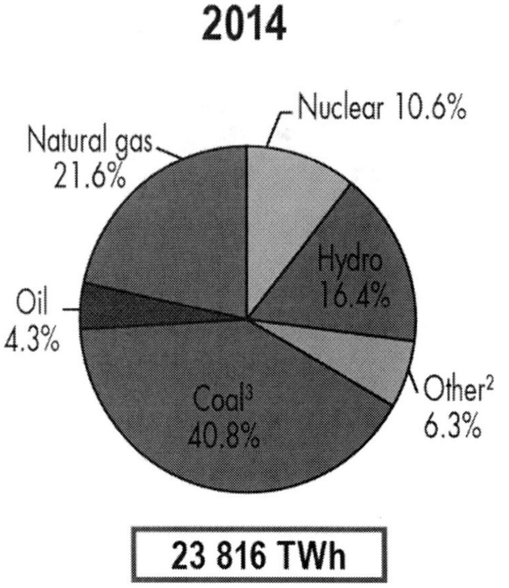

Fig. 9. – World electricity generation (2014) [6], © OECD/IEA 2016.

As far as carbon and other greenhouse gas emissions are concerned, for nuclear energy from fission an accurate estimate of the CO_2 equivalent emissions should include all activities in the whole plant lifetime, from fuel mining/extraction to the final decommissioning of the plant. Such a detailed study has been performed, for instance, in [9], showing that a nuclear plant produces an estimated 66 gCO_2 equiv/kWh, typical renewable energy installations range between 9 and 32, fossile fuels are all above 400. Considering the capability to deploy a large quantity of electric power per plant (typically an actual power of around 1 GWe with high availability) and the relatively low associate emission level per kWh, it is clear why nuclear energy is still considered one of the players in meeting greenhouse gas reduction targets. Indeed, both in the so-called New Policies Scenario and in the more aggressive "450" Scenario (not exceeding 450 ppm of CO_2 equivalent in the atmosphere) [6], nuclear power production is expected to grow.

4·3. *Safety*. – Safety is a very important aspect in nuclear power production. The core of a nuclear reactor produces radiation both while in operation and after shutdown, because of the high content of radioactive materials in the fuel and because of the neutron irradiation of materials comprised by the core and surrounding the core, which produces radioactive nuclei starting from stable ones. Therefore, the safety culture that permeates modern industry is an even more essential aspect of nuclear industry. Attention to safety is present from the design stage down to operation, shutdown and decommissioning of the plants and particular care is devoted to the analysis of possible accidental scenarios. A fundamental principle is the *defence-in-depth*, which means that a series of barriers, starting from the solid fuel form itself up to the reactor containment building, is put in place to contain radioactivity as much as possible at all times and in all instances. Obviously, in this respect, the role of independent regulatory authorities is crucial in overseeing all stages of a plant lifetime.

4·4. *Radioactive waste.* – Among many important safety aspects of fission reactors, a particularly relevant one is the production of radioactive waste. There are two mechanisms for the production of radioactive species in addition to those already present in the fuel (*e.g.*, the two uranium isotopes are radioactive in the first place): the first mechanism is fission itself, which is producing many radioactive fission products (or fragments), *i.e.*, unstable nuclei that decay by β-decay at a later time, with the β-decay being followed very often by emission of gamma rays. The radioactive half-lives of fission products for β-decay is enormously variable, ranging from fractions of a second to million years. Fission products are continuously produced in the fuel material inside a reactor and can be found in the spent fuel once it is removed from the reactor at the end of a power production cycle. Fission products with a half-life until several days are responsible for the large amount of decay heat that is produced within the core immediately after reactor shutdown, *i.e.*, after stopping the chain reaction. However, their radioactivity dies away in a few days or weeks, which means that in case of accidental release of radioactive products they pose a threat only for such a limited amount of time and that when the spent fuel is stored in the cooling pools they do not produce anymore heat after a short amount of time. Other Short-Lived Fission Products (SLFP) with half-lives up to a few years can instead be a safety concern also after an accordingly longer period of time, as are Medium-Lived Fission Products with half-lives up to about 100 years (*e.g.*, ^{137}Cs and ^{90}Sr, with half-life of about 30 years). Such isotopes contribute to heat production in the spent fuel and can be a safety concern if accidentally released into the environment, for decades. Finally, Long-Lived Fission Products (LLFP) have half-lives over about 200000 years (no fission product has half-life between about 100 years and 200000 years). There is a second mechanism producing radioactive isotopes. Radiative capture, which leads to the production of plutonium, also produces a series of heavier (transuranic) nuclides, americium, curium, etc., with increasing atomic number $Z > 92$. Capture on ^{235}U leads also to the formation of ^{237}Np. Neptunium, americium, curium and other similar isotopes are called *Minor Actinides, MA*. Many of these MA decay predominantly by α-decay and have half-lives ranging from a few hundreds of years up to above 2 million years (^{237}Np).

The typical inventory from a 1 GWe PWR fuelled with about 27 tons of U (3.5% ^{235}U) at discharge (when the spent fuel is removed from the core) is approximately 280 kg of ^{235}U, 266 kg of Pu of which 56 kg of fissile ^{239}Pu, 20 kg of MA, 13 kg of ^{90}Sr and 30 kg of ^{137}Cs (both with 30 years half-life), and 63 kg of LLFP. The very long decay time of both LLFP and MA is the reason why disposal of spent fuel and similar waste requires very long-term storage, for which one possible solution envisaged is storage deep underground in the so-called geological repositories. Some countries adopt the so-called reprocessing of the spent fuel, where the plutonium is extracted from the fuel rods and used to produce new fuel containing both uranium and plutonium.

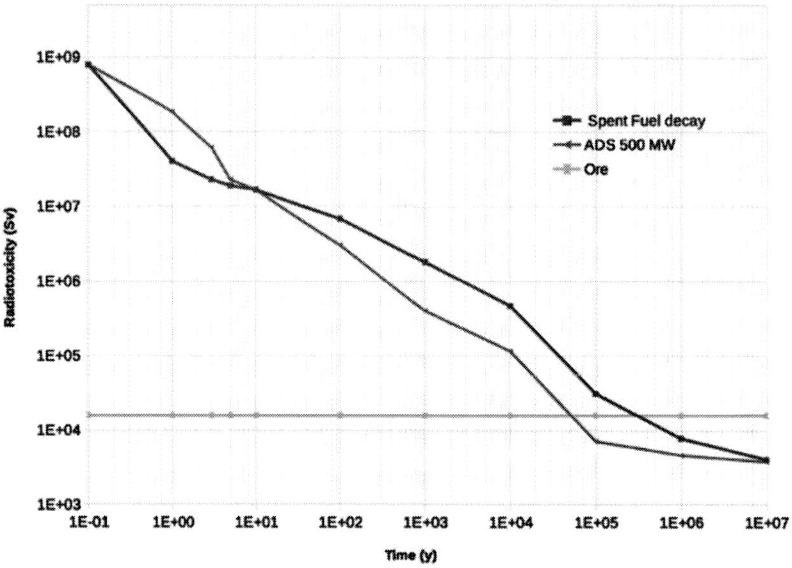

Fig. 10. – Radiotoxicity in two different fuel cycle scenarios. Blue curve: natural decay in the spent fuel; red curve: uranium and fission products separated and stored, plutonium and MA mixed in new fuel and irradiated for 3 years in a 500 MWth ADS.

4·5. *Fast reactors, subcritical reactors and fuel breeding*. – In the effort to improve the safety, security and efficiency of nuclear plants, new concepts of reactors have been developed with goals including the minimization of the production of MA, a better and more efficient use of the fuel, a better thermodynamic efficiency, the possible production of hydrogen at high temperatures and finally improved safety features to minimize the risk of accidents. All these various concepts are considered within the so-called Generation-IV reactors, which are the subject of an international initiative [10]. As an example, we saw that in a fast reactor, by using, *e.g.*, a liquid metal or gas as coolant, the energy spectrum of the neutrons is harder. In such a reactor core, it becomes possible to burn not only fissile elements like ^{235}U and ^{239}Pu, but also fissionable elements like ^{238}U for which fission occurs significantly only above a certain energy threshold around 0.5 MeV, as can be seen in fig. 3. Although given the much lower fission cross-section typically the fuel has to be richer in fissile content, this means that also ^{238}U can be to some extent considered to be a fuel as well, which has obvious implications in terms of how long uranium resources will last. Another important aspect in fast-neutron systems is that, for energies, say, above 0.5 MeV, fission becomes dominant over capture, as again seen in fig. 3. This means that the MA production, which occurs namely via capture, is suppressed. On the other hand, in a fast reactor the MA produced can be partly destroyed in the reactor itself by fission. An important physics aspect when considering fission of MA is that the amount of delayed neutrons produced upon their fission is significantly less than for ^{235}U and ^{239}Pu. This means that, for instance, fueling a reactor purely with MA is not possible, as one would lose the crucial safety feature of the reactor response times being dominated by the relatively long delayed neutron emission times. However, for ADS based on a subcritical reactor core that requires an external neutron source, the role of delayed neutrons is not so crucial and, therefore, there is the possibility to mix a larger quantity of MA into the fuel, thereby offering the possibility to use the apparatus to perform transmutation of the MA into fission products and reduce the radiotoxicity of the final materials in the fuel. This is called *P&T, partitioning and transmutation* approach, meaning that first some nuclides are separated from the spent fuel (partitioning) and then they are irradiated in a special core to be either destroyed or transformed into stable elements (transmutation). Figure 10 shows the radiotoxicity of the material originated from the spent fuel, *i.e.*, its potential danger to human health, estimated according to two scenarios of fuel cycle in a very simple model [11]. In the open cycle without reprocessing, the radiotoxicity of the spent fuel obtained from 1 ton of natural uranium (ore) reaches the level of the ore in about 300000 years. In the closed fuel cycle where MA are separated and burnt together with plutonium for three years in a high-power ADS (here uranium and FP are left aside), the radiotoxicity could be reduced to the ore level in an order of magnitude less years. More accurate models of the closed cycle show that by P&T a reduction to the ore level in a few hundred years can be achieved [12].

REFERENCES

[1] Ongena J., *Magnetic Fusion Status and Outlook* (these proceedings).
[2] *Java-based Nuclear Data Information System*, `http://www.oecd-nea.org/janis/`.
[3] Nifenecker H., Meplan O. and David S., *Accelerator Driven Subcritical Reactors* (Institute of Physics Publishing, Bristol, Philadelphia) 2003.
[4] *Nuclear Energy Today*, second edition, © OECD 2012, NEA No. 6885, 2012, p. 9, `http://www.oecd-nea.org/pub/nuclearenergytoday/6885-nuclear-energy-today.pdf`.
[5] *IAEA Power Reactor Information System (PRIS)*, Copyright © 2017 IAEA, `https://www.iaea.org/PRIS/home.aspx`.
[6] © OECD/IEA 2016 *Key World Energy Statistics* (IEA Publishing) 2016, p. 24, `https://www.iea.org/publications/freepublications/publication/KeyWorld2016.pdf`, License: `http://www.iea.org/t&c`.
[7] *Uranium 2014: Resources, Production and Demand*, © OECD 2014, NEA No. 7209, 2014, `http://www.oecd-nea.org/ndd/pubs/2014/7209-uranium-2014.pdf`.
[8] © OECD/IEA and OECD/NEA 2015 *Projected Costs of Generating Electricity*, 2015 edition, pp. 14–15, `https://www.oecd-nea.org/ndd/pubs/2015/7057-proj-costs-electricity-2015.pdf`.
[9] Sovacool B. K., *Energy Policy*, **36** (2008) 2940.
[10] *Generation IV International Forum*, `https://www.gen-4.org/`.
[11] Panza F., private communication.
[12] *Implications of partitioning and transmutation in radioactive waste management*, Technical Reports Series No. 435, Copyright © 2004 IAEA, `http://www-pub.iaea.org/MTCD/Publications/PDF/TRS435_web.pdf`.

EPJ Web of Conferences **189**, 00014 (2018) https://doi.org/10.1051/epjconf/201818900014
LNES 2017

Radioactive wastes and disposal options

Bernhard Kienzler(*) and Horst Geckeis

Institute for Nuclear Waste Disposal (INE) Karlsruhe Institute of Technology (KIT)
Hermann-von-Helmholtz Platz 1, 76344 Eggenstein-Leopoldshafen, Germany

Summary. — This paper provides for a summary of facts which are indispensable for nuclear waste disposal. Information is presented on types of radioactive wastes and the origin, the waste treatment procedures and some characteristics of the waste forms. Finally the various host rocks and the international disposal concepts are discussed and the procedures for safety analyses are shortly described.

1. – Introduction

In 2016, 450 nuclear power plant units with an installed electric net capacity of about 392 GW_{electric} were in operation worldwide. Additional 60 plants with an installed capacity of 60 GW were under construction in 16 countries (`https://www.euronuclear.org/info/encyclopedia/n/nuclear-power-plant-world-wide.htm` June 2017). In Europe, France has 58 units in operation, producing 63 GW_{el}. The efficiency of a nuclear power plant is in the range of 35%, and the average annual availability is about 80%. A combined gas-steam power plant has an efficiency of about 60%. Nevertheless, the nuclear power plants in France save annually the emission of 130 million metric tons of

(*) E-mail: `Bernhard.kienzler@kit.edu`

CO_2 in comparison with the production by combined gas-steam power plant or about 280 million metric tons of CO_2 produced by lignite fired power plants (data taken from `https://www.eia.gov/tools/faqs/faq.php?id=73&t=11`, June 2017). For this reason, nuclear power plays still an important role in preventing CO_2 emissions. However, the energy production by nuclear fission produces radioactive wastes. The mass and volume of the nuclear waste is relatively small but due to its high radiotoxicity it needs to be treated and disposed safely for very long periods of times.

2. – Origin, type and amount of radioactive wastes

Today, most of the nuclear power plant reactors are pressure water (PWR) or boiling water (BWR) reactors. Both reactor types are light-water nuclear reactor. The main difference between a BWR and PWR is that in a BWR, the reactor core heats water, which turns to steam and then drives a steam turbine. In a PWR, the reactor core heats also water, under elevated pressure. This hot water goes through steam generators heating a lower-pressure steam system, which drives the turbine. A typical PWR (*e.g.* Gösgen, CH) with a net electrical output 1010 MW needs about 77 metric tons of fuel in the core of the reactor. The enrichment of UO_2 is in the range of 4.6–4.95% ^{235}U. Every year, parts of the fuel elements are replaced by fresh fuel and the other fuel elements are transferred to different locations in the reactor core according to their burn-up. The burn-up is a measure for the energy produced per mass of initial fuel in gigawatt-days/metric ton of heavy metal (GWd/t_{HM}),

Typically, fuel elements remain in the nuclear power plant (NPP) for 4 cycles (`https://www.kkg.ch/upload/cms/user/KKG_TB_englisch_2016.pdf`, June 2017). A NPP generates almost 20 t of spent nuclear fuel (SNF) per year which need to be stored in a water pool for several years. Figure 1 shows the design of a fuel rod, a photo of a PWR fuel element (assembly) and pictures of the UO_2 fuel after discharge from the NPP.

The energy in a NPP is produced by fission of fissile isotopes, such as the natural isotope ^{235}U. Only 0.7% of the natural uranium consist of the isotope ^{235}U. The effective energy release by fission of ^{235}U is about 205.4 MeV per fission. This energy is distributed between 2 fission products, recoil nucleus and 2 to 3 neutrons. The energy of the neutrons is in the range between 0.1 and 10 MeV which is too fast for keeping up a chain reaction. In a LWR, fission is caused by neutrons in the thermal energy range. By interactions with light atoms (hydrogen of the water) the fission neutrons are moderated to thermal energies of 0.025 eV. It needs to be mentioned that fission can be generated also by high-energy neutrons for example in fast reactors. Details of the neutron interactions with matter, the interaction coefficients for scattering, capturing and fission processes as function of the neutron energy are given in the numerous text books on nuclear physics.

The mass distribution of the fission products resulting from the fissile isotopes ^{233}U, ^{235}U and ^{239}Pu are shown in fig. 2. For a burn-up of 50 GWd/t_{HM}, 5.5% of the initial ^{235}U atoms are affected. The formation of the fission products cause disturbances in the crystal structure of the UO_2 fuel, resulting in swelling and distortion of the UO_2

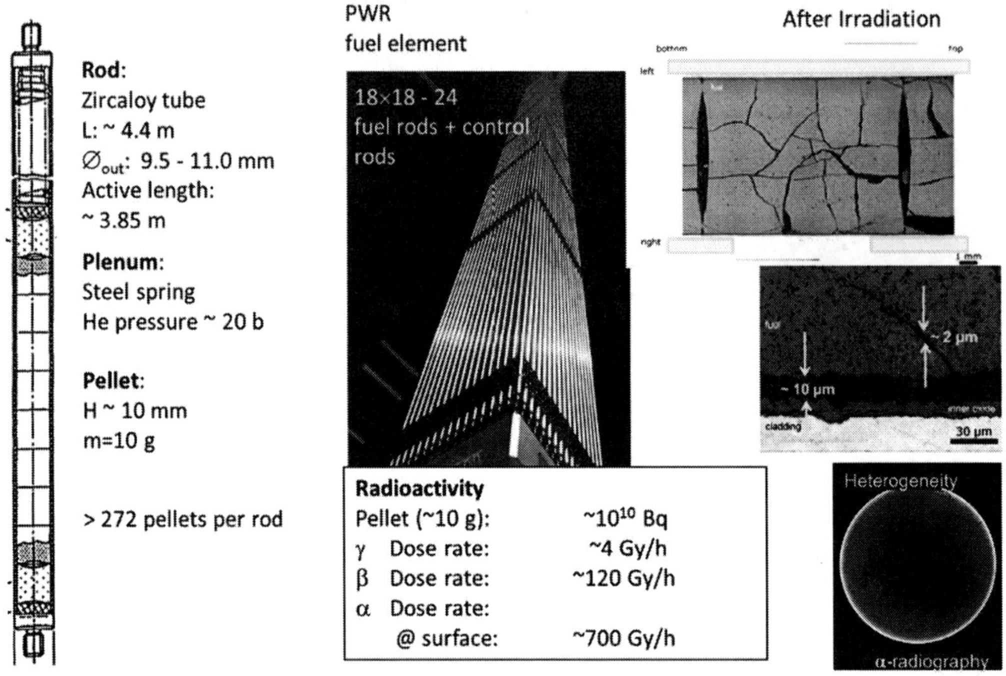

Fig. 1. – Design of a fuel rod (left), photo of a PWR fuel element assembled from fuel rods (middle), fuel after discharge (right).

lattice, oxygen redistribution, reduction of heat conductivity, and changes in specific heat, mechanical properties, formation of fractures and pores. The fission products according fig. 2 are characterized as follows:

– Noble gases (Kr and Xe) and other volatile fission products (FP).

– FPs dissolved as oxides in the UO_2 matrix with adequate oxidized valence state.

– FPs forming metallic precipitates.

– FPs forming oxide precipitates

– Actinides dissolved as oxides in the fuel matrix.

If a thermal or epithermal (0.55–10 eV) neutron hits one of the majority uranium atoms ^{238}U, it can be captured forming ^{239}U. This isotope is unstable and undergoes beta decay (half-life 23.47 min) forming ^{239}Np. Also this isotope undergoes beta decay ($t_{1/2} = 2.355$ days) forming ^{239}Pu. This plutonium isotope has a half-life of 24 114 years and is also fissile. During the 4 years of the fuel in a NPP, ^{239}Pu is formed and partly burned. By neutron capture, also higher actinides are formed.

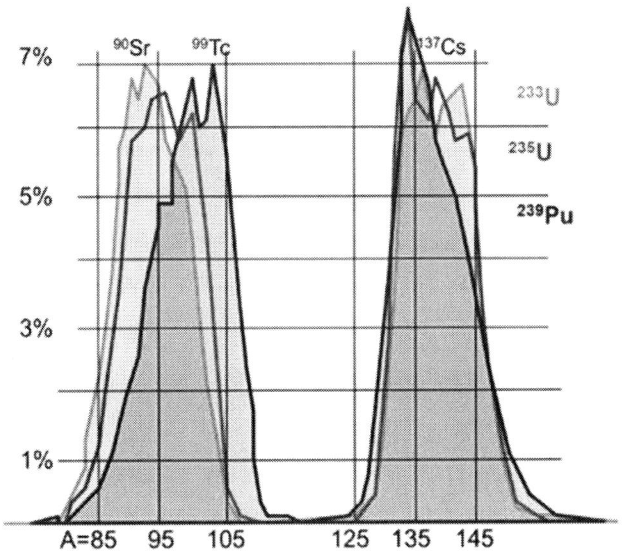

Fig. 2. – Distribution of the main fission products as a function of the atomic number for 3 fissile isotopes.

From the chemical point of view, the spent nuclear fuel is one of the most complex materials with respect to the elemental composition, the chemical state of the elements and the heterogeneity. Information on the radioactivity of spent nuclear fuel is given in fig. 1.

3. – Reprocessing or direct disposal of spent nuclear fuel

After discharge of the UO_2 fuel from a NPP with a burn-up of 50 GWd/t_{HM}, the spent fuel (SNF) contains still 0.8% ^{235}U and about 1% ^{239}Pu fissile isotopes. Especially in the 1960/1970ies when the total worldwide uranium resources were significantly underestimated, many states developed plans to retrieve the fissile materials for further use in NPPs. The objective was a closed fuel cycle with multiple reprocessing and reuse of plutonium (see fig. 3).

The reprocessing processing process is a complicated chemical procedure, dissolving the UO_2 fuel in hot, highly concentrated nitric acid. Afterwards, the acid solution which contains the fission products, as well as some activation products, the actinides and the complete mass of uranium is treated by the PUREX process. This is a chemical method used to purify fuel for nuclear reactors or nuclear weapons. It is an acronym standing for Plutonium Uranium Redox EXtraction. PUREX is the *de facto* standard aqueous nuclear reprocessing method for the recovery of uranium and plutonium from used ("spent", or "depleted") nuclear fuel. It is based on liquid-liquid extraction ion-exchange.

Reprocessing and re-fabrication of nuclear fuel with the remaining ^{235}U and the plutonium (closed fuel cycle) was considered as the most appropriate method. In the atomic

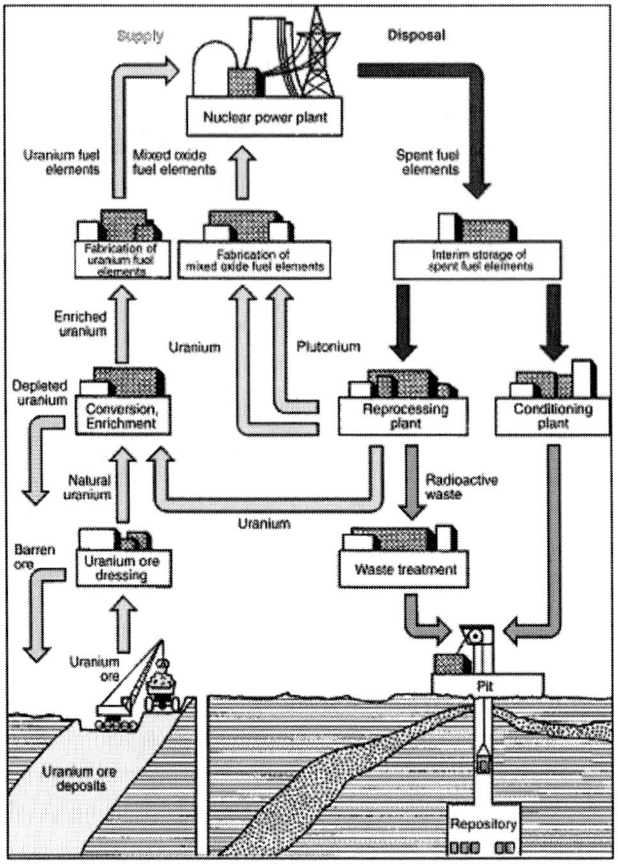

Fig. 3. – Schematic view of the open and closed fuel cycles. (Picture taken from ENS, European Nuclear Society, `https://www.euronuclear.org/info/encyclopedia/n/nuclear-fuel-cycle.htm`.)

law in Germany, until 1994 reprocessing of used fuel was promoted and direct disposal was forbidden.

During the reprocessing of UO_2 fuels different waste streams are generated. These comprise

- Fission gases,
- zircaloy cladding (hulls) of the fuel rods including the foot and end-pieces (solids);
- high-level radioactive waste in HNO_3 solution (liquid);
- intermediate-level radioactive wastes mainly in acidic solutions (liquid);
- low-level radioactive wastes, HTO-bearing liquids, decontamination detergents (liquid);
- solid wastes, *e.g.* filters, etc.

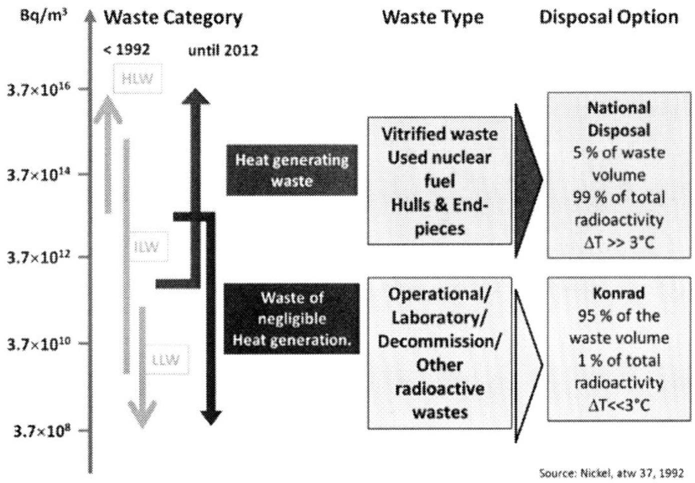

Fig. 4. – International (left) and German classification of radioactive wastes [1].

In the closed fuel cycle, the separated plutonium would be applied to produce mixed uranium plutonium fuels (MOX) to be used in LWRs or in fast reactors.

4. – Classification of wastes

Radioactive wastes do not only originate from reprocessing, but also during the operation and decommissioning of NPP, from industrial use (sterilization, radiographic techniques, etc.), from research and from medical applications. The nuclear waste classification applied In Germany is shown in fig. 4, the radioactive wastes are classified with respect to the heat release in a repository. Heat-producing waste comprises the solidified high-level waste from reprocessing, hulls and end-pieces as well as SNF.

All wastes need to be transformed in a solid form before disposing.

4·1. *High-level wastes.* – For liquid high-level radioactive waste (HLW) from reprocessing, the vitrification of the liquid is presently the method of choice. In Europe, vitrified HLW is/was produced by AREVA in La Hague, France, in Sellafield, UK and also at the pilot reprocessing plant WAK in Germany. The molten glass is poured in stainless-steel canisters of 150 l. The outcome of vitrification is a solid waste form with a content of 15–20 wt.% of fission product oxides in a borosilicate matrix. Guarantee values of HLW glass canisters are 3×10^{16} Bq of β/γ activity, 1.4×10^{14} Bq of α activity and a surface dose rate of 1800 Gy/h.

Zircaloy rod cladding (hulls), spacers etc. as well as foot and end-pieces of the fuel elements are compacted and stored in the same type of canisters as the vitrified waste.

In the European countries, different package concepts for spent fuel elements were developed. Well-known is the Swedish KBS-3 concept, where 12 complete fuel elements are inserted in steel canisters which are surrounded by a 5 cm thick copper mantel. The

Fig. 5. – Swedish KBS-3 concept: Disposal of a Cu-coated steel canister in a vertical drill hole into the granitic host rock surrounded by clay buffer. (Picture taken from Wikipedia, https://commons.wikimedia.org/wiki/File:SKB_KBS-3.jpg#file.)

Cu layer is considered to be stable under the Scandinavian granitic rock conditions for several 100 000 years (see fig. 5).

4·2. *Intermediate- and low-level wastes.* – Almost all countries distinguish between long-lived and short-lived Intermediate- (ILW) and low-level waste (LLW). Liquid ILW and LLW from reprocessing are concentrated by evaporation and the concentrates mainly mixed with cement in order to form solid waste forms. Typically these waste forms contain about 10 wt.% of $NaNO_3$ salts. In some facilities, the liquid ILW was incorporated in bitumen. At the reprocessing plant in La Hague, France, the bituminization process is replaced by a vitrification process using a "cold crucible" technology [2].

4·3. *Operational wastes.* – In LWRs, different waste streams have to be treated. In Germany, about 300 m^3 of radioactive wastes accumulate annually. The wastes include solids, such as scrap metals, debris, organic materials (paper, plastic, etc.) and liquids like oils, slurries, evaporator and filter concentrates and ion exchanger resins. The various processing techniques aim at volume reduction, producing solid, compact and dry waste products. The specific techniques and the resulting waste forms are shown in fig. 6. For combustible wastes a volume reduction by a factor of 50 is achieved from incineration. In the case of compressible raw waste, a volume reduction by a factor of 2 to 5 can be achieved [3]. Cementation is the preferred method for fixation of radioactive liquids.

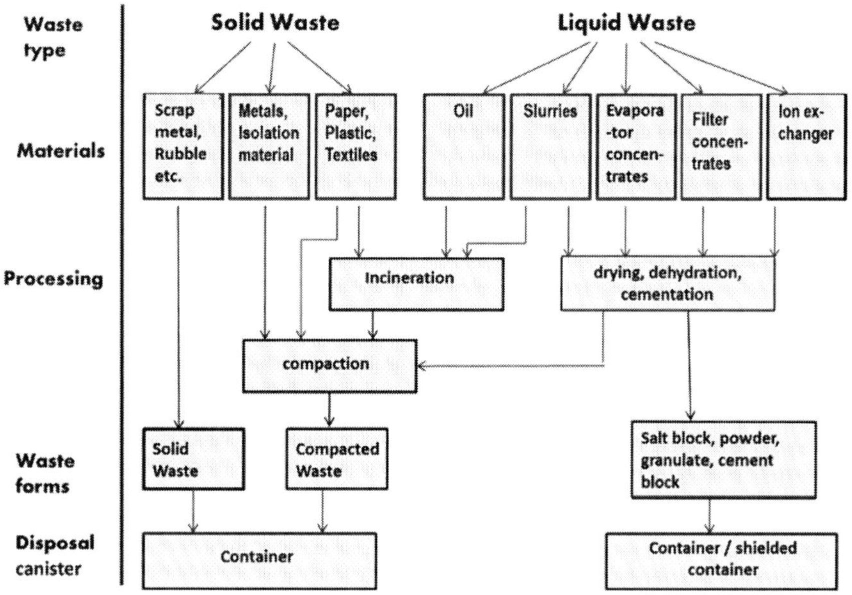

Fig. 6. – Waste types and specific treatment techniques for LWR operational wastes.

4'4. *Decommissioning wastes.* – Decommissioning wastes represent a huge amount of mass. The total mass of the radiation-controlled area of a large nuclear power plant with a PWR is estimated to 156 500 t. About 143 000 t cover concrete structures. This material can be almost completely conventionally recycled following removal of any surface contamination. Only approximately 600 t of the concrete requires final disposal as radioactive waste. The mechanical installations —essentially piping and components— including the entire steel construction (*e.g.* platforms and mountings) form a mass of around 13 500 t in the radiation-controlled area of a PWR. Among these materials, only around 3 000 t require disposal as radioactive waste and around 500 t as radioactive secondary waste (including from decontamination). The remaining 9800 t can likewise be directly released or recycled after decontamination or melting-down [3]. During the decommissioning activities each material is analyzed with respect to radiation dose. In the case that the dose is below a certain limit (which is expressed in Bq/g or Bq/cm^3) for the different materials and structures such as debris, surfaces, buildings and scrap metal, the materials can be disposed in conventional surface disposal or fed to re-use processes [4].

However, the different European countries follow different disposal concepts. In France, all materials arising from decommissioning of nuclear facilities need to go to specific disposal (very low waste disposal).

5. – Disposal – International disposal projects

An international consensus exists that long-lived highly radioactive and heat-producing wastes should be disposed in a deep repository. However, finding a well-suited

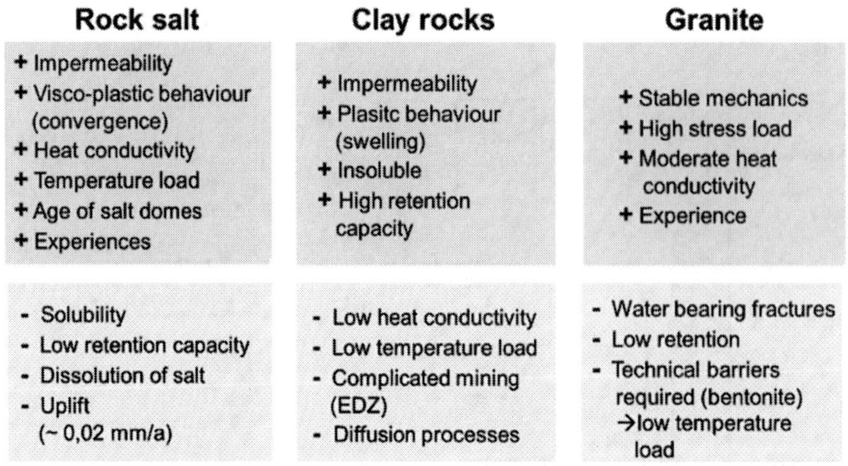

Fig. 7. – Comparison of important characteristics of potential host rocks.

site is a challenge. The siting process has to take into account a multitude of different aspects, scientific, social, economic and also ethical aspects. In most countries, the import and export of radioactive wastes are not allowed. For this reason all countries using nuclear power need to take decisions. A summary of the international approaches for deep geological disposal of nuclear waste and the geological challenges in radioactive waste isolation was published recently by Lawrence Berkeley and Sandia National Laboratories in the USA (LBNL-1006984, 2016) [5]. It covers the disposal plans of 23 countries worldwide.

Before disposing HLW in a deep underground facility, the heat capacity has to decay below a certain level. This is achieved by interim storage, in many cases for periods of minimum 30 years. Of course, prolonged interim storage might be considered for example in The Netherlands. The Dutch waste management organization COVRA has a license for an interim storage facility for a period of 100 years.

Presently, three host rock types are under consideration for HLW and SNF. These are rock salt which has been investigated in Germany and USA since decades. Crystalline rocks, such as granite are available in Scandinavia and therefore this host rock was selected for Swedish and Finnish disposal concepts. It is also under investigation in Korea and China. In Belgium, an indurated clay rock formation (Boom clay) is under investigation, whereas in Switzerland the Opalinus clay and in France the Callovo argillite is selected. These two clay rocks consist of a mineral assemblage made of illite and calcite as main components. In fig. 7, some important features of the different host rocks are presented.

Figure 7 shows pros and cons of some general properties which, however, might vary over the selected host rock formation. An intensive exploration of the site is required in any case. Safety analyses performed for the different host rocks and the related designs

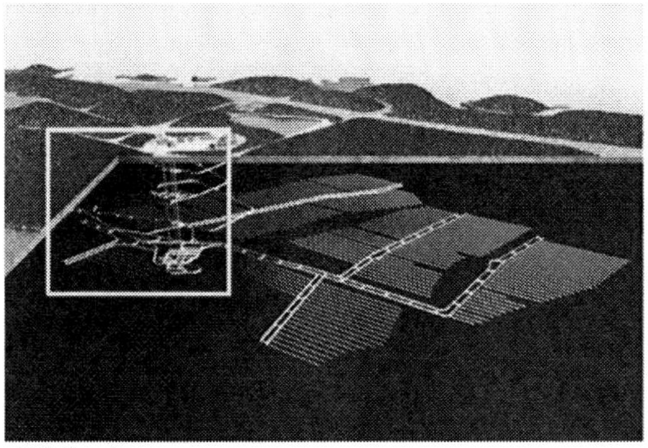

a) Finland
(http://www.posiva.fi/en/final_disposal/final_disposal_facility/repository#.WcDzaU1rymQ)

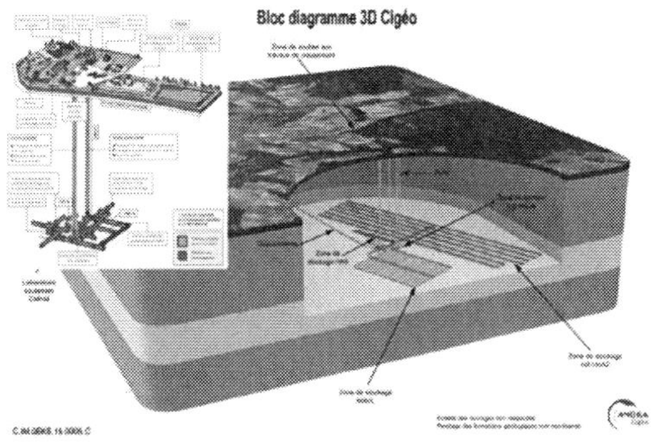

b) France
(http://www.usinenouvelle.com/article/l-asn-a-encore-des-reserves-sur-le-projet-cigeo-d-enfouissement-de-dechets-nucleaires-a-bure.N573223)

c) Sweden, Forsmark site
(http://www.skb.com/future-projects/the-spent-fuel-repository/)

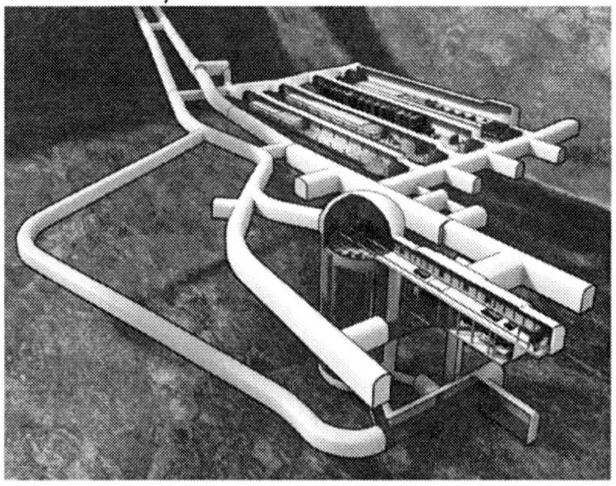

d) Sweden, LLW and ILW repository SFR
Sweden, LLW and ILW repository SFR
(http://www.skb.com/our-operations/sfr/)

Fig. 8. – European disposal concepts. (Pictures taken from the public web pages of the respective waste management organizations: a) Finland, picture taken from POSIVA OY, Finland, `http://www.posiva.fi/en/final_disposal/final_disposal_facility/repository#.WcDzaU1rymQ`; b) France, picture taken from ANDRA, Agence nationale pour la gestion des déchets radioactifs, France, `http://www.usinenouvelle.com/article/l-asn-a-encore-des-reserves-sur-le-projet-cigeo-d-enfouissement-de-dechets-nucleaires-a-bure.N573223`; c) Sweden, Forsmark site, picture taken from SKB, Svensk Kärnbränslehantering AB, `http://www.skb.com/future-projects/the-spent-fuel-repository/`; d) Sweden, LLW and ILW repository SFR Sweden, picture taken from SKB, Svensk Kärnbränslehantering AB, `http://www.skb.com/our-operations/sfr/`.)

of the repositories have shown that a safe disposal is feasible in all types of host rocks. However, the specific multi-barrier system consisting of technical, geo-engineered and natural barriers as well as the impact of each barrier on the disposal safety is different for the different host rocks.

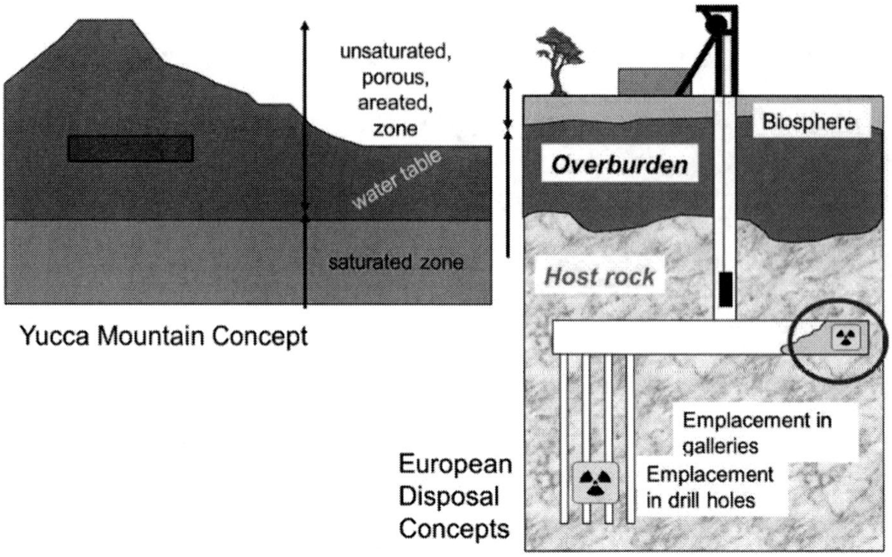

Fig. 9. – Layout and general design of a HLW/SNF repository.

All reasonably discussed disposal concepts for HLW and SNF are based on mines in the host rock. For disposal of heavy disposal casks access tunnels or shafts are foreseen. Some disposal concepts need to demonstrate the possibility of retrieving the wastes for a certain period of time (*e.g.* France, Germany), whereas other disposal concepts promote disposal "in an irreversible manner" (Finland). It is clear that if the tunnels of a mine have to be accessible over decades, specific layout and constructive measures have to be considered.

Figure 8a shows the planned Finnish HLW repository, start of operation is planned for 2025. The yellow square shows the disposal research mine ONKOLO, where site-specific investigations are running. Figure 8b presents the French disposal concept Cigéo which will be located in the Callovo-Oxfordinan clay formation. The license for construction is expected in 2018, the operation could start after 2025. In the left corner of fig. 8b, the research mine at Bure is shown.

Figure 8c shows a photo of the selected site at Forsmark in Sweden, where the Swedish waste management organization Svensk Kärnbränslehantering AB (SKB) will construct the HLW repository. It is planned to start operation in 2030. Close to this location, SKB operates the LLW/ILW disposal facility called SFR since 1988. In contrast to the deep disposal concept for HLW, the SFR is situated 50 metres below the bottom of the Baltic and comprises four 160-metre long rock vaults and a chamber in the bedrock with a 50-metre high concrete silo for the most radioactive waste.

Figure 9 shows the design and general layout of repositories. The disposal mine is generally located deep below the groundwater table which guarantees slow groundwater movements and reducing conditions. The only exemption was the US disposal site at Yucca Mountain in Nevada. This site is located in a desert with low precipitation rates which caused problems with corrosion of canister materials.

Fig. 10. – Final layout of the Belgian LLW surface disposal at Desel. (Picture taken from ONDRAF·NIRAS, Belgium, `https://www.ondraf.be/stockage-en-surface-%C3%A0-dessel`.)

Today, deep borehole disposal concepts are under discussion in some countries (5000 m). The concepts are not sufficiently developed and will not be further addressed in this presentation.

For ILW and LLW, disposal in deep geological formations is considered in some countries only (*e.g.* Germany and Switzerland). Most countries use shallow surface disposals (France, Belgium, Spain, ...) for short-lived radioactive wastes. Figure 10 shows the layout of the Belgian LLW disposal at Dessel after final closure of the site. The wastes are stored in drums which are packed in concrete structures. These are placed in buildings which are finally covered by clay layers. This type of disposal should hold several hundred years until the decay of the most abundant fission products ^{90}Sr and ^{137}Cs.

As already mentioned, the safety of repositories relay not only on the hostrock, but also on a multi-barrier system. The barriers consist of constructions and materials with the following properties:

- waste packages, cansisters, etc.;
- barrieres against penetration of groundwater;
- barriers against migration of radionuclides (sorption properties);
- chemical "buffering" (pH, redox, composition of solutions ...);
- mechanical stability;
- heat conductivity (for HLW/SNF);
- decoupling of canister and hostrock (in the case of crystalline hostrocks).

The barrier materials need to be designed according to the hostrock, groundwater composition and the expected long-term evolution of the system.

6. – Safety analysis, data and models

Operating of an underground disposal requires a series of measures to guarantee the orperational safety. One important aspect accounts for non-reactive waste forms (WIPP accident 2014 [6, 7]). The disposal of radioactive wastes in deep underground repositories aims at a multitude long-term safety relevant features. These comprise:

- concentration of radiotoxic materials,
- isolation from biosphere,
- protection of environment and population,
- protection against intended and unintended access,
- safeguards (nonproliferation of fissile materials).

Additionally, ethical and societal features are of high importance:

- passive safety,
- unattended (maintenance free),
- no burdens for future generations.

Due to the long half-lives of the actinides and some fission products (^{135}Cs, ^{99}Tc, ^{129}I), these features have to be maintained over long periods of time. In Germany, safety for 1 million years is claimed by law.

In the underground, below the groundwater table, water is omnipresent and contact of water with the waste forms cannot be excluded completely. It is clear, that the different host rocks provide different probabilities of water penetration to the wastes. From a thermodynamic viewpoint, the components disposed of are not inert in contact with water, but tend to react over long periods of time. The reactions include interactions with backfill/buffer materials, corrosion of canister materials, and leaching and corrosion of the waste itself. Safety assessment is based on models which simulate the whole suite of possible processes in a disposal including corrosion, waste form behavior, radionuclide mobilization and retention. The reactions and reaction rates depend on the availability of the relevant reactants, *inter alia* of water and groundwater constituents. Both, the transport of water and of radionuclides is affected by reactions causing alterations of the mineralogy and consequently of the porosity and permeability of near-field barriers. It is clear that the models require reasonable understanding and databases.

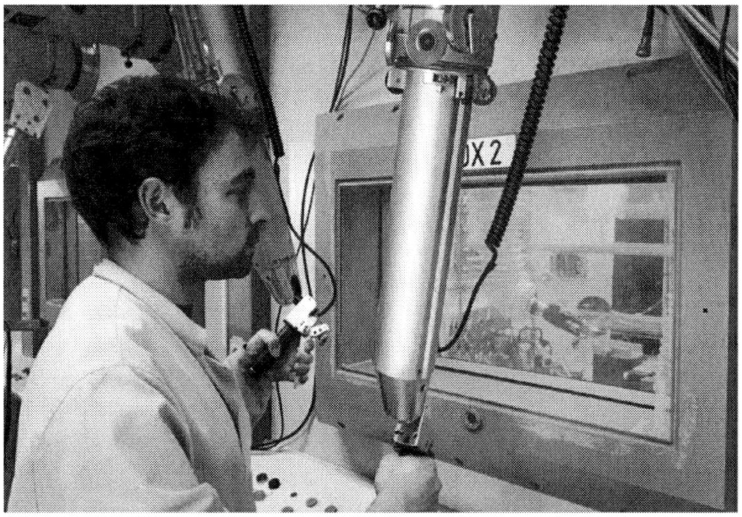

Fig. 11. – Experiments with SNF in the shielded boxes at KIT-INE.

6·1. *Alteration/Corrosion of waste forms.*

HLW glass

The mechanism of glass dissolution is primarily driven by two main reactions that are opposed to one another: These are the parallel reactions of "matrix dissolution" of the glass and "water diffusion/hydration of the glass matrix/ion exchange reactions" [8]. The fastest process is the rate-dominating process in parallel reactions, whereas in sequential reactions, the slowest reaction governs the rate. The combination of the processes lead to consecutive regimes of glass alteration having different reaction rates. First, the formation of the gel layers and/or the accumulation of glass constituents in solution, later the formation of secondary phases take place and the glass corrosion rate decreases. with time. Radionuclides distributed homogeneously in the glass matrix are released. Soluble elements remain dissolved in the attacking water, less soluble elements react according to their thermodynamic potentials forming solids or being sorbed onto solid phases.

SNF

In contrast to HLW glass, radionuclides in the SNF are not distributed homogeneously. There exist segregated phases containing easily soluble elements such as Cs, and I, metallic phases with noble metals and Mo and the oxide phases which contain oxidized plutonium and rare-earth elements. The release behavior of the segregated elements has been investigated, *e.g.* [9-13]. It needs to be mentioned that these investigations require heavy shielding and radiation protection (fig. 11). A summary of the long-term radionuclide release form SNF under reducing conditions in deep repository is given by Metz *et al.* [14]. The long-term radionuclide release is affected by α-radiolytical processes, the availability of hydrogen and the presence of dissolved elements interacting

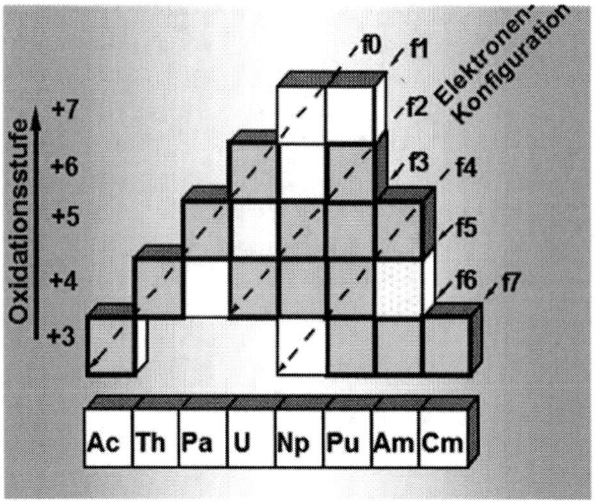

Fig. 12. – Oxidation states and electron configuration of the disposal relevant actinide elements.

with radiolytical compounds (Cl^-, Br^-). The α-radiolytical processes depend on the specific α-activity of the SNF which decays with time [15].

Cemented waste forms

Cemented waste forms are produced using relatively high water to cement ratios which are representative for the pore system of the cement matrix. Groundwater penetrates into the pore system and starts leaching of soluble components such as the portlandite ($Ca(OH)_2$). Depending on flow-through processes, the whole portlandite might be dissolved causing a decrease of the pH in the cement pores. As a consequence, the main constituents of the hardened cement the calcium-silicate phases are no longer stable and decompose [16]. In the case of contact of cement products with magnesium chloride brines, the decomposition rate is no longer controlled by diffusion but by Ca-Mg exchange reactions. This reaction shows a much faster degradation of cemented waste forms [17]. Soluble elements (Cs) are released quickly, in cement incorporated actinides are dissolved according their thermodynamic properties [18].

6˙2. *Solubility and complexation of radionuclides.* – As already mentioned, actinides show low solubility under the prevailing pH and redox conditions. Figure 12 shows the potential oxidation states and electron configuration of some actinide elements. Each of these valence states has own solid phases and complexation properties with groundwater constituents. For prediction of the actual radionuclide concentrations in the specific groundwater, thermodynamic data are required. Besides the measurement of the concentrations, the kind of complexes, their charge, pH and ionic strength dependence need to be quantified exactly. An important tool is the time resolved laser fluorescence spectrometry TRLFS (see fig. 13). The solubility controlling solid phases must be characterized

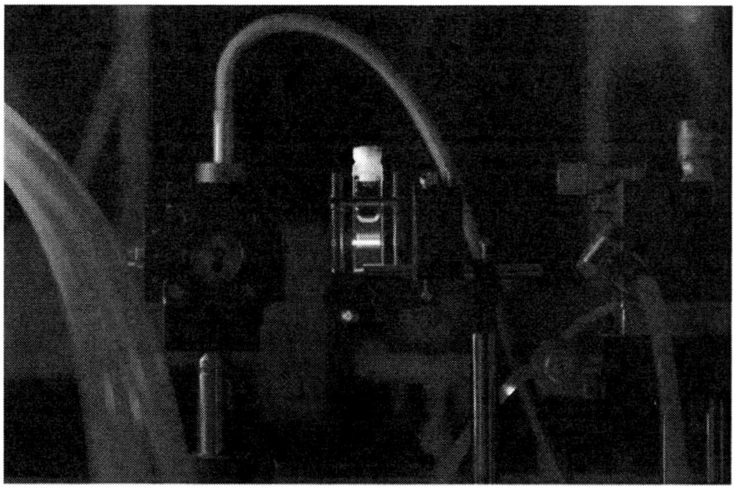

Fig. 13. – Laser spectroscopic characterization of dissolved and sorbed species at KIT-INE.

be different methods such as X-ray diffractometry or extended X-ray absorption fine structure measurements (EXAFS) and related methods.

Due to the strong sorption capacity of the actinides, the elements might sorb onto groundwater colloids or form eigen-colloids. Specific techniques have been developed for detection and quantification of the colloids (*e.g.* Laser-induced breakdown detection LIBD).

6'3. *Sorption onto artificial and natural barrier materials.* – Radionuclides can be sorbed onto the different solid materials of the barriers. The sorption processes comprise

- ion-exchange reactions,
- surface complexation reactions (outer and inner surface complexes),
- surface precipitation by co-precipitation and by redox controlled processes.

The sorption coefficients are strongly influenced by the *p*H of the groundwater and the dissolved ligands such as carbonate. It is also clear that sorption takes place onto the surfaces of the solids. The sizes and properties of the surfaces play an important role. For this reason, the sorption data bases (*e.g.* Vandergraaf [19]) can hardly be transferred to different systems. Sorption data need to be determined site specifically.

For quantification of the sorption processes and characterization of the relevant surface complexes highly sophisticated methods are required. These techniques include X-ray photoelectron spectroscopy, XPS, TRLFS and EXAFS/XANES.

6'4. *Transport processes.* – The migration of radionuclides from deep repositories to the biosphere is facilitated by transport in water. Transport of gaseous radionuclides is possible only under very specific conditions. Diffusion of radionuclides in water is certainly the slowest process. Advectively driven processes are significantly faster. The following mechanism might trigger the water/radionuclide transport:

- thermal convection,
- density driven convective flow,
- forced flow as consequence of gas formation,
- forced flow as consequence of volume changes.

One can expect that the design and the barrier system of a HLW repository would prevent thermal and density driven flow processes, also the changes of open volumes by swelling (clay) or convergence (rock salt) are relatively slow processes and can be evaluated reasonably.

As the repository system evolves, gases may be produced, such as hydrogen from the corrosion of metals and from the radiolysis of water, and radon from the radioactive decay of some of the waste. If present, biodegradable wastes can also produce carbon dioxide and methane. Understanding how these gases move in a repository setting was a topic of the Euratom FORGE project, which studied key gas migration issues in repository performance assessment (`https://www.bgs.ac.uk/forge/`).

6˙5. *Models and safety analysis.* – Demonstration of the safety of a nuclear waste disposal requires numerical model calculations. However, in this context models results are representative only for the aspect for which the model has been developed. Extrapolations beyond the parameter range on which the model is based are not possible. This is also true for the parameter used in the calculations. Besides reasonable models one has also to deal with natural processes and unpredictive human activities. Further the models as well as the data have uncertainties which need to be addressed.

Deterministic safety analysis

For safety analysis specific tools are used. As the future is unknown, safety analyses are build-up on the concept of scenarios. This concept representing the deterministic safety analysis is demonstrated in fig. 14. In a first step, a selection of the features, events, and processes (FEP) is required. The FEP should cover all conceivable facts and figures of the disposal system including wastes, barriers and hostrock. Relevant FEP can be selected from FEP databases or from site specific information. A scenario is constructed by combination of the selected FEPs. A common definition is as follows: "A scenario is specified by a combination of some FEPs which characterize one potential evolution of the disposal system". Especially in the case of an incidence scenario, the initiating FEP is of high importance.

After the combination of the FEPs to form a scenario, a model is developed, mainly as a numerical code. Initial and boundary conditions as well as model parameters are derived from site-investigation or from experimental data with wastes, barriers, etc. This process delivers results for a multitude of scenarios and realizations with different sets of parameters. Figure 14 shows the different steps which are required for deducing the long-term performance and the resulting dose rates from FEPs and scenarios.

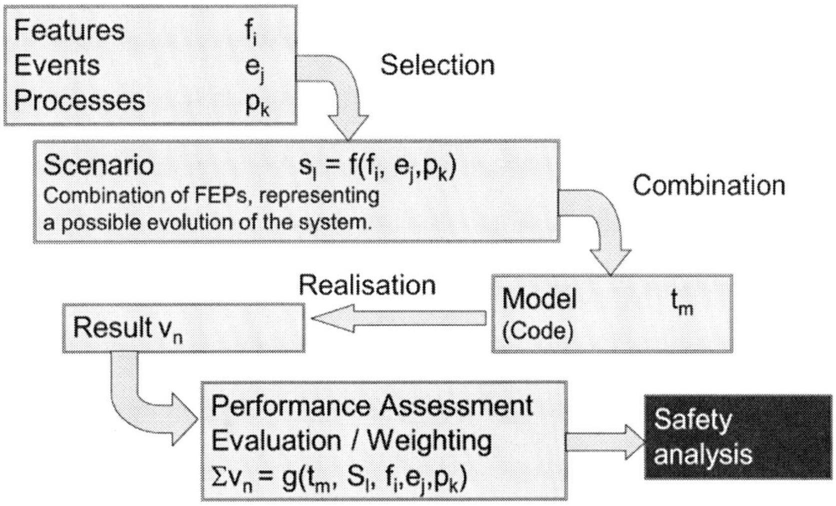

Fig. 14. – Scenario: From FEP to performance assessment and safety analysis.

One important question is concerned with the weighting of the results to be used for safety analysis. This can be done by probabilistic tools.

Probabilistic Safety Analysis

Probabilistic safety assessment (PSA) is an established technique to numerically quantify risk measures in nuclear power plants. It is also applied for disposal safety analysis. PSA comprises a model of the disposal, in which all safety relevant systems are modelled and logically linked together to determine the overall likelihood of radionuclide release and the dose to man([1]). System failures are identified and quantified by models like Fault Trees (FT) which deduce logical combinations of the FEPs. At the lowest level, the FEPs of the fault trees are assigned probability distributions. These probability distributions are propagated up through the tree logic to reach a probability distribution of the top event. The PSA methodology is a logical, deductive technique which specifies an undesired top event and uses fault trees and event trees to model the various parallel and sequential combinations of processes that might lead to an undesired event such as radionuclide release. The number of combinations that might lead to this incident increases exponentially with the number of modelled components. Consequently, effective computer codes and quantification techniques are necessary to solve any large scale problem. The PSA models are traditionally built from logical Boolean expressions. In PSA such logical expressions are typically based on coupled event tree models. Event trees (ET) depict the potential event sequences from the initiating event to the associ-

([1]) The detriment adjusted nominal risk coefficient for cancer and hereditary effects by radiation is 0.057 per Sv [20].

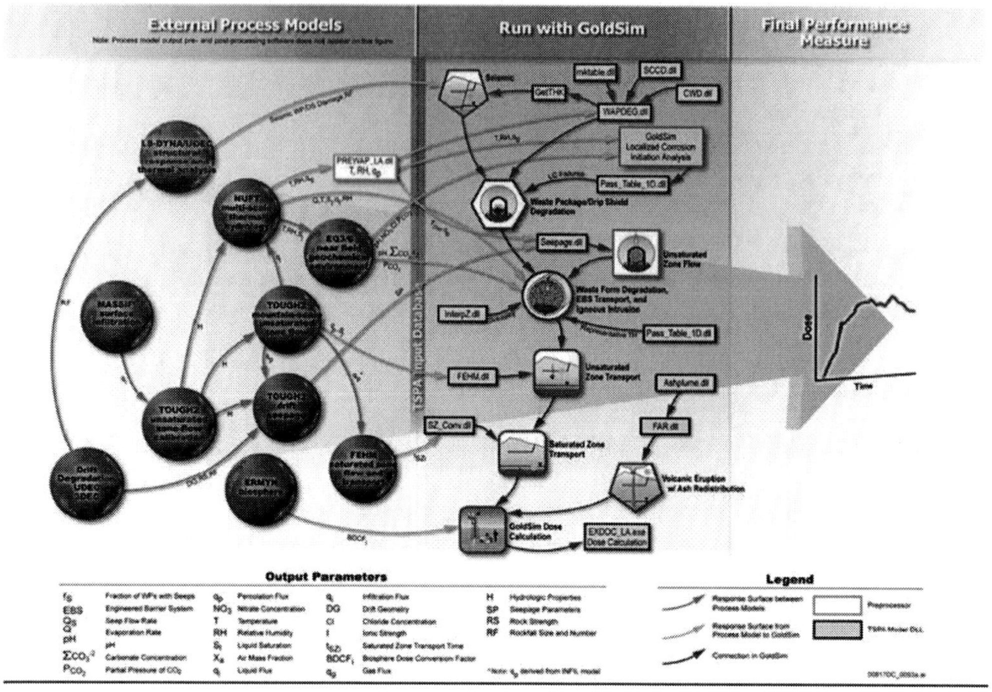

Fig. 15. – TSPA Model Architecture. (Picture taken from C. W. Hansen *et al.*, Sandia National Laboratories, USA, `http://energy.sandia.gov/wp-content//gallery/uploads/SAND2010-3754C.pdf`.)

ated consequences. The ET begins with the initiating event. An initiating event (IE) could be failure of a barrier. For the Total System Performance Assessment (TSPA) of the Yucca Mountain, four questions were underlying the probabilistic safety analysis:

1. What events and processes can take place at the facility?
2. How likely are these events or processes?
3. What are the consequences of these events or processes?
4. How certain are the answers to the first 3 questions?

The complexity of the TSPA model architecture is shown in fig. 15. (GoldSim is a dynamic, probabilistic simulation software developed by GoldSim Technology Group.)

7. – Conclusions

It was shown that various types of radioactive wastes originating from the nuclear fuel cycle, direct disposal of used fuel elements, or from industry, research and medicine can be treated in a way that the final products are suited for disposal. Depending on the classification of the wastes, different disposal strategies have been or will be implemented in the European countries. For the high-level radioactive wastes consensus exist to dispose of these wastes in deep underground facilities. Due to the different geologic settings of the countries, different host rocks can be selected. In few countries, such as in Germany, several suited host rock formations are available.

Development and demonstration of safe disposal of radioactive wastes is a multi-disciplinary task requiring experiences from nuclear physicists, chemists, geologists, mining engineers and many more scientific disciplines. However, even if a disposal concept sounds scientifically and technically reasonable and safe, it must be accepted by the society and by the population living close to the site.

The safe isolation of the wastes from the environment over the long periods of time can be demonstrated by model calculations and safety analyses. Deterministic and probabilistic procedures for safety analyses are described. Scientists may accept the methodologies, can review the models and databases as well as evaluate the conclusions drawn from the experimental findings on which the data were determined. Of course, different opinions of scientists exist. These different opinions can be disputed on a scientific level. However, it is necessary to communicate the concepts, analyses and results to the society including politicians, the population at the site and action groups. The communication of scientific facts and data is extremely complicated: Breaking facts and figures down to easily understandable statements for the public could be misinterpreted by the scientific community. In the context of safe disposal of nuclear wastes the communication to the public is indispensable and procedures resolving this deficit need to be developed.

REFERENCES

[1] Nickel H., "The spent fuel and waste management concept of German nuclear power plants" *Atomwirtsch.-Atomtech.*, **37** (1992) 368.
[2] Boen R., "L'elaboration de verres nucleaires en creuset froid" *Clefs CEA*, **59** (2010) 17.
[3] Working Panel Waste management of VGB PowerTech e.V., "Waste disposal for nuclear power plants", VGB PowerTech e.V., Essen, 2012.
[4] StrlSchV, "Verordnung über den Schutz vor Schäden durch ionisierende Strahlen" *BGBl I 2001, 1714, (2002, 1459) vom 20. Juli 2001. Geändert durch Art. 2 V v. 18. 6.2002 I 1869*, 2001.
[5] Faybishenko B., Birkholzer J., Sassani D. and Swift P., *International Approaches for Deep Geological Disposal of Nuclear Waste: Geological Challenges in Radioactive Waste Isolation, Fifth Worldwide Review* (Lawrence Berkeley National Laboratory, Sandia National Laboratories) LBNL-1006984, 2016.
[6] DOE U. S., *Accident Investigation Report: Underground Salt Haul Truck Fire at the Waste Isolation Pilot Plant February 5, 2014* (U.S. Department of Energy, Office of Environmental Management) 2014.
[7] DOE U. S., *Accident Investigation Report: Phase 1 Radiological Release Event at the Waste Isolation Pilot Plant on February 14, 2014* (U.S. Department of Energy, Office of Environmental Management) 2014.
[8] Grambow B., "Coupled modelling of the source-term for radionuclide release from nuclear waste forms in the near-field" *Geochimica Et Cosmochimica Acta*, **69** (2005) A408.
[9] Kienzler B., Duro L., Lemmens K., Metz V., Pablo J. d., Valls A., Wegen D. H., Johnson L. and Spahiu K., "Summary of the Euratom Collaborative Project FIRST-Nuclides and Conclusions for the Safety Case" *Nucl. Technol.*, **198** (2017) 260.

[10] Johnson L. H., Garisto N. and Stroes-Gascoyne S., "Used-fuel dissolution sudies in Canada" in *Waste Management 1985 Tucson, Arizona, March 24-28, 1985* (Arizona Board of Regents, Tucson, AZ) 1985, pp. 479–482.

[11] Ferry C., Lovera P., Poinssont C. and Johnson L., "Quantitative assessment of the instant release fraction (IRF) for fission gases and volatile elements as a function of burnup and time under geological disposal conditions", in *Scientific Basis for Nuclear Waste Management XXVII, Kalmar, Sweden, 2003* (Materials Research Society, Warrendale, PA, USA) 2003, pp. 35–40.

[12] Johnson L., Günther-Leopold I., Kobler Waldis J., Linder H. P., Low J., Cui D., Ekeroth E., Spahiu K. and Evins L. Z., "Rapid aqueous release of fission products from high burn-up LWR fuel: Experimental results and correlations with fission gas release" *J. Nucl. Mater.*, **420** (2012) 54.

[13] González-Robles Corrales E., "Study of radionuclide release in commercial UO_2 Spent Nuclear Fuel - Effect of burn-up and high burn-up structure", PhD, Universitat Politècnica de Catalunya, Barcelona, 2011.

[14] Metz V., Geckeis H., González-Robles E., Loida A., Bube C. and Kienzler B., "Radionuclide behaviour in the near-field of a geological repository for spent nuclear fuel" *Radiochim. Acta*, **100** (2012) 699.

[15] Spahiu K., Devoy J., Cui D. Q. and Lundstrom M., "The reduction of U(VI) by near field hydrogen in the presence of UO2(S)" *Radiochim. Acta*, **92** (2004) 597.

[16] Berner U. R., "Evolution of Pore Water Chemistry during Degradation of Cement in a Radioactive Waste Repository Environment" *Waste Management*, **12** (1992) 201.

[17] Kienzler B., Borkel C., Metz V. and Schlieker M., *Long-Term Interactions of Full-Scale Cemented Waste Simulates with Salt Brines* (Karlsruhe Institute of Technology) KIT SR 7721, 2016.

[18] Kienzler B., Metz V., Brendebach B., Finck N., Plaschke M., Rabung T., Rothe J. and Schild D., "Chemical status of U(VI) in cemented waste forms under saline conditions" *Radiochim. Acta*, **98** (2010) 675.

[19] Vandergraaf T. T., Ticknor K. V. and Melnyk T. W., "The selection of a Sorption Data Base for the Geosphere Model in the Canadian Nuclear Fuel Waste Management Program" *J. Contaminant Hydrol.*, **13** (1993) 327.

[20] International Commission on Radiological Protection (ICRP), *Biological and Epidemiological Information on Health Risks Attributable to Ionising Radiation: A Summary of Judgements for the Purposes of Radiological Protection of Humans, Committee 1 Task Group Report: C1 Foundation Document (Annex A of Main Recommendations)* Vol. 12/ 180/06, 2006.

EPJ Web of Conferences **189**, 00015 (2018) https://doi.org/10.1051/epjconf/201818900015
LNES 2017

Fusion: a true challenge for an enormous reward

J. Ongena
Plasma Physics Laboratory, Royal Military Academy
Renaissancelaan 30, 1000 Brussels, Belgium

Summary. — Nuclear physics shows that energy can be released from both fission of heavy nuclei and fusion of light nuclei. Steady progress shows that fusion — an important additional option for energy production in the future — promises to be a clean and safe solution for mankind's long-term energy needs with minimal environmental impact. A source of energy which would be inexhaustible, inherently safe and environmentally friendly, is this not a marvellous prospect? Nuclear fusion, a possible candidate for this role, has been the energy source of our Sun and the stars in the universe for billions of years. This process requires temperatures of tens of millions of degrees, so extremely high and foreign to our daily experience that it seems out of reach. Nevertheless, these extremely high temperatures are routinely realised in several laboratories all over the world, and since the early 1990s, tens of MW fusion power have been released from fusion reactions. We are witnessing the birth of a new technology destined to meet the gigantic future energy needs of mankind with minimal impact on the environment.

1. – Fusion reactions in our Sun and on Earth

The dominant fusion reaction in our Sun converting hydrogen into helium is the p-p fusion chain, the other being the carbon-nitrogen-oxygen or CNO cycle. The CNO cycle is dominant in stars that are about 1.3 times heavier than our Sun; only about 1.7% of the ^{4}He nuclei in our Sun are estimated to originate from the CNO cycle.

At first glance it seems easy to overcome the repulsion of the positively charged nuclei involved in fusion reactions. Why not simply use a particle accelerator to provide the nuclei with sufficient energy and make them collide? Unfortunately, the probability for a fusion reaction is extremely small. Rather than fuse, the nuclei are mostly scattered in all directions and only few fusion reactions take place.

Our Sun offers a more promising solution: its whole mass of gas is at a high temperature enabling a much larger number of fusion reactions to occur per unit of time. Gravity is responsible for confinement and heating of the fusing protons. An estimate of the temperature in the centre of the Sun can be found from the observation that the protons need to be sufficiently fast (*i.e.* have sufficient energy and thus be sufficiently hot) to overcome the compressional forces from gravitation. This means that kinetic energy of a proton in the centre = potential energy from gravity, or

$$\frac{3}{2}kT_{\text{proton in centre}} = \frac{Gm_{\text{proton}}M_{\text{Sun}}}{R_{\text{Sun}}}, \tag{1}$$

$$T_{\text{proton in centre}} = \frac{2Gm_{\text{proton}}M_{\text{Sun}}}{3k\,R_{\text{Sun}}}, \tag{2}$$

with $k = 1.38 \times 10^{-23}$ J/K Boltzmann's constant, $G = 6.67 \times 10^{-11}\ \text{Nm}^2/\text{kg}^2$ the gravitational constant, $M_{\text{Sun}} = 1.989 \times 10^{30}$ kg and $R_{\text{Sun}} = 695\,500$ km $= 6.955 \times 10^8$ m (more than 2 times the distance Earth-Moon ...) we obtain

$$T_{\text{protons in centre}} = 1.56 \times 10^7\ \text{K} = 15.6\ \text{MK}. \tag{3}$$

Fusion of heavier nuclei in stars will occur at higher temperatures. From the formula above, it is immediately clear that a contraction of the star under its own gravity will further raise the central temperature. Fusion in heavy stars will continue to keep the star alive as long as the reactions are exothermic and thus are able to help the star defend itself against a full collapse under its own gravity. The mass defect curve (fig. 1) shows that energy can be gained from fusion reactions up to ^{62}Ni, which is the tightest bound nucleus [1]. Once the star tries to fuse heavier nuclei, gravity finally wins, and a collapse of the star follows. This ends in a spectacular supernova, leading to a neutron star or for the heaviest stars, the formation of a black hole.

The p-p and CNO fusion reactions in our Sun have a much too low reaction rate to be useful on Earth. One of the underlying reasons is that for the formation of a ^{4}He nucleus two protons have to be converted into two neutrons, involving a slow beta-plus decay reaction. Despite this fact, every seconds about 700 million tonnes of H are converted in our Sun into 695 million tonnes of ^{4}He; it thus means that every second about 5 million tonnes of the Sun's mass disappear in the form of radiation [1]!

Thus, on Earth we have to use nuclear fusion reactions that avoid the conversion from protons to neutrons, *i.e.* use as reaction products light atomic nuclei that already contain neutrons, needed for the formation of ^{4}He. Thus, as a substitute for the p-p fusion cycle in the Sun, we need to make use of the isotopes D and T or possibly also

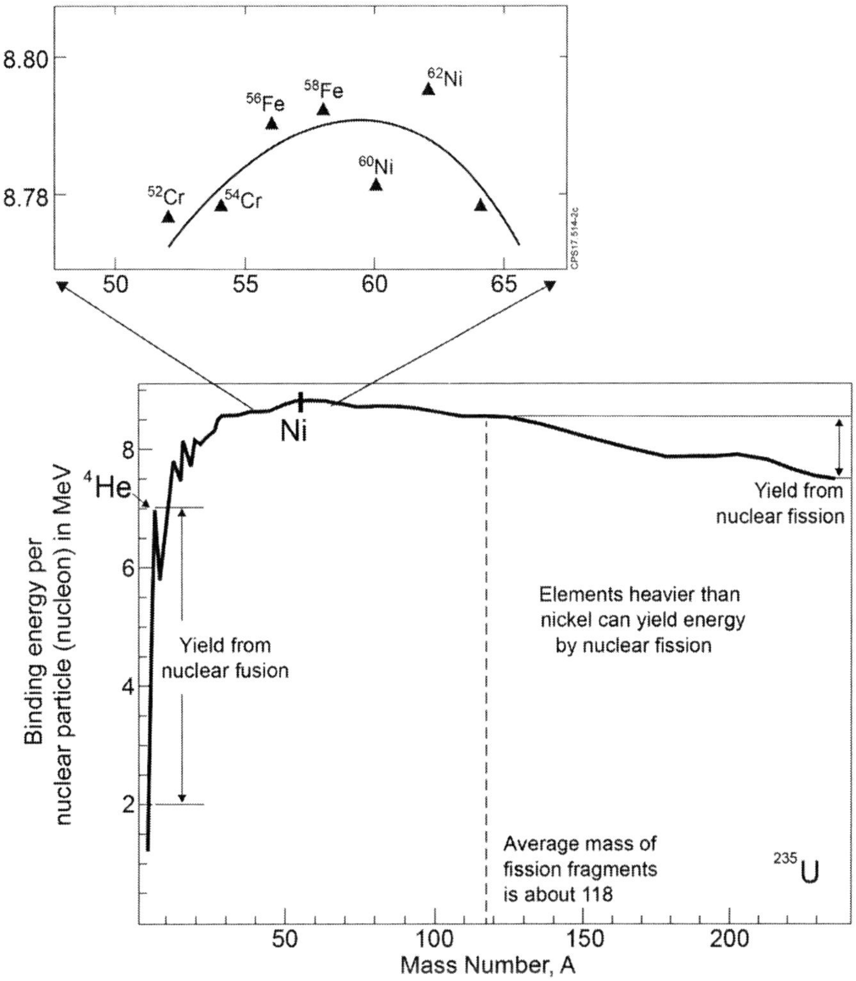

Fig. 1. – Binding energy per nucleon *versus* atomic mass number showing the maximum in the curve reached for ^{62}Ni. Note the exceptionally large binding energy of the ^{4}He nucleus among the light nuclei. An approximate indication is given for the difference in binding energy per nucleon in fusion reactions of light nuclides with ^{4}He as one of the reaction products ($\sim 5\,\mathrm{MeV/nucleon}$) and fission of ^{235}U ($\sim 0.9\,\mathrm{MeV/nucleon}$). The exact numbers for the mass defects of selected nuclides with mass numbers up to 20 can be found in the appendix.

the rare isotope ^{3}He. Here D symbolizes deuterium (the stable isotope of hydrogen with a nucleus consisting of one proton and one neutron) and T is the symbol for tritium (the radioactive hydrogen isotope with a nucleus of 2 neutrons and 1 proton). A number of fusion reactions [2] together with their cross-sections and energy released are listed in table I. Note that the energy released is the largest in reactions where a ^{4}He nucleus is formed because of its exceptionally large binding energy (fig. 1). The fusion reaction

TABLE I. – *Fusion reactions.*

Reaction	σ at 10 keV (barn)	σ_{max} (barn)	Center-of-mass energy (keV) for σ_{max}	Energy released (MeV)
$D + T \to {}^4He + n$	2.72×10^{-2}	5.0	64	17.59
$D + D \to T + p$	2.81×10^{-4}	0.096	1250	4.04
$D + D \to {}^3He + n$	2.78×10^{-4}	0.11	1750	3.27
$T + T \to {}^4He + 2n$	7.90×10^{-4}	0.16	1000	11.33
$D + {}^3He \to {}^4He + p$	2.2×10^{-7}	0.9	250	18.35
$p + {}^6Li \to {}^4He + {}^3He$	6×10^{-10}	0.22	1500	4.02
$p + {}^{11}B \to 3\,{}^4He$	4.6×10^{-17}	1.2	550	8.68
$p + p \to D + e^+ + \nu$	3.6×10^{-26}	–	–	$1.44 + 0.27(\nu)$
$p + {}^{12}C \to {}^{13}N + \gamma$	1.9×10^{-26}	1.0×10^{-4}	400	1.94

that results in the formation of ^{4}He and has the largest cross-section is the so-called D-T reaction:

$$D + T \to {}^4He(3.5\,MeV) + n(14.1\,MeV). \tag{4}$$

In this fusion reaction, "only" a rearrangement of the nuclides is needed to form ^{4}He, which is much easier to realize: the cross-section σ for this reaction at *e.g.* 10 keV is about 10^{24} times larger than that for the p-p reaction.

The reaction products are an α-particle (^{4}He nucleus) and a neutron. Twenty percent of the reaction energy is carried by the α-particles and 80% by the much lighter neutron. In magnetic confinement systems the neutron does not feel the presence of the magnetic field (because it is not charged) and escapes immediately from the reactor volume, while the charged α-particle remains confined by the magnetic field. The kinetic energy of these escaping fast neutrons will be converted into heat in a blanket and then into electricity using conventional technology (steam cycle). About one million times more energy is released from a fusion reaction in comparison with a chemical one (MeV's instead of eV's for the latter). This is the reason why so little fuel can produce so much energy: when burnt in a fusion reactor, the deuterium contained in 1 litre of water (about 33 mg) will produce as much energy as burning 260 litres of gasoline.

Other possible fusion reactions of interest between isotopes of hydrogen and helium are

$$D + D \to {}^3He(0.82\,MeV) + n(2.45\,MeV), \tag{5}$$

$$D + D \to T(1.01\,MeV) + H(3.02\,MeV), \tag{6}$$

$$D + {}^3He \to {}^4He(3.6\,MeV) + H(14.7\,MeV). \tag{7}$$

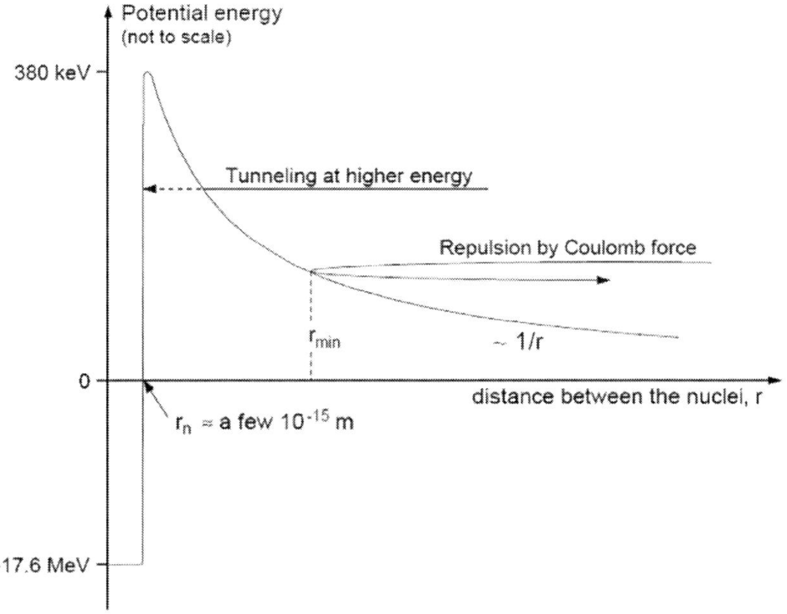

Fig. 2. – Atomic particles can overcome the Coulomb barrier (electrostatic repulsion) at much lower energies than the maximum by "tunnelling" through it. This increases the probability for fusion.

These are more difficult to achieve and have a lower power density than the D-T reaction [3, 4] but show even more benign environmental features. The D-D reaction would eliminate the need for tritium and produce neutrons with lower energies which are therefore easier to absorb and shield. A reactor based on the D-^{3}He reaction would proceed with low neutron production (some neutrons would be produced in competing but much less occurring D-D reactions) and lead to much less induced radioactivity in the reactor structures. However, the prospects for these "advanced" fuels are still too speculative and only the D-T reaction has immediate future prospects. For the interested reader, the appendix provides the masses and binding energies per nucleon for a selection of lighter nuclides, including those involved in reactions (4) to (7).

In order to undergo fusion reactions, the reacting particles have to approach each other to within short distances. A first estimate of the energy necessary to realise fusion reactions can be obtained from the potential energy of one nucleus in the presence of the electric field of another one at a distance such that, in the classical representation, the nuclei just touch each other. The maximum of the Coulomb barrier between two nuclei with radius R_1 and R_2 and charges qZ_1 and qZ_2 can be found from

$$V_C[\mathrm{J}] = \frac{q^2 Z_1 Z_2}{4\pi\varepsilon_0(R_1 + R_2)}, \tag{8}$$

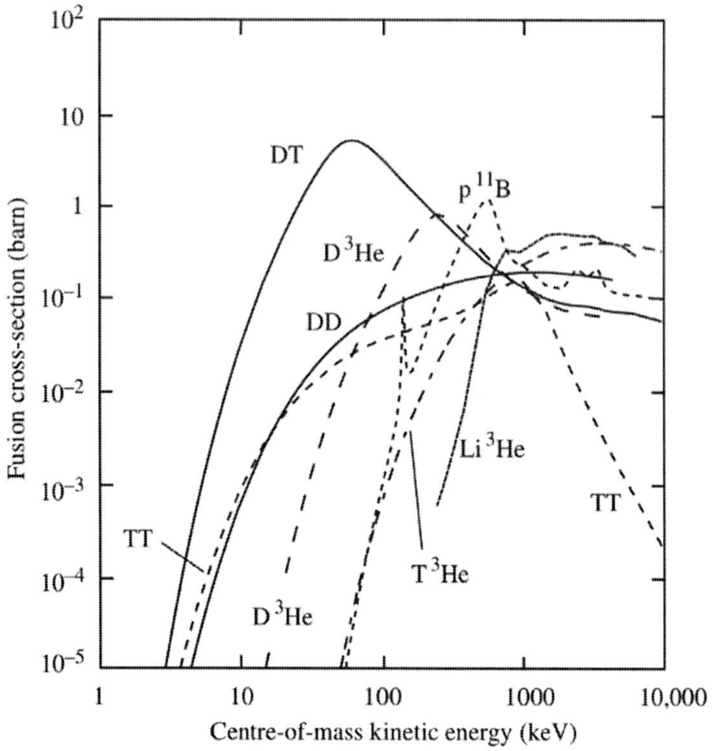

Fig. 3. – Cross-sections (in barn) as a function of the centre-of-mass energies of the reacting particles for various fusion reactions.

or

$$V_C[\mathrm{MeV}] = 1.44 \frac{Z_1 Z_2}{1.4(A_1^{1/3} + A_2^{1/3})}, \tag{9}$$

where we made use of the formula for the radius R of the nuclei [5]:

$$R = 1.4A^{1/3} \times 10^{-15}\,\mathrm{m}. \tag{10}$$

For the reaction between deuterium and tritium we find 0.38 MeV. In the classical picture, nuclei should therefore at least acquire this energy before any fusion reaction can take place. The equivalent temperature of a plasma gas with this value as average energy amounts to more than 4 billion degrees.

A reduction of this huge temperature is made possible by the tunnel effect. Owing to the wave character of matter, nuclei have a finite probability to "tunnel" through the Coulomb barrier and thus can fuse at energies much lower than those given by the maximum of the barrier (fig. 2). As we shall see further, this reduction is substantial, and the optimum temperature for a magnetically confined D-T plasma is about 13 keV.

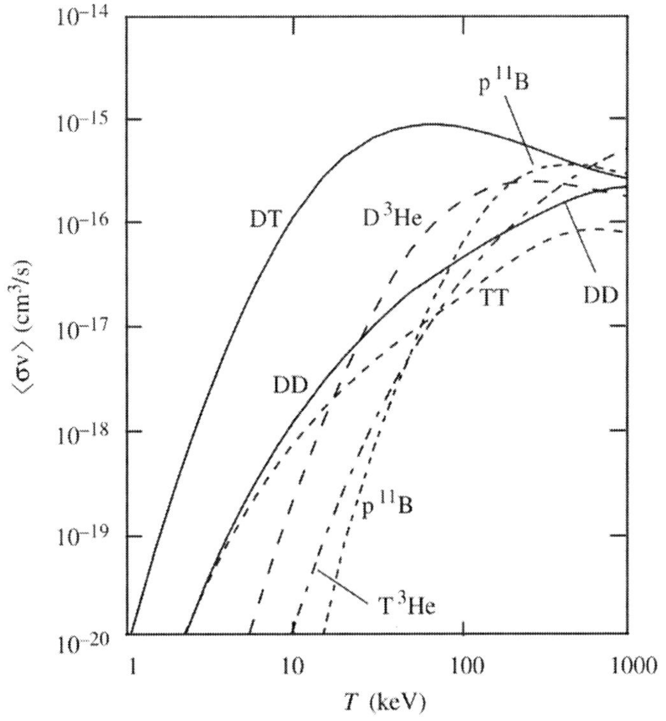

Fig. 4. – Reactivity averaged over a Maxwell-Boltzmann distribution as a function of temperature for the fusion reactions of fig. 3.

Cross-sections [2] as a function of the centre-of-mass energies of the reacting particles for the reactions (4) to (7) above are shown in fig. 3. The largest cross-section is found for the D-T reaction (about 5 barn) at the centre-of-mass energy of the colliding particles $1/2\,\mu(v_1 - v_2)^2$ (where $\mu = 1/2(m_1 m_2/(m_1 + m_2))$ of about 60 keV. However in a gas one has to take into account that the velocities of the colliding particles are described by a Maxwell-Boltzmann distribution and values for the reactivity $\langle \sigma v \rangle$ averaged over this distribution as a function of the temperature are shown in fig. 4 for various fusion reactions. The maximum is now much broader but still around energies of about 60 keV for the D-T reaction. Thus one is inclined to conclude that the plasma has to be heated to temperatures of over 600 million degrees to obtain the maximum fusion rate. However, one has to recall that in a magnetic fusion device, tokamak or stellarator, the operating pressure $p = nkT$ is constant, and that therefore there is an inverse relation between temperature and density. The fusion reaction rate $R = \frac{n^2}{4}\langle \sigma v \rangle$ can thus also be written as $R = \frac{1}{4}\frac{\langle \sigma v \rangle}{T^2}$ and the operating temperature of a reactor will be determined by the maximum of this function [6] as shown in fig. 5. It is clear that for the D-T reaction the peak is located at about 13 keV. For the D-D or D-^{3}He reactions, the temperature is higher or the power density is smaller at the maximum of the $\langle \sigma v \rangle$ curve. For the D-T reaction

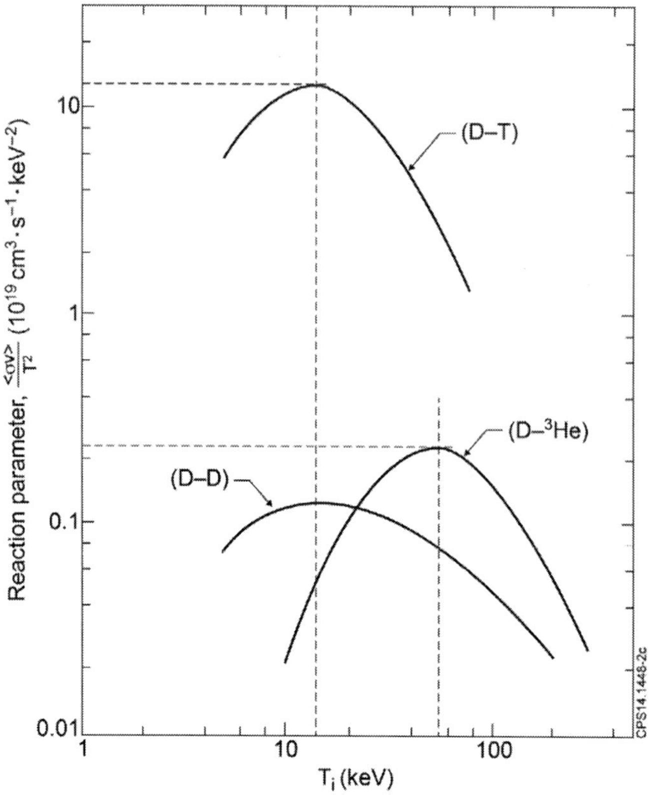

Fig. 5. – Maxwellian averaged values for $\frac{\langle\sigma v\rangle}{T^2}$ as a function of temperature for various fusion reactions. (Adapted from [6].)

we have thus to realize a temperature of about 150 million degrees (the temperature of a Maxwellian gas with an average energy of 1 eV corresponds to 11600 K), which is 10 times larger than in the Sun's core. In other words, the realisation of nuclear fusion on Earth requires the creation of a small but very hot Sun. Two questions then immediately arise: how does one realise such extreme temperatures and — equally important — how can one confine a gas at such high temperatures in a safe way?

2. – Plasma confinement

Two radically different approaches have been followed towards the realisation of nuclear fusion on Earth: inertial and magnetic confinement. In the first approach, lasers or particle beams are used to rapidly compress the nuclear fuel in order to achieve the conditions required for fusion. To this end, a small sphere with a diameter of a few hundred μm is filled with equal amounts of deuterium and tritium. Powerful laser or particle beams irradiate the surface of this little sphere as uniformly as possible, for about a billionth of a second. The outermost layers of the sphere vaporise nearly imme-

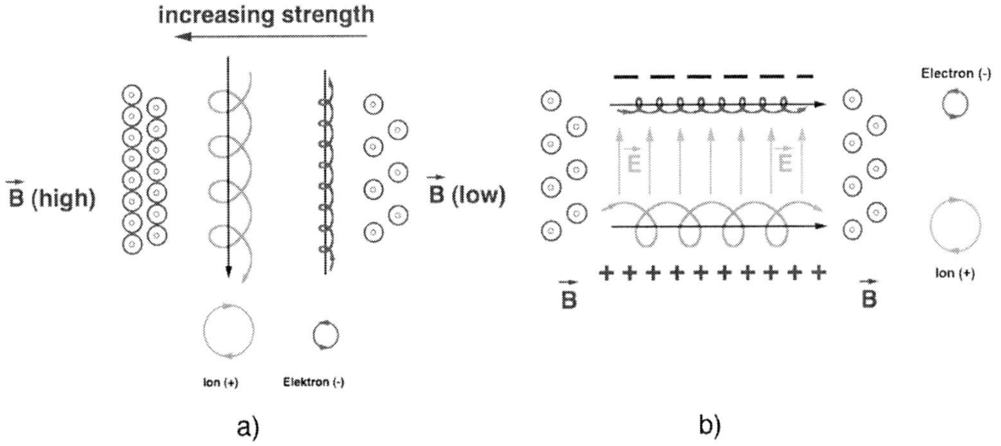

Fig. 6. – Illustration of the effect of the inhomogeneity of the toroidal magnetic field on plasma confinement. The effect can be easily explained in two steps considering the formula for the Larmor radius: $\rho = \frac{mv_\perp}{qB}$. For the first step see a) at the left: formation of the electric field due to the inhomogeneity of the toroidal field. The figure shows a cross-section of the plasma; the orbits of ions and electrons are no longer perfectly circular, but the particles spiral vertically up and down: indeed from the formula follows that ρ decreases as B increases and increases as B decreases. As ions and electrons rotate in opposite directions, they move vertically in opposite directions, creating an electric field in the plasma. Second step, see b) at the right: effect of the created electric field on the plasma particles. We consider a small section of the previous figure, such that B is to a good approximation homogeneous over the section, and thus the effect of E alone can be studied. As B is constant over this section, the only parameter influencing the movement of the particles is the velocity. Positive ions are accelerated towards the negative layer of charges and decelerated towards the positive layer of charges, leading to an increase resp. decrease of the velocity of the particle. From the formula follows that with increasing velocity the Larmor radius decreases and decreases with increasing velocity: thus the particle spirals vertically outwards. A similar reasoning for the electrons leads to exactly the same result. Both electrons and ions move thus horizontally outward: the plasma is unstable. To solve this problem, the formation of the electric field has thus to be avoided. Adding a poloidal component to the toroidal field is a way to do this.

diately, generating an inward-propagating spherical shock wave, which compresses the sphere's contents enormously. At the end of this short compression phase, a density of a thousand times that of water is reached at a temperature of some tens of millions of degrees centigrade. Under these conditions, the deuterium and tritium nuclei start to fuse. The energy released by these first fusion reactions further heats the rest of the strongly compressed fuel, allowing fusion reactions to spread throughout the sphere.

The second approach is by magnetic confinement. Since this has proven to be the most successful method to date, magnetic fusion research is under development in laboratories all over Europe and throughout the world.

At the high temperatures necessary for nuclear fusion, atoms break up into their constituents, electrons and nuclei, and the fuel becomes fully ionised, a plasma. To

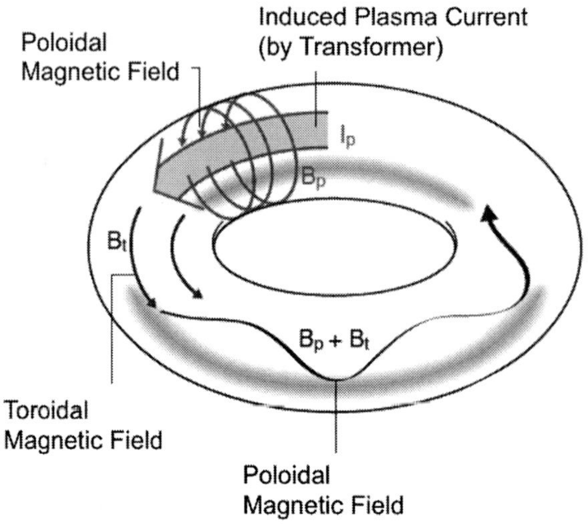

Fig. 7. – Principle of the tokamak: A large current induced in the plasma ring generates a poloidal magnetic field. The combination of this poloidal field with the main toroidal field results in a helical magnetic field, necessary for a stable confinement of the plasma in a toroidal configuration.

confine a hot plasma one can make use of magnetic fields, since charged particles will follow a helical path around the field lines owing to the Lorentz force. Possible movements perpendicular to the field are in this way highly restricted.

The most obvious way to realise this is a purely cylindrical configuration, which, however, suffers from plasma losses at both ends. These losses can be avoided by bending the magnetic field lines onto themselves, leading to a doughnut-shaped or toroidal magnetic field configuration. A toroidal magnetic field, however, is not homogeneous. Its curvature destroys most of the nice confinement properties of a purely cylindrical configuration: the magnetic field is largest near the axis of the torus, and decreases towards the outboard side. As a result, ions and electrons move in opposite vertical directions (fig. 6), giving rise to an electric field which, when combined with the toroidal magnetic field, expels all particles to the outside, irrespective of their charge. Without special measures, it is thus impossible to confine a toroidal plasma in a stable way. However, if the magnetic field can be given an extra twist, vertical drifts can be cancelled and a suitable trap can be made for fusion plasmas. This extra twist can be produced by means of electrical currents flowing inside or outside of the plasma ring. The *tokamak* concept relies on currents produced inside the plasma, whereas the *stellarator* concept relies on currents external to the plasma.

In a tokamak, a set of coils placed around the doughnut-shaped plasma chamber produces the main toroidal magnetic field. The electrical current necessary for the stability of the plasma flows in the plasma ring, which serves as the (only) secondary winding of an enormous transformer. The induced current in the plasma ring generates a poloidal

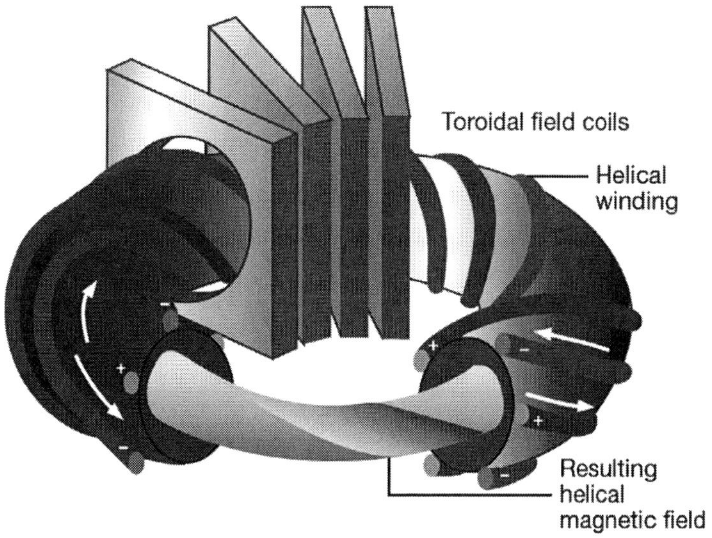

Fig. 8. – In a stellarator, helical coils external to the plasma ring create a supplementary magnetic field that, in combination with the main toroidal magnetic field, confines the plasma.

magnetic field. The combination of this poloidal field with the main toroidal field results in a helical magnetic field (fig. 7). The magnetic structure thus generated consists of an infinite set of nested toroidal magnetic surfaces, each with a slightly different twist, reducing further the leakage of particles and heat from the plasma. On each of these surfaces, the plasma pressure is constant. Tokamak research started in the Kurchatov Institute in Moscow under the leadership of Lev Artsimovich. The success of this configuration became at once clear at the Novosibirsk IAEA conference of 1968 where he announced that much higher temperatures could be obtained in a tokamak than all other magnetic configurations under study at that time.

The tokamak is however a pulsed device, since the transformer needs a steadily increasing current in its primary winding to induce the necessary plasma current.

Continuous operation of a fusion device can be obtained if the need for at transformer to induce the plasma current could be avoided. The stellarator is such a device, and it relies on currents external to the plasma to create the needed helical magnetic configuration. In its basic configuration, extra helical coils around the toroidal plasma provide the necessary additional twist to the toroidal magnetic field generated by the main field coils (fig. 8). These helical windings around the plasma ring, however, complicate the construction. In addition such "classical" stellarator configurations lack good confinement properties. Modern stellarators have optimised confinement properties, and are equipped with a set of coils with a complex shape that is determined numerically (fig. 9). Several devices of the stellarator type are in operation or in construction at this moment all over the world. The largest stellarator in the world that recently started operations (2016) is Wendelstein 7-X, located in Greifswald, Germany (fig. 10). The first experimental cam-

Fig. 9. – An illustration of the complex shapes of the coils of a modern stellarator, which combine the functions of the separate coil systems of fig. 8.

Fig. 10. – A picture of the stellarator Wendelstein 7-X, at the time of construction in November 2011, and now in operation since 2016 in Greifswald, Germany.

paign showed encouraging results [7]. The device will be further upgraded in the coming years and in its final configuration will be able to run plasma pulses of 30 minutes. The largest stellarator-like device outside Europe is the LHD (Large Helical Device) of the National Institute for Fusion Science (NIFS, near Nagoya, Japan) that started operations in March 1998. Also this device has led to several rather important contributions [8].

3. – Plasma heating

The plasma current, needed to create the helical magnetic field providing stable confinement at the same time heats the plasma. Since the heating results from the finite resistance of the plasma, just as in an electrical heating element, this is referred to as ohmic heating. The plasma resistance, however, decreases with increasing plasma temperature, and at some millions of degrees, the efficiency of ohmic heating becomes too low to be useful for further heating. In fact, at these temperatures the plasma becomes a better conductor than copper at room temperature. That is why auxiliary heating methods have to be used to reach the temperatures required for fusion.

One additional heating method consists of injecting energetic particle beams of neutral hydrogen or deuterium into the plasma. This is possible because only neutral particles can penetrate the magnetic fields needed to confine the plasma. To this end, ions are created in a plasma source, accelerated with voltages up to 150 000 volts and sent through a cloud of neutral gas. The accelerated ions "steal" electrons from atoms in this neutral gas cloud, and become energetic neutral particles, which in turn then enter the hot tokamak plasma without impediment from the confining magnetic fields of the fusion device. Once in the hot plasma, they are immediately ionised again, thus confined, and deposit their energy via collisions to the rest of the plasma particles. Powers of up to several million watts per neutral injector can be delivered to the plasma using this technique.

A second heating method sends electromagnetic waves into the plasma, by antennas or wave guides at the plasma edge. The principle underlying this method is similar to that of a microwave oven: the energy from the waves is most easily absorbed if the frequency used is equal to a "natural" frequency of the particles to be heated. Using the cyclotron frequency with which the charged plasma particles gyrate around the magnetic field lines, is an evident choice. Thus, two different heating systems exist: Ion Cyclotron Resonance Heating (ICRH) and Electron Cyclotron Resonance Heating (ECRH), depending on whether ions or electrons are heated. Ion cyclotron frequencies are in the MHz range (20 MHz and upwards), while electron cyclotron frequencies are approximately a 1000 times higher (at frequencies in the range of 100–200 GHz), due to the smaller mass of the electrons. Heating powers for high-frequency systems range from 100 kW to several tens of MW.

4. – Characterizing the fusion reactivity of plasmas

The power amplification factor Q is defined as the ratio of the power produced by fusion reactions to the total heating power supplied externally. Two important milestones for the value of Q are customarily used in fusion research. The first, *breakeven*, is reached when the heating power is equal to the power produced from fusion reactions, corresponding to a Q value of one. The second, *ignition*, is reached when the additional heating systems can be switched off and the heat of the fusion reactions alone is sufficient to maintain the high temperatures needed for fusion. This corresponds to an infinite value for Q.

5. – Progress in recent years

The progress obtained in fusion research in the past three decades is clearly visible from the successively obtained values for the fusion product. Compared to the first fusion experiments, this value has increased by a factor of more than 10 000! The results obtained on the largest tokamak in the world, JET (Joint European Torus, located at Culham Labs, close to Oxford, UK), have contributed significantly to this progress, and to date, values close to breakeven have been obtained in D-T experiments on this device. An important milestone was reached in 1991 with the first production of MWs of fusion power from controlled fusion reactions, as tritium was used for the first time as fuel in a tokamak [9]. These experiments, obtained with a mixture of 90% D and 10% T, generated fusion powers in the megawatt range for nearly 2 seconds, with a maximum of about 1.7 MW, corresponding to a Q value of about 0.15.

Further successes were obtained early in 1994 in the American tokamak TFTR (Tokamak Fusion Test Reactor) at the Plasma Physics Laboratory of Princeton University. In plasmas consisting of a mixture of 50% deuterium and 50% tritium, multi-megawatt level fusion powers were generated for about one second, with a maximum of 6.3 MW [10]. Plasma temperatures in excess of 300 million degrees were reached in the plasma centre, 20 to 30 times hotter than in the centre of the Sun! In November 1994, fusion powers of more than 10 MW were generated, corresponding to Q values of about 0.27 [11].

Other record values were reached in 1996 on the Japanese Tokamak JT-60U of the Naka Fusion Research Establishment, located 150 km north of Tokyo, a division of the Japanese Atomic Energy Research Institute (JAERI). This machine demonstrated temperatures in excess of 520 million degrees [12], the highest temperature ever realised by man on Earth. Even more important, a record value for the fusion triple product was obtained in pure deuterium plasmas. If the same conditions had been realised in deuterium-tritium plasmas, it would have resulted in a Q value of about 1.25, *i.e.* conditions better than breakeven [13].

The most impressive results in fusion research up to now were obtained in the JET, in October and November 1997. Experiments in 50% D-50% T plasmas resulted in over 16 MW of fusion power during about 1 second, with Q values in excess of 0.65 [14]. These are the highest fusion powers and Q values ever reached in D-T reactions, thereby effectively resulting in the first demonstration of breakeven in reactor grade D-T fusion plasmas. A quasi steady-state generation of fusion power has also been demonstrated: over 4 MW of fusion power were produced for time intervals of more than 5 seconds [15], a duration only limited by the actual technical constraints of JET.

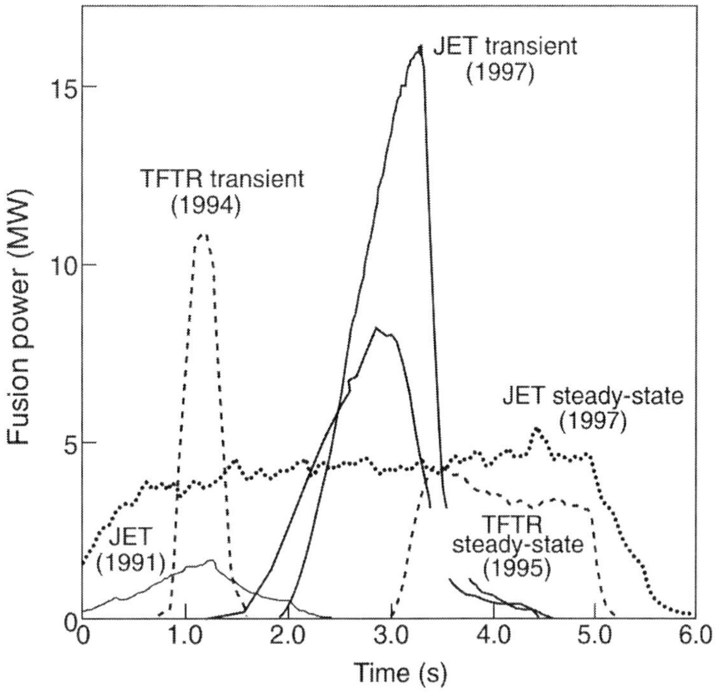

Fig. 11. – Time traces of the fusion power released in different high-performance deuterium-tritium fusion experiments in TFTR and JET. (Adapted from [15].)

A summary of the different high performance D-T pulses obtained on JET and TFTR, as discussed above, is presented in fig. 11.

The success of the D-T experiments on JET and TFTR proves the scientific feasibility of controlled nuclear fusion. In addition, those experiments also allowed to get a first glimpse of the effects on the plasma of fast alpha particles.

6. – Conditions for a fusion reactor: Lawson criterion

The Lawson criterion determines conditions for an operating reactor. In its simplest form it can be defined as follows: the sum of the power densities from auxiliary heating P_{heat} and α particles has to be at least equal to the power densities lost by radiation $P_{\mathrm{radiation}}$ and transport, $P_{\mathrm{transport}}$ (convection, conduction). This can be written as

$$P_{\mathrm{heat}} + P_{\alpha} \geq P_{\mathrm{transport}} + P_{\mathrm{radiation}}. \tag{11}$$

To simplify the calculations, we consider in what follows a pure D-T plasma without impurities (*i.e.* effective charge of the plasma $Z_{\mathrm{eff}} = 1$); this implies also that we neglect the presence of He ash from the reaction. We assume only Bremsstrahlung losses for $P_{\mathrm{radiation}}$ (given by the formula P_{Brems} (in W/m^3) $= C_B T^{1/2} n^2$ if T is expressed in keV, and n in m^{-3} with $C_B = 5.35 \times 10^{-37}\,\mathrm{W\,m^3\,keV^{-1/2}}$ [16]) and we characterize all transport-related energy losses by the term $P_{\mathrm{transport}}$, making use of the confinement time τ_{E}, defined by the relation $\mathrm{d}W_{\mathrm{p}}/\mathrm{d}t = W_{\mathrm{p}}/\tau_{\mathrm{E}}$, where W_{p} is the energy density in the plasma. We also take into account that

$$P_{\mathrm{heat}} Q = P_{\mathrm{fusion}} = 5P_\alpha. \quad (12)$$

And that the total plasma energy density is the sum of the ion and electron energy densities:

$$W_p = \frac{3}{2}(n_e k T_e + n_i k T_i) \quad (13)$$

with T_i and T_e in K, n_e and n_i in m^{-3}, and $k = 1.38 \times 10^{-23}$ J/K the Boltzmann constant.

In a neutral D-T plasma electron and ion densities are the same $n_e = n_i = n$; we also assume that the plasma is in thermal equilibrium *i.e.* there is no difference between electron and ion temperatures $T_e = T_i = T$. Therefore

$$W_p = 3nkT. \quad (14)$$

Assuming a pure 50% D-50% T plasma leads also to $n_T = n_D = 1/2n_e = 1/2n$. The expression for the power density delivered to the plasma by the α particles, can thus be written as

$$P_\alpha = \frac{1}{4} n_D n_T E_\alpha \langle \sigma v \rangle = \frac{1}{4} n^2 E_\alpha \langle \sigma v \rangle. \quad (15)$$

Expressing E_α and T_k in keV, (11) can be written as

$$\frac{1}{4} n^2 E_\alpha \langle \sigma v \rangle \left(\frac{Q+5}{Q} \right) \geq \frac{3nT_k}{\tau_E} + C_B^* Z_{\mathrm{eff}} T_k^{1/2} n^2. \quad (16)$$

And finally

$$n\tau_E \geq \frac{3T_k}{\frac{\langle \sigma v \rangle E_\alpha (Q+5)}{4Q} - C_B^* Z_{\mathrm{eff}} T_k^{1/2}}, \quad (17)$$

where $C_B^* = C_B/1.602 \times 10^{-16} = 3.34 \times 10^{-21}\,\mathrm{W\,m^3\,keV^{-1/2}}$.

The right-hand side formula shows a minimum for about $T_k = 25\,\mathrm{keV}$. Approximating $\langle \sigma v \rangle$ [in $\mathrm{m^{-3}\,s}$] $= 1.1 \times 10^{-24} T_k^2$ (valid to within 15% in the interval $10\,\mathrm{keV} < T_k < 25\,\mathrm{keV}$) and setting $Z_{\mathrm{eff}} = 1$ (pure D-T plasma) we find

$$\text{for breakeven } (Q = 1):\ n\tau_E \geq 2.5 \times 10^{19}\,\mathrm{m^{-3}\,s} \quad (18)$$

and

$$\text{for ignition } (Q = \infty)\text{: } n\tau_E \geq 1.5 \times 10^{20}\,\mathrm{m}^{-3}\,\mathrm{s}. \tag{19}$$

An alternative formulation of the Lawson criterion is in the form of the so-called fusion triple product. Considering again only the temperature range where the minimum of eq. (17) occurs ($kT_e \sim 10$–$25\,\mathrm{keV}$), we can neglect to a first approximation the corrections for the Bremsstrahlung losses. Thus we find, with the same conventions for the various physical quantities as in eq. (17):

$$n\tau_E T_k \geq \frac{12QT_k^2}{(Q+5)\langle\sigma v\rangle E_\alpha}\,. \tag{20}$$

The minimum for the right-hand side is obtained at the maximum for $\langle\sigma v\rangle/T_k^2$, as shown above, about $T_k = 13\,\mathrm{keV}$. We thus find

$$\text{for breakeven } (Q = 1)\text{: } n\tau_E T_k \geq 5 \times 10^{20}\,\mathrm{m}^{-3}\,\mathrm{s\,keV} \tag{21}$$

and

$$\text{for ignition } (Q = \infty)\text{: } n\tau_E T_k \geq 3 \times 10^{21}\,\mathrm{m}^{-3}\,\mathrm{s\,keV}. \tag{22}$$

These simplified expressions should give a good idea of the physics behind the famous Lawson criterion. For a more detailed treatment, taking into account radiation, impurities etc. we refer to the interesting paper by Rebhan and Van Oost [17].

7. – Future prospects: JET as a testbed for ITER, JT-60SA, ITER and DEMO

7·1. *JET as a testbed for ITER.* – JET is currently focused on providing experimental evidence to questions raised by the ITER team. Because of its worldwide unique capabilities (able to operate with both beryllium as wall material and tritium) and its large size, JET is ideally placed to do so. JET's "Programme in Support of ITER" has been elaborated along three main axes: i) experimentation with an ITER-like wall (combination of W and Be as a first wall); ii) development of plasma configurations and parameters towards the most ITER-relevant conditions achievable today and iii) integrated experimentation in deuterium-tritium. The JET auxiliary heating systems are upgraded with currently 45 MW of installed heating power, allowing access to ITER-relevant operational conditions. This will provide access to conditions of melt layer formation both on the beryllium first wall and the tungsten divertor and, help to make further progress in hybrid and advanced scenarios for ITER, which require full or partial current profile control, thereby making use of new dedicated diagnostics. Results from the first years of

TABLE II. – *Main parameters of important tokamaks.*

	T-10	DIII-D	JT-60SA	JET	ITER
Land/Organisation	Russian Federation	USA	Japan	GB/ EURATOM	International
Plasma shape	circular	elliptical (D)	elliptical (D)	elliptical (D)	elliptical (D)
Minor radius (m)	0.3	0.67 (hor) 1.74 (vert)	1.18 (hor) 2.30 (vert)	1.25 (hor) 2.1 (vert)	2.0 (hor) 3.7 (vert)
Major Radius (m)	1.5	1.67	2.96	2.96	6.2
Toroidal Magnetic field (T)	2.5	2.2	2.25	3.5	5.3
Plasma Current (MA)	0.7	3.0	5.5	5.0	15 (17)
Pulse length (s)	4	10	100	60	300-500
Injection of Neutral particle beams (MW)	–	20	34	30	73 (130) in total
Injection of electromagnetic waves (MW)	1.5	8	7	38	

operations with the ITER like wall are summarized in [18]. Preparations for a new D-T campaign are currently underway in JET.

7·2. *JT-60SA.* – In February 2007, Euratom and Japan signed the Broader Approach agreement. This aims to complement the ITER Project and to accelerate the realisation of fusion energy by carrying out R&D and developing some advanced technologies for future demonstration fusion power reactors (DEMO). Within the Broader Approach, three main projects are being implemented: i) the construction in Japan of the large superconducting tokamak JT-60SA aimed at developing operational scenarios relevant for ITER (International Thermonuclear Experimental Reactor) and DEMO; ii) the construction of facilities to test components to be adopted in the research machine IFMIF (International Fusion Material Irradiation Facility) aimed to study the material damage caused by high-energy neutron flux generated in fusion reactions, the construction of a prototype of the liquid lithium IFMIF target; iii) the creation of the International Fusion Energy Research Centre (IFERC) including a centre for advanced materials development and a supercomputing centre [19].

Parameters for JT-60SA are summarized in table II. JT-60SA is a fusion experiment designed to support the operation of ITER and to investigate how best to optimise the operation of fusion power plants that are built after ITER. It is a joint international research and development project involving Japan and Europe, and is to be built in Naka, Japan using infrastructure of the former JT-60 Upgrade experiment, for which we cited several important experimental results in sect. **5** above. SA stands for "super, advanced", since the experiment will have superconducting coils and study advanced modes of plasma operation. According to the current planning, first plasmas on JT-60SA are foreseen in 2019. More information on this interesting future tokamak can be found in [20].

7·3. *ITER*. – The results summarized in sect. **5** provide crucial information for the design of a next large tokamak, aimed at demonstrating the technical feasibility for large-scale energy production. This next step is ITER, short for International Thermonuclear Experimental Reactor, currently under construction in Cadarache, France as a combined effort between Europe, Japan, the Russian Federation, South Korea, India, China and the United States. This device will thus for the first time in history allow mankind to produce hundreds of MW of fusion power from nuclear fusion reactions at temperatures over 100 million degrees. ITER is expected to generate fusion powers of the order of 500 MW in pulses of 300–500 seconds. Specifications for ITER (and a few other major tokamaks) are summarised in table II. After ITER, the construction of a demonstration reactor, currently termed DEMO, is foreseen, which should show not only the technical, but also the economical feasibility of fusion.

7·4. *DEMO*. – Before DEMO can become a practical demonstration of electricity generation on a power-plant scale that satisfies various socio-economic goals, major challenges must be resolved. Although ITER will bring significant advances, there remains a large gap between ITER and DEMO. Main differences between ITER and DEMO are summarized in table III.

DEMO is currently based on the tokamak, as this is the most advanced fusion concept to date. Reactor studies are also being developed for Helical Devices (see, *e.g.*, [21-24]). However, a decision on a next step stellarator/helical device can only take place when the main results of the current large helical devices in operation or construction have been obtained.

To lead a coordinated effort in the EU towards DEMO, the Power Plant Physics and Technology Department (PPP&T) has been established in 2011 [25]. The aims of the DEMO studies in Europe are: i) to quantify key physics and technology prerequisites for DEMO; ii) to identify the most urgent technical issues that need to be solved in physics and technology and iii) to plan and implement supporting physics and technology R&D.

Two DEMO design options are currently being investigated by PPP&T. (See table IV for main characteristics):

i) DEMO Model 1: A "conservative baseline design" that could be delivered in the short to medium term, based on the expected performance of ITER with reasonable

TABLE III. – *Main differences between ITER and DEMO.*

ITER	DEMO
Experimental Device	Close to commercial plant
400s pulses Long interpulse time	Long pulses, high duty cycle or steady state
Many diagnostics	Minimum set of diagnostics only needed for operations
Many H&CD systems	Reduced set of H&CD systems
No T breeding required	Self-sufficient T breeding
316 SS structural material	Reduced activation structural material
Low n-fluence, low dpa Low material damage	High n-fluence, high dpa Significant material damage

TABLE IV. – *Main parameters of the early DEMO 1 and more advanced DEMO 2 model currently under investigation by the PPP&T Department of EFDA. Shown are the thermal output power (P_{th}), the net electrical power to the grid (P_{net}), the recirculating power (P_{rec}), the auxiliary heating power (P_{aux}), major radius (R_0) and minor radius (a) of the device, plasma current (I_p) and toroidal magnetic field on axis (B_t).*

Device Operation Mode	DEMO 1 Pulsed	DEMO 2 Steady State
P_{th} (MW)	2200	2700
P_{net} (MW)	500	500
P_{rec} (MW)	594	600
P_{aux} (MW)	50	350
R_0 (m)	9.0	8.15
a (m)	2.25	3.0
I_p (MA)	14.1	19.8
B_t (T) on axis	6.8	5.0

improvements in science and technology *i.e.* a large, modest power density, long-pulse inductively supported plasma in a conventional plasma scenario.

ii) DEMO Model 2: an "optimistic design" based upon more advanced assumptions which are at the upper limit of what may be achieved, leading to a steady state plasma scenario where a large fraction of the plasma current is induced non-inductively, *i.e.* without making use of the transformer. This is currently a rather speculative option.

From table III there follows immediately a minimum set of research topics that need to be (much) further developed: heating systems capable of heating the large plasmas foreseen for DEMO, diagnostic systems that can sustain the heavy neutron loading, tritium-generating modules (so-called tritium blankets) to foresee tritium self-sufficiency, and new materials that will be compatible with the high heat and neutron loads. This implies also that new laboratories will have to be built to be able to explore possibilities and further develop existing systems. Among others, a material research facility to study the behaviour of various candidate materials under a flux of 14.1 MeV neutrons (preparations for such a facility, IFMIF, are already underway in Japan and Europe [26]), a facility to develop, *e.g.*, Neutral Beam Heating systems (if these are viable at all for DEMO) with acceleration energies between 1 and 2 MeV, based on negative ion sources (in contrast to the positive ion sources used currently) etc. A first step in this direction is the PRIMA facility being built in Padova, Italy for the development of the ITER NBI systems [27]. But more efforts in various other fields will have to follow, and this will unavoidably take time.

The European Fusion Development Agreement (EFDA) has released in November 2012 a roadmap for the realization of fusion electricity to the grid by 2050 [28]. This roadmap covers three periods: i) the upcoming European Research Framework Programme, Horizon 2020, ii) the years 2021-2030 and iii) the period 2031-2050. ITER is the key facility of the roadmap as it is expected to achieve most of the important milestones on the path to fusion power. The second period is focused on maximizing ITER exploitation and on preparing the construction of DEMO. Building and operating DEMO is the subject of the last roadmap phase (time horizon about 2050).

8. – Will fusion be a safe, clean and inexhaustible energy source?

The concept of a future fusion power device based on a tokamak is outlined in fig. 12. Due to the high reactivity of the D-T fusion reaction, deuterium and tritium will most likely be selected as fuel for the first generation of fusion power stations. The electrically charged α-particles from fusion reactions are confined by the magnetic field. These high-energy helium nuclei collide with the background plasma particles, *i.e.* with the D and T ions in the plasma and maintain the temperature rendering additional heating systems quasi unnecessary as soon as ignition is reached.

In contrast to the α-particles, the neutrons generated by the D-T reaction have no electrical charge and escape immediately from the confining magnetic field in the plasma chamber. These neutrons are not lost, but serve a dual purpose in the reactor. First,

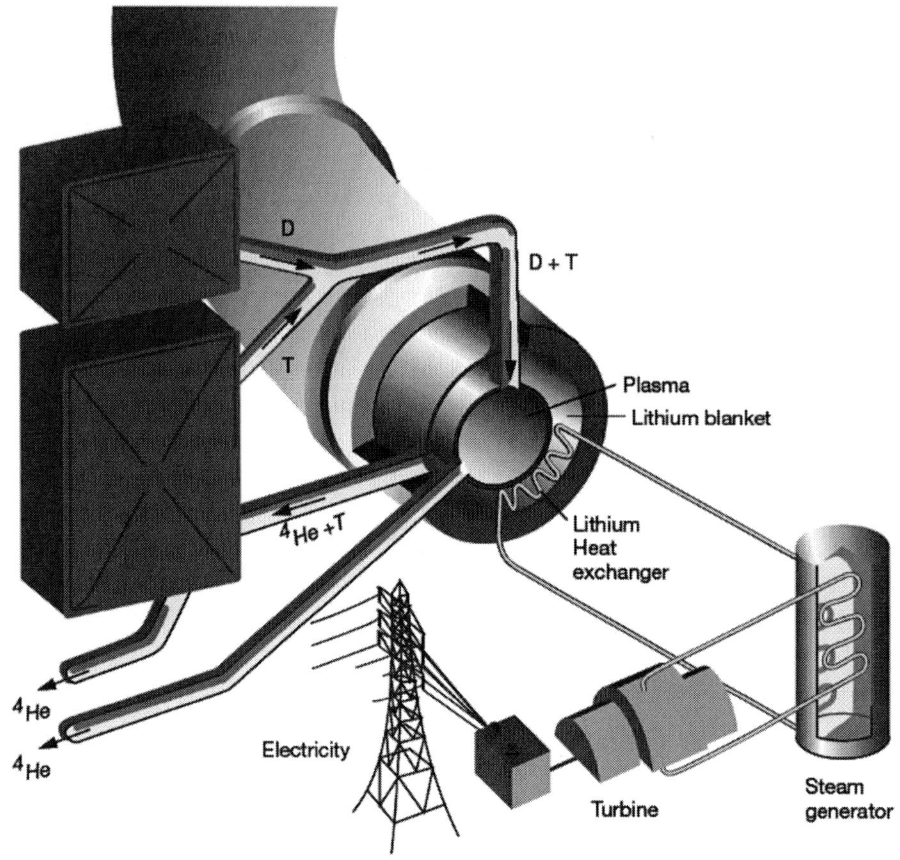

Fig. 12. – Many obstacles still have to be overcome before the first nuclear fusion power plant will be put into operation. Eventually, heat generated in fusion reactions will be used to generate steam to drive a turbine, with electrical power as final result.

they produce tritium in a blanket surrounding the reactor containing lithium or lithium compounds, according to the reactions given in table V. This is necessary to refuel the plasma with tritium, consumed in fusion reactions. Second, the energy of the neutrons deposited in the blanket is converted into heat, which is in turn is transported to the exterior by the cooling system and used to generate steam and drive classical turbines for the production of electricity.

The main advantage of fusion is the absence of any long-lived radioactive waste and its inherent safety. The quantity of fuel in the reaction chamber at any moment is only a few grams, in contrast to the presence of several tons of fuel in the core of a fission reactor. Furthermore, the fusion process is not susceptible to an uncontrolled release of energy, as it is in essence a “gas burner” and thus can be halted in a very short time by cutting the supply of fusion fuel. There is thus no risk of a runaway reaction, and Chernobyl-like accidents are absolutely excluded.

TABLE V. – *Tritium production from Lithium.*

$^{7}Li + n \rightarrow {}^{4}He + T + n - 2.47\,MeV$
$^{6}Li + n \rightarrow {}^{4}He(2.05\,MeV) + T\ (2.73\,MeV)$

As previously mentioned, the fuels for the most accessible fusion reactions are deuterium and tritium. Deuterium can be extracted in large quantities from ordinary water found in rivers and oceans, rendering each country less dependent on fuel supply from other parts of the planet. Moreover, a minimal amount of fuel is necessary to cover the lifetime electricity needs of a European citizen: about 10 g of D and 15 g of T (and about twice this amount for a citizen of the United States). As explained above, tritium will be produced *in situ* from the irradiation of Li with the neutrons from the fusion reaction. Thus the real combustibles are in fact D and Li, both of which are non-radioactive. World reserves of Li are sufficient for many thousand years of operation of D-T fusion power plants. Studies are underway to also extract it from seawater [29]. If at a later stage the D-D reaction could be used (requiring, however, even much higher plasma temperatures), we could have at our disposal a virtually unlimited energy source, as reserves of D are sufficient for several million years of energy production. In both cases, the reaction product is helium, which is chemically inert and non-radioactive. With nuclear fusion there is thus no release of gases which contribute to the greenhouse effect, acid rain or the depletion of the ozone layer. Moreover, helium is the only gas still liquid at temperatures down to absolute zero, thus offering the only possibility for cooling low temperature superconductors (as *e.g.* in modern MRI devices). It is thus not a "waste product" but in fact is essential for all applications involving ultra-low temperatures. This is especially important since helium is rather scarce on Earth.

The high-energy neutrons will activate the metallic reactor structure and hence make it radioactive. Low activation materials have to be selected to minimise this effect. In this way, the total radioactive inventory of a fusion reactor could be made one million times smaller than that of existing fission reactors. The activation problem could be drastically reduced in a distant future by the use of other fusion reactions, which liberate fewer neutrons. The radioactivity induced by the D-^{3}He reaction for example, would be about 50 times lower.

The various steps that remain to be taken by thermonuclear fusion research indicate that some decades will still be needed to realise a practically useable energy source. The most recent results support the attractiveness of the fusion concept for energy production. A sustained effort — in laboratories worldwide — brings the vision of an energy supply by thermonuclear fusion ever closer to realisation. We all know that fusion is a challenging undertaking and that patience will be needed, but it is more than worth the effort given the difficulties we are facing in the future with our current energy supply and its suspected influence on climate. It will evidently be up to you, young researchers, to tackle these interesting problems. If successful, this will be your important contribution to the benefit of all people on Earth. A true challenge for an immense reward!

Appendix A.

Atomic mass and binding energy per nucleon [30], stability and natural abundance for light nuclides. Note the remarkable absence of stable nuclides with mass number 5 and 8.

Nucleus	Stable/ Unstable (half-life)	Natural Abundance	Binding Energy per nucleon (keV)	Atomic Mass (u)
^{1}H	s	99.985%	n.a	1.0078250
n	u (12 min)	n.a.	n.a	1.0086649
^{2}H = D	s	0.015%	1112.283	2.0141018
^{3}H = T	u (12.32 yr)	n.a.	2827.265	3.0160493
^{3}He	s	0.000137%	2572.680	3.0160293
^{4}He	s	99.999863%	7073.915	4.0026033
^{6}Li	s	92.41%	5332.331	6.0151229
^{7}Li	s	7.59%	5606.439	7.0160034
^{9}Be	s	100%	6462.668	9.0121831
^{10}B	s	19.8%	6475.083	10.0129369
^{11}B	s	80.2%	6927.732	11.0093052
^{12}C	s	98.89%	7680.144	12.0000000
^{13}C	s	1.11%	7469.849	13.0033548
^{14}N	s	99.634%	7475.614	14.0030740
^{15}N	s	0.366%	7699.460	15.0001089
^{16}O	s	99.762%	7976.206	15.9949146
^{17}O	s	0.038%	7750.728	16.9991318
^{18}O	s	0.2%	7767.097	17.9991596
^{19}F	s	100%	7779.018	18.9984032
^{20}Ne	s	90.48%	8032.240	19.9924402
^{21}Ne	s	0.27%	7971.713	20.9938467
^{22}Ne	s	9.25%	8080.465	21.9913851

Further Reading

A recent overview of the status of fusion research (magnetic and inertial) can be found in the "Insight Section" (66 pages) of the May 2016 volume of *Nature Physics* (`http://www.nature.com/nphys/insight/nuclear-fusion/index.html`).

REFERENCES

[1] LANG and KENNETH R., in *The Sun from space*, 2nd edition (Springer Verlag, Berlin) 2009, ISBN-13: 978-3540769521, Chapt. 3.
[2] NACRE (Nuclear Astrophysics Compilation of Reaction Rates), `http://pntpm.ulb.ac.be/nacre.htm`; ANGULO C. *et al.*, *Nucl. Phys. A*, **656** (1999) 3.
[3] RAEDER J. *et al.*, in *Controlled Nuclear Fusion: Fundamentals of its Utilization for Energy Supply* (John Wiley & Sons, New York) 1986, ISBN-13: 978-047110312, Chapt. 1.
[4] REBHAN E., in *Heisser als das Sonnenfeuer* (Piper Verlag, München and Zürich) 1992, ISBN-13: 978-3492031097, Chapt. 5.
[5] ATZENI S. and MEYER-TER-VEHN J., in *The Physics of Inertial Fusion: Beam-Plasma Interaction, Hydrodynamics, Hot Dense Matter* (Oxford University Press, Oxford), 2004, ISBN-13: 978-0198562641, Chapt. 1.
[6] EMMERT G. A., EL-GUEBALY L. and KULCINSKI G. L. *et al.*, *Nucl. Fusion*, **29** (1989) 1427.
[7] WOLF R. *et al.*, *Nucl. Fusion*, **57** (2017) 102020.
[8] YAMADA H. *et al.*, *Nucl. Fusion*, **51** (2011) 094021.
[9] The JET team, *Nucl. Fusion*, **32** (1992) 187.
[10] STRACHAN J., ADLER H. and ALLING P. *et al.*, *Phys. Rev. Lett.*, **72** (1994) 3526; HAWRYLUK R. J., ADLER H. and ALLING P. *et al.*, *Phys. Rev. Lett.*, **72** (1994) 3530.
[11] MCGUIRE K. M., ADLER H. and ALLING P. *et al.*, *Phys. Plasmas*, **2** (1995) 2176.
[12] ISHIDA S. *et al.*, in *Proceedings of the 16th IAEA Conference on Fusion Energy, Montreal* Vol. **1** (IAEA, Vienna) 1997, p. 315.
[13] FUJITA T., KAMADA Y. and ISHIDA S. *et al.*, *Nucl. Fusion*, **39** (1999) 1637.
[14] KEILHACKER M., GIBSON P. and GORMEZANO C. *et al.*, *Nucl. Fusion*, **39** (1999) 209.
[15] JACQUINOT J. and the JET Team, "Deuterium-tritium operation in magnetic confinement experiments: results and underlying physics", *Plasma Phys. Control. Fusion*, **41** (1999) A13.
[16] WESSON J., in *Tokamaks*, 3rd edition (Oxford University Press, Oxford, UK) 2004, ISBN 0-19-850922-7.
[17] REBHAN E. and VAN OOST G., in *Proceedings of the Sixth Carolus Magnus Euro-Summer School On Plasma And Fusion Energy Physics Brussels, Belgium*, *Fusion Sci. Technol.*, **45** (2004) 15.
[18] LITAUDON X. *et al.*, *Nucl. Fusion*, **57** (2017) 102001.
[19] TSUNEMATSU T., *Fusion Eng. Des.*, **84** (2009) 122.
[20] BARABASCHI P., KAMADA Y. and ISHIDA S., in *Proceedings of the 25th IAEA Conference on Fusion Energy, St-Petersburg, Russian Federation, 2014* (IAEA) 2014, paper OV/3.
[21] BEIDLER C. D. *et al.*, *Plasma Fusion Research Series*, **5** (2002) 149.
[22] WOBIG H. and ANDREEVA T. *et al.*, in *Proceedings of the 36th EPS Plasma Physics Conference Sofia, Bulgaria* (European Conference Abstracts) 2009, Paper P4.192.
[23] SAGARA A. *et al.*, *Fusion Eng. Des.*, **81** (2006) 2703.
[24] SAGARA A. *et al.*, *Nucl. Fusion*, **45** (2005) 258.

[25] FEDERICI G. *et al.*, in *Proceedings of the 25th Symposium Fusion Engineering (SOFE), 10–14 June 2013, San Francisco, USA* (SOFE) 2013, paper WO1-1.
[26] KNASTER J. *et al.*, *Nucl. Fusion*, **53** (2013) 116001.
[27] SONATO P. *et al.*, in *Proceedings of the 3rd International Symposium on Negative Ions, Beams and Sources, (NIBS), Jyväskylä, Finland, AIP Conf. Proc.*, **1515** (2013) 549.
[28] ROMANELLI F. *et al.*, in *Fusion Electricity: A roadmap to the realization of fusion energy* (EFDA, Garching bei München, Germany) 2012, ISBN-13: 978-3000407208.
[29] HOSHINO T., *Desalination*, **359** (2015) 59.
[30] WANG M. *et al.*, *Chin. Phys. C*, **41** (2017) 030003.

EPJ Web of Conferences **189**, 00016 (2018) https://doi.org/10.1051/epjconf/201818900016
LNES 2017

Energy intensitites, EROI (energy returned on invested), for electric energy sources

D. WEISSBACH, F. HERRMANN, G. RUPRECHT and A. HUKE
Institut für Festkörper-Kernphysik gGmbH - Leistikowstr. 2, 14050 Berlin, Germany

K. CZERSKI
Institut für Festkörper-Kernphysik gGmbH - Leistikowstr. 2, 14050 Berlin, Germany
Instytut Fizyki, Wydział Matematyczno-Fizyczny, Uniwersytet Szczeciński
ul. Wielkopolska 15, 70-451, Szczecin, Poland

S. GOTTLIEB
Institut für Festkörper-Kernphysik gGmbH - Leistikowstr. 2, 14050 Berlin, Germany
Fachhochschule Dortmund - Sonnenstraße 96, 44139 Dortmund, Germany

A. HUSSEIN
Department of Physics, University of Northern British Columbia
3333 University Way, Prince George, BC, V6P 3S6, Canada

Summary. — The EROI (Energy Returned On Invested) has been calculated for various systems, especially power plants, applying several different methodologies. This quantity is often used as an argument in discussions about economics to emphasize the performance of a certain technology. Contrary to that, the publications often answer different questions, *e.g.* analyses of emissions or material flows, rendering the EROI unphysical. The present work introduces a physical approach based on exergy flows for EROI calculation to pinpoint and compare the efficiency of different power plant technologies.

1. – Introduction

The economic strength of a society depends heavily (or even relies predominantly) on two factors: how efficiently it can provide valuable (or better: usable) work — hereafter called *exergy* — and how proficient it is at using it for the production of goods and services. Measuring this efficiency by investigating the flows of money, goods and services is a complex task, due to unreproducible and arbitrary changes caused by societal processes such as political decisions. Therefore, it is expedient to introduce quantities that are invariant against these issues — physical quantities.

Note that in the present work, the term "society" is defined as: A large group of humans living together over several generations, while operating an economic and industrial system to provide goods and services for themselves and possibly others. Thus, all modern nations fulfill our definition of society. A small set of millionaires living on an island and importing all necessities from the outside is *not* a society meeting our criteria.

While it is difficult (yet possible) to investigate the physical benefit of goods and services, evaluating a society's capability of gaining usable work is far more intuitive. In order to provide exergy, a certain amount of exergy has to be invested. A motor which delivers mechanical work or electricity has to be built first, and the motor factory will require some work and electricity to do so. Thus, one crucial question arises: Which fraction of the available usable work (or exergy) has to be invested to provide it? The answer lies in a single quantity: the EROI, Energy Returned On Invested, which has been determined for several technologies used to deliver exergy.

These systems are heat engines, because only they are able to provide more exergy than they consume. Keep in mind that exergy is different from energy in general, as the latter encompasses both diffuse (heat, noise) and ordered energy (electricity, mechanical work, lasers, etc.), while the term *exergy* refers exclusively to the second type. For a technological society, the most important heat engines are vehicle motors (piston and turbine/jet engines) on the one hand and electrical power plants on the other: the roles of both kinds are highly intertwined, so that the failure of one would cause immediate social chaos. As automation progresses with computers and robots taking over increasingly large parts of the economy, the importance of electrical power plants will grow proportionally.

In fact, heat engines can be seen as economic amplifiers, with the EROI, or R for short, as the amplification factor. Man provides 1 unit of usable work — and technology amplifies it R-fold by delivering R units. Using a more vivid metaphor, the EROI tells us how much work machines can do for us. Thus, the historical change of R of society's power generation is the most important proxy for industrial progress and wealth evolution, hovering around 1 for most of human history, until industrialization — the large-scale application of heat engines — began boosting it upwards starting in the late 18th century, reaching values of multiples of 10 for OECD-like societies of today.

The term EROI has been used in many ways to answer questions different from the one above. It is often applied inconsistently, inhibiting the comparison of the economic efficiency of power plant technologies. This lecture, based on the respective publication [1] of one of the lecture authors and his colleagues, introduces a physical approach for obtaining the EROI that not only potentially eliminates these flaws but may also provide answers to the question of how certain types of power plants affect the economic efficiency of a society.

2. – Description

It is crucial to treat the EROI (or R just for practical mathematical notation) as a physical quantity. From an industrial standpoint, mutability is the most important attribute of a certain form of energy, allowing its adaptation to any purpose. Disordered energy (also called *anergy*, that is, heat, noise or diffuse radiation) can only partially be converted to exergy, while exergy can always be converted completely to anergy. This is what gives exergy provided by heat engines its central importance: It is *versatile*, it can be transformed and transported over great distances as electrical current. Heat has its uses too, *e.g.* enabling proper factory operation, but it is not able to do any useful work alone. In order to create goods and services, coherent, ordered energy is needed — exergy. The maximum amount of exergy that can be extracted in a closed system is limited, but from the point of view of a society, it can be generated when and where needed, provided that fuel supply is not cut off.

Speaking in terms of statistical physics, only the application of exergy is able to reduce the entropy of an (open) system. For a more thorough explanation of these terms, the reader is referred to introductory literature on engineering, thermodynamics and statistical physics, thus in this work extensive explanations will be skipped in order to avoid disturbing the reading flow.

The EROI R of a power plant can be calculated from the following simple equation:

$$R := \frac{E_{\text{out}}}{E_{\text{in}}} \,.$$

with the total electricity output E_{out} and the cumulative demand or input E_{in}, which both, ideally, should be exergies.

The electricity output is, by definition, pure exergy, but the input has to be carefully evaluated, since the data often contains no information about the energy type, making it difficult to extract or convert to exergy. Nevertheless, it is consistent to exclusively use the same pure physical quantity, which can be uniformly applied to all thermodynamic machines, both for output and input.

The methodological conditions for obtaining the output and the input should further ensure that all relevant inputs (fuel extraction, construction, buffering, maintenance, etc.) are considered to achieve comparability (see below).

3. – Methodology

Strict exergy usage only: In order to maintain comparability, at least the output of the power plant should not contain any non-exergetical parts like waste heat, since some plant types (wind, photovoltaics) provide electricity exclusively. As mentioned above, only exergy (here: electricity) is usable for industrial processes, *e.g.* construction and operation of power plants, while heat is far less versatile. Most of all, its thermodynamic quality depends on its temperature — lower temperatures meaning lower quality — which is not uniform among power plant technologies.

No quality weighting of output: Several works present an equivalent of the electricity output of renewables, but not the output itself, to express the amount of fuel (coal, natural gas, ...) not used in a fossil-fueled plant and thus saved. The "EROI" obtained by this approach answers a different question: how much more fuel can be saved by further expanding a certain renewable technology. This addresses resource stockpile problems, but has little to do with the topic at hand; therefore, the "EROI" derived in these works is not an EROI from the standpoint of physics. For the physical approach, this issue is obviously completely irrelevant, thus only the unweighted electricity output must be considered.

No quality weighting of input: As mentioned above, input data often contain several energy fractions of different quality. If possible, only the unweighted amount of exergy used should be considered. In contrast, in most works, all mechanical work or electricity is converted to its primary energy equivalent by multiplying it with the inverse of the mean thermal efficiency of all power plants operated by the society. This is quite arbitrary, since the turbine efficiency might change and the conversion factor rather represents a market value difference. Low-quality heat alone is not sufficient to build and operate power plants, this renders the weighting approach unphysical.

Nevertheless, in the author's work, gathering sufficient data to use pure exergy inputs could not be achieved either. Thus, an unweighted addition of different energy types as well as fuel energy inventories was considered the best practical approach. This immediately causes comparability issues, since data of power technologies show different fractions of exergy usage. To mitigate this problem, all data and the calculations were collected in a table open to the public. See [2].

Considering input as-is: Some works analyzing exergy flows only use theoretical values derived from thermodynamics and chemistry, *e.g.* work amounts and chemical processes without friction or losses, assuming an efficiency of 100%. That does not describe the actual economics of the society, but reflects the theoretical limits of a given technology which cannot be exploited due to physical reasons. Instead, the real used inputs are considered in this analysis.

Discarding fuel energy content: Coal, uranium or any other fuel can be considered a "thermodynamic black box", in the sense that its heat content is in itself not relevant for the society using it — only the usable work extracted from it is relevant for this approach. Thus, the exergy expended to mine fuel, and that gained by burning it and running the combustion heat through a thermodynamic engine enter our calculations, not the raw energy content of the fuel itself. Putting it more clearly: Humans never injected the primary energy content into fossil fuel stockpiles, they were created and made available by natural processes. Contrary to this, many analyses, *e.g.* Raugei *et al.*, include the fuel's calorific value in the input, which thus becomes dominant for fuel-consuming power plants. This inherently causes the EROI to be roughly equal to the thermal efficiency of the plant, instead of signifying exergy amplification.

Usage of physical inputs only: As shown above, for the purposes of this analysis, only exergy flows should be considered. This is done by adding up all such relevant inputs from all relevant fields and processes, which is also known as a bottom-up analysis.

So-called top-down investigations were performed in other works, where non-physical efforts like money transfer or labor costs are converted into energy. For instance, this is often done by linking the energy consumption to the gross-domestic product of a country. Since the conversion ratio is highly volatile and strongly depends on the location, it is hardly possibly to treat this as a physical quantity in the strict sense. Further details are mentioned in sect. **6**.

Treat recycling as it is: In many works, recycling has been taken into account by completely subtracting material inventories which could theoretically be reused, assuming they will be fully recovered during the operation of the plant. In reality, this is not the case. Materials are only partially recycled — processing them to avoid downcycling consumes energy. Also, the recycling procedure does not necessarily take place during decommission: should we link the recycled material to the present power plant or a future one? To be consistent in this issue, our analysis uses the present material inventories which already include recycled fractions while omitting such assumptions as mentioned above.

4. – System borders and auxiliary facilities

As stated above, all relevant inputs have to be included, even, for example, the facilities that produce machines which are then used to extract fuel. To be more precise, the system has to be enlarged to encompass the infrastructure that created these facilities and so on. Obviously, this gets very complex when delving deeper into the infrastructure and it is necessary to perform an appropriate analysis truncation: system borders need to be defined. To this end, the following condition is used: The contribution of the remaining infrastructure outside of the limits should be lower than the error of the entire input estimate. Thus, auxiliary facilities such as storage units, mining sites as well as material inventory and exergy flows regarding the power plant are included. All infrastructure outside, such as production of mining machinery or lubricants, contributes negligibly and need not be taken into account.

Figure 1 shows a schematic of the industrial infrastructure around a power plant.

All input sources which fulfill the aforementioned conditions are shown. They are located outside the analyzed system, the power generating system including auxiliary facilities, as defined by the system borders. Note that some energy flows enter the system from the outside while others occur within it (*e.g.* electricity produced at the power plant to directly supply the plant's auxiliary systems). In the latter case, the demand is subtracted from the output instead of adding it to the input, as it remains within the analyzed system — it is not coming from the (consuming) society outside — and therefore, from the "investor's point of view", it can be seen as internal friction.

A power plant not only has to produce electricity, but must also ensure that its output satisfies society's demand, so unpredictable volatility has to be eliminated via buffering (*e.g.* for wind farms). The following assumptions are made in this analysis:

- The output (as well as society's demand) is treated as constant base load. Maintenance outages of all power plants are mostly predictable and therefore neglected.

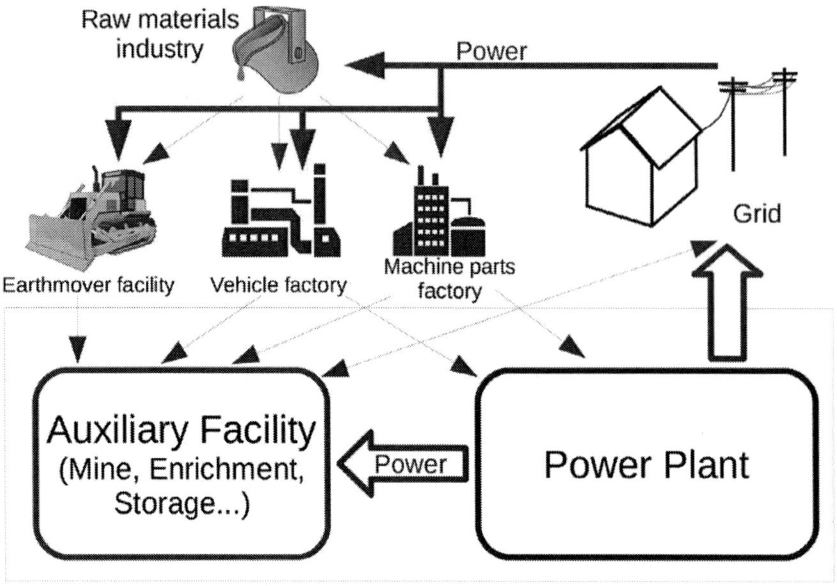

Fig. 1. – System borders for a typical power plant. Whether electricity for a auxiliary facility like a coal mine is taken from the grid (thin red line) or from the power plant does make a difference. Note that if the auxiliary facility is an energy buffer, it also feeds back to the grid.

This issue is already taken into account by the average annual load that technically can be achieved.

- The actual power technology has to fully supply the society by itself. Combinations with other technology mask the characteristics of the analyzed system and hampers comparability.
- The output has to follow load, not vice versa, as often proposed by advocates of smart grids. To this end, buffering for volatile techniques like wind and solar are required — though not for fueled plants as their fuel acts as the storage. Instead, fueled plants need mines or wells as an auxiliary facility, see fig. 1.

The buffering efforts can be estimated by using the data from Popp [3] and the calculation is explained in the authors' publication [1]. This intuitive approach — including buffering for volatile technologies instead of fuel for fueled plants — has so far never been used in any EROI works known to the authors today. Of course, one might argue that the society's demand is never base-load-like so non-volatile power plants have to adjust for that, but these predictable changes can be managed through optimized operation of such plants, not significantly reducing their EROI. Only for niche applications like air condition during the day in hot regions, unbuffered solar power is suitable, but this is by far not a full-supply scenario.

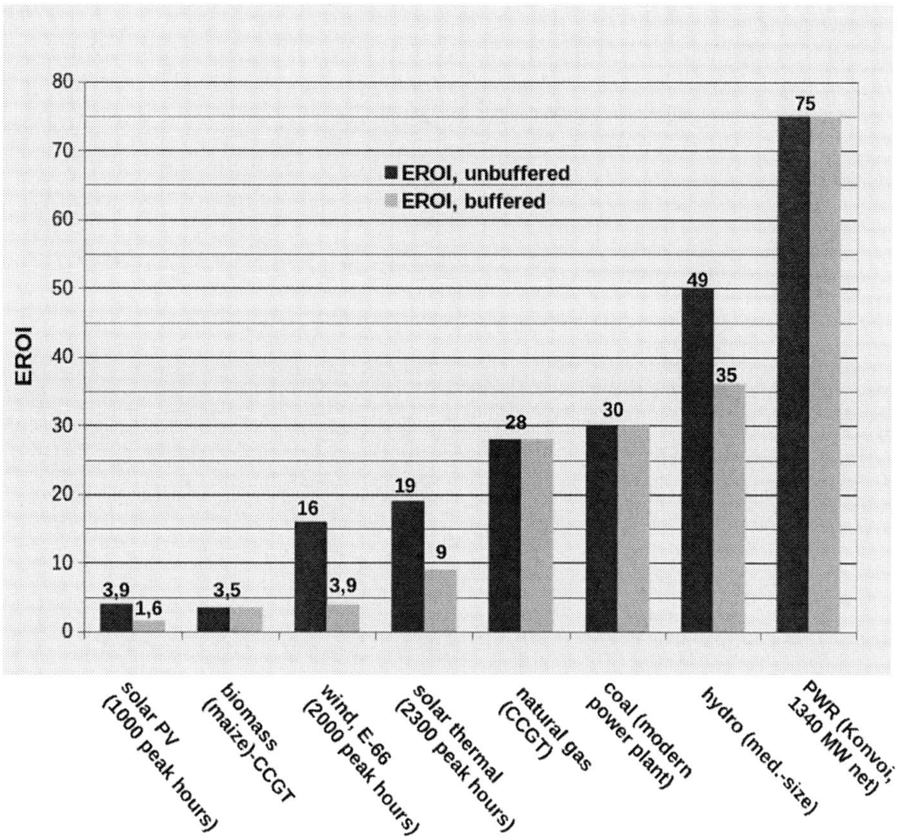

Fig. 2. – EROIs of several electricity generation technologies.

5. – Results and nuclear power as example

The diagram in fig. 2, taken from the authors' publication [1], shows the EROI of some power plant technologies. The respective parameters, also mentioned in this publication, are adapted to German conditions.

Unbuffered EROIs for the volatile technologies (solar, wind, etc.) are shown only for completeness' sake — electricity cannot be stored without converting it to some form of potential energy first, and as the EROI by definition compares exergies, unbuffered EROIs do not exist for fluctuating sources in a strict sense. Additional buffering efforts for all technologies due to load-following are very small compared to compensation for volatility, so they are neglected here.

The economies of all OECD nations rest on fossil burning. This means that in order to achieve what is perceived as prosperity in the 21st century, EROIs around 30 or larger are needed. More generally speaking, the EROI of non-mechanized human work (a single-digit value close to 1) should be exceeded by at least one order of magnitude, putting

TABLE I. – *EROIs and key figures of the reference nuclear power plant (*100% *centrifuge enrichment in brackets), based on the authors' publication* [1]. *The values outside the brackets (before year* 2000*) correspond to a composition of* 83% *centrifuge and* 17% *diffusion enrichment where the latter consumes almost* 50 *times more electricity than the former.*

Installed capacity (net)	1 340 MW
Full-load hours	8 000
Lifetime	60 a
Output	**2 315 000 TJ**
Construction energy demand	4 050 TJ, thereof 35% electrical
Decommissioning energy demand	1 150 TJ, thereof 40% electrical
Maintenance energy demand	6 900 TJ, thereof 68% electrical
Fuel extraction	4 200 TJ, thereof 50% electrical
Fuel element fabrication, conversion	2 480 TJ, thereof 15% electrical
Fuel enrichment	10 740 TJ (1 560 TJ), thereof 100% electrical
Fuel disposal	1 380 TJ, thereof 13% electrical
Sum energy demand	**30 900 TJ (21 750 TJ), thereof 60% (50%) electrical**
EROI	**75 (105)**

the economic threshold at a value of around 10. Physically, this means that low-EROI technologies cannot stand on their own without hampering the present economic wealth in industrialized countries: they would collapse quickly if they had to deliver the energy needed for their own construction, maintenance and decommission. Current endeavors in (even unbuffered) solar and wind are critically dependent on fossil or nuclear power driving the production of units.

It is obvious that volatile technologies or power plants with low power density show the smallest EROI values. Wind energy (unbuffered) achieves relatively high values due to the mostly cheap (in terms of energy) materials, but its high fluctuations need substantial buffering. Coal and Combined-Cycle Gas Turbine (CCGT) show similar EROI values, but CCGT inputs are dominated to a far higher degree by fuel extraction expenses when compared with coal, while extraction and enrichment are the main input contributors of nuclear, see table I.

When looking at the distribution of monetary costs, remarkable differences to the energy input fractions can be observed. For instance, coal extraction dominates the monetary as well as the energy investment, but for nuclear power, fuel extraction plus enrichment absorbs the largest share of energy input, while costs are dominated by plant construction. This demonstrates the necessity of evaluating the physical economic efficiency of a given power plant — coal seems to be managed quite efficiently, but the technological potential of the Light Water Reactor (LWR) is not fully exhausted, as the electricity costs show: they are similar for coal and nuclear power plants despite the substantial EROI difference. In fact, the theoretical potential of nuclear power in general is exploited to a far smaller degree than that of all other existing technologies.

The caloric value of actinides is about 3 million times higher than that of chemical fuels. Even when accounting for the 1000 (mines today) to 30000 (mean concentration in Earth's crust) times lower abundance in nature, far higher EROIs than those of current-technology plants, surpassing fossil plants by only a factor of 2.5, should be achievable. LWRs are not capable of unleashing the full potential of the Strong Force. This is caused in part by their very low fuel usage (less than 1%), as well as, relative to the amount, high fuel processing demand. Additionally, LWRs are complex facilities. Improving these flaws, together with a better economic exploitation of the technology, would enable nuclear to boost society to a completely new level of energy flux densities and productivity.

6. – Comparison with other results

The following example, shown in table II, demonstrates how the same EROI for solar photovoltaics can be obtained using different methodologies — see the analysis of Prieto [4] as compared to the present work. Detailed numbers can be found in his book [5]. The main difference is the extensive usage of additional top-down parts by Prieto *vs.* the higher buffering fraction assumed by Weißbach *et al.* — Prieto included lower backup efforts provided by the surrounding power plants.

Prieto used to operate large photovoltaic field installations in Spain, collecting extensive economic data on the whole production infrastructure and gaining comprehensive maintenance experience. His absolute input energy values for the modules are roughly the same as the ones from the author, the relative difference being caused simply by the different plant load. Comparing the output data of Prieto and Weißbach in their respective works shows that the electric efficiency as well as the output reduction due to weather, inverter efficiency etc., or the so-called performance factor is roughly the same. All these amounts are real used energies, but Prieto adds additional money-equivalent energy inputs.

About 60% of the entire input calculated by Prieto is derived from monetary costs and further 10% are calculated from the primary energy consumption per capita for the whole labor involved. This methodology shows some serious issues when calculating the EROI as a physical quantity. While it is accepted that money and labor are important proxys for evaluating the economic efficiency of power plant technologies, it heavily masks its physical efficiency in several ways.

First, the energy-to-money ratio as well as the *per capita* primary energy consumption depend heavily on the country the analyzed plant is located in or on societal processes in the said country. Although the plant itself is physically the same no matter which location is considered, the physical efficiency, the EROI, changes significantly — a gas turbine in a less developed country would have a higher EROI than in an industrialized country. This is caused simply by the different living standards, which brings us to the

TABLE II. – *Values are given as fractions from the output, so their sum corresponds to the energy intensity* $1/R$.

	Prieto (2016)	**Weißbach (2013)**	**Comments**
Manufacturing	12%	21%	difference corresponds to insolation
Installation, transports	3.5%	4.1%	Prieto: mostly top-down
Backup fraction	4%	18%	Prieto: top-down (CCGT backup), Weißbach: Pump hydro
direct/indirect labor	5.5%	0%	Prieto: 90 MWh/year per worker (primary energy)
Operation/maintenance	7.7%	0%	Prieto: top-down, Weißbach: no energy data known
surveillance, taxes, fees, insurance, PR	3.5%	0%	Prieto: top-down, money-to-energy conversion 2 kWh/euro
losses, production overcapacities, grid fraction	8% (3.5% grid-related)	0.8% (grid connection only)	Prieto: top-down, fraction of whole grid included
EROI, buffered	**2–3**	**2.3**	**Poly-Si, field, Pietro: 1730, Weißbach: 1000 full-load hours per year**

second issue. The living standards, which are now included in the input, are the result of the society's economic efficiency, which itself, at first, should be a consequence of the EROI. Such feedback loops in the EROI calculations lower the EROI for power plant technologies with high monetary flows (*e.g.* nuclear, see sect. **5**) or which necessitate a lot of labor (*e.g.* solar, see above), considerably obfuscating the achievable physical efficiency and hampering comparability.

For this reason, such efforts are not taken into account in the present work. The solar PV system analyzed by Prieto thus would have an EROI of about 4 when assuming reasonable buffering. This value shows the maximum physical efficiency of the actual system with the available technology and a certain insolation, invariant against societal processes. Regarding other analyses, the manifold methodologies applied may cause EROI variations by a factor of 50 — mainly due to input or output weighting, top-down calculations and outdated numbers, best seen in the analysis of nuclear power plants by Leeuwen *et al.* [6], with an EROI of roughly 2.

Prieto's analysis show a similar EROI value for solar PV compared to the work by Weißbach *et al.* indicating the systems having a similar economic performance. Nevertheless, the main difference is the methodology — Prieto used energy inputs derived from monetary flows, rendering the EROI more unphysical, making it not invariant against non-physical quantities thus less comparable to other technologies. While it is easily possible to adjust for the different insolation (1750 and 1000 full-load hours annually in Spain and Germany, respectively) it is not possible to compare both PV systems because of arbitrary societal influence — the administrative and salary situation in Spain differs from that in Germany. The results just seem to be accidentally identical because there are several derivations in both directions.

7. – Conclusion

In the large field of life cycle analysis, until now attention focused mostly on emissions, material and energy flows and also monetary quantities, answering questions about economy, environmental impacts and resource consumption. When used, the EROI was usually not treated as a physical quantity, but rather as a proxy for these issues. Thus, its value for a given system varied substantially from one publication to the other — it is actually hard to find two publications with the same result. The methodologies applied were neither consistent nor uniform among each other, and societal properties (*e.g.* monetary flows) were often included indirectly in the analysis. This approach not only disguises economical properties of the system as physical quantities (*i.e.* especially its physical efficiency), but also inhibits comparability among analyzed systems.

The present work introduces a consistent method for calculating the EROI, emphasizing the inherent physical characteristics of power plants, which are among the most important infrastructure elements for technologically advanced societies. It was found that only *real used exergy flows* should be included in the input and output, but the lack of sufficient data makes it hard to do so. Nevertheless, with the raw data presented in a transparent way, EROI values for several power plant technologies have been calculated using a consistent methodology. Comparing them gives strong clues to their benefit for industrialized societies. It comes as no surprise that technologies with the highest power and energy density as well as the lowest volatility show the highest EROI values.

There are numerous parameters which can also be discussed, such as fuel resources, "death toll" or land use; but these are independent problems and should be treated separately (though fuel scarcity has a certain influence on the EROI). Combining all these aspects will hopefully allow for a sound discussion about present and future energy supply.

REFERENCES

[1] Weissbach D., Ruprecht G., Huke A., Czerski K., Gottlieb S. and A. Hussein, *Energy*, **52** (2013) 210, `http://dx.doi.org/10.1016/j.energy.2013.01.029`.
[2] Google+ table this work is based on: `https://docs.google.com/spreadsheets/d/11BK3pntKdd3bo8oAAvjnpQvYaLZp1G-ieuS5GA5NGV4`.
[3] Popp M., *Speicherbedarf bei einer Stromversorgung mit erneuerbaren Energien* (Springer, Berlin) 2010, pp. 135–137, `http://www.springer.com/engineering/energy+technology/book/978-3-642-01926-5`.
[4] P. Prieto, *Science & Energy, Course 3 at Ecole de Physique des Houches*, presentation, Les Houches 2016, `http://science-and-energy.org/wp-content/uploads/2016/03/20160307-Des-Houches-Case-Study-for-Solar-PV.pdf`.
[5] Pedro A. Prieto and Charles Hall, *Spain's Photovoltaic Revolution* (Springer, New York) 2013, `http://www.springer.com/gb/book/9781441994363`.
[6] Storm van Leeuwen J. W., *Nuclear power – the energy balance*, website, Part D, Ceedata Consultancy 2007. Visited 2017/07/18, `http://www.stormsmith.nl/np-ebalance.html`.

EPJ Web of Conferences **189**, 00017 (2018) https://doi.org/10.1051/epjconf/201818900017
LNES 2017

Power system and market integration of renewable electricity

Georg Erdmann(*)
Energy Systems of TU Berlin - Einsteinufer 25, D-10587 Berlin, Germany

These lecture notes were originally published in *EPJ Web of Conferences*, **148** (2017) DOI: 10.1051/epjconf/201714800005 as a contribution to the Proceedings of the 5th course of the MRS-EMRS "Materials for Energy and Sustainability" and 3rd course of the Joint EPS-SIF International School on Energy held in 2016. They are here reproduced as a good summary of the material presented in July 2017 during the course "Advances in Basic Energy Issues".

(*) E-mail: `Georg.erdmann@tu-berlin.de`

EPJ Web of Conferences **189**, 00018 (2018) https://doi.org/10.1051/epjconf/201818900018
LNES 2017

Modelling the material and energy costs of the transition to low-carbon energy

OLIVIER VIDAL(*), HUGO LE BOULZEC and CYRIL FRANÇOIS

Isterre, CNRS, Université Grenoble-Alpes - France

Summary. — The evolution from 2000 to 2050 of the needs in concrete, steel, aluminium and copper to build the infrastructure of electricity generation is modelled for the scenarios of García-Olivares *et al.* (2012), Ecofys-WWF (2012) and the blue map scenario of the IEA (2010). A simple dynamic model is used to estimate the primary production, recycling and lost flows as well as the cumulative stocks of material to be produced, to go into the infrastructure and to be lost. The energy of material production is also estimated. When compared with the expected evolution of global material and energy demands, the modelling results suggest that i) the transition to low-carbon energies implies a substantial increase of raw materials and energy consumption, ii) the shorter lifetime of wind and solar facilities and the loss of recycling implies that the total amount of metal to be produced during the deployment of the infrastructure of energy is significantly higher than their amount stocked in the infrastructure, and iii) the needs in materials and energy will occur in a period of expected increase of primary metal consumption at the world scale and limited potential of recycling.

Introduction

Meeting a growing demand for energy while reducing greenhouse gas emissions is a crucial challenge of the 21st century. The European Commission aims to reach a share of renewable energy of at least 27% of EU energy consumption by 2030 and the COP21 Paris agreement on climate change targets reaching the "carbon neutrality" worldwide

(*) E-mail: `Olivier.Vidal@univ-grenoble-alpes.fr`

by 2050. To achieve these objectives, it is necessary to transform within 40 years the existing fossil-fuels–based system of energy generation, transportation, storage, distribution and use. Large infrastructures of energy generation with a material intensity significantly larger than that of fossil fuels power plants are required to capture renewable but diluted energy [1-6]. The latest generations of > 5 MW off-shore wind turbines with their foundation contain about 1000 t of steel and they require wiring network and grounding systems that consume up to 9.6 tons of copper per MW [7]. This corresponds to steel intensities 2 to 6 times larger than fossil fuel power plants and copper intensities 4 to 13 times larger [4]. The same observation is made for many other mineral resources including cement, aggregates, aluminium, and a variety of rare metals and other "critical" raw materials like REE, platinium group metals, etc. [8]. The production of all these materials is not neutral and it requires high amounts of energy. At present, about 10% of the world energy demand is used for cement and steel productions alone. It appears thus that the transition to renewable energy will be a source of energy and material overconsumption during the next 40 years. An estimation of the needs must be made to evaluate the best deployment options in regard to the actual and forecasted global supply of material and energy for the next decades.

In the following, we propose a quite simple dynamic modelling of material flows and stocks and energy flows. It is used to estimate the needs in primary and recycled material for the future infrastructure of electricity generation for three contrasted energy scenarios. The results are then compared to the needs of matter and energy at the global scale. Our study focuses on the case of steel, cement, aluminium and copper, which are the major and non-substitutable materials carrying the most important part of embodied energy in the energy production facilities. Moreover, the historical consumption and production of these raw materials are well documented since several decades, and their future needs can be estimated in a relatively reliable way as it poorly depends on technological innovation (in contrast to rare metals used in high technologies metals).

1. – The considered scenarios of energy

Many scenarios of the future energy demand have been published during the last decades, which foresee contrasted evolutions in the shares of conventional and renewables capacities. In the following, the IEA blue map scenario (BM) from [9] was selected because it is rather conservative in anticipating that about 40% of the global electricity (36500 TWh/yr in 2050) will be produced from renewable sources. The Ecofys-WWF scenario (WWF) from [10] is more ambitious. It assumes a 100% renewable electricity production in 2050, produced with a mix of solar PV and concentrated solar power (CSP) technologies, wind turbines and hydropower (fig. 1). The WWF scenario is based on the sobriety of use and assumes a strong reduction of global energy consumption down to 260 EJ in 2050 (393 EJ in 2015; see [11]). Similarly, it assumes a limited increase of electricity consumption to 29000 TWh/yr in 2050 (36500 TWh/yr for BM). The last considered scenario (GO) proposed by García-Olivares *et al.* [2] assumes that the unique source of energy in 2030 is electricity produced by wind turbines, CSP and hydropower.

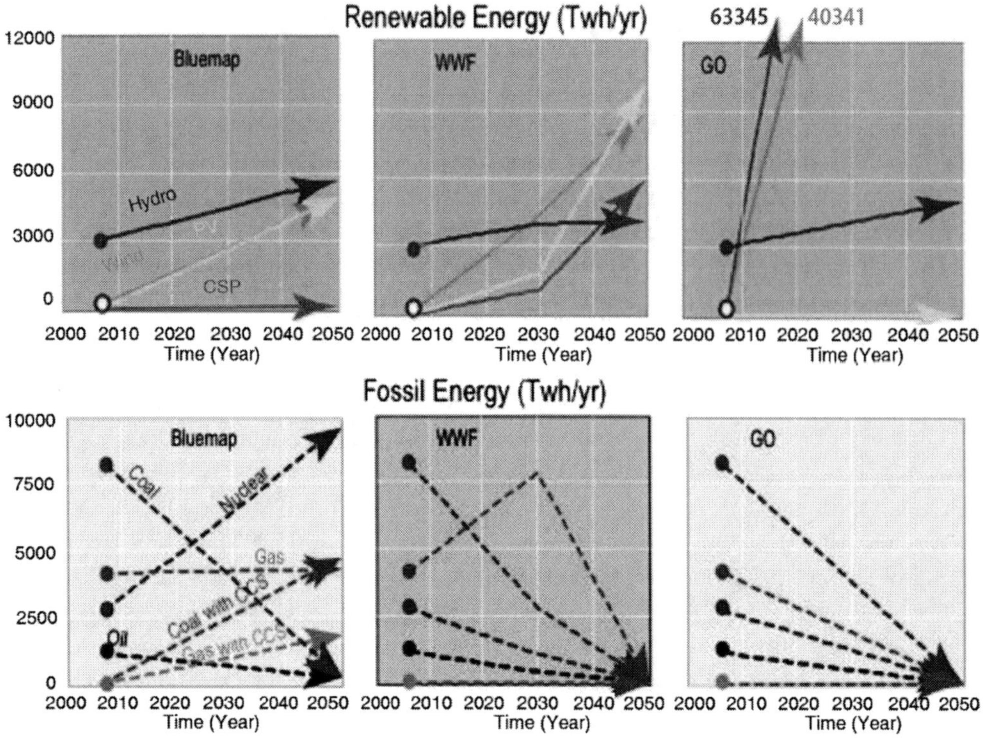

Fig. 1. – Evolution of the energy mix for the blue map [9], Ecofys-WWF [10] and García-Olivares *et al.* [2] (GO) scenarios.

The global demand in electricity will reach the enormous value of 108000 TWh/yr in 2030. Such an evolution is clearly unrealistic, but the GO scenario remains interesting because it provides an upper limit in raw materials demand for the transition to low-carbon energy. In order to be more realistic, we have considered that the scenario is achieved in 2050 instead of in 2030.

2. – The dynamic model

The value chain from primary extraction, use, end-of-life and recycling has been modelled as shown in fig. 2. Three types of power generation infrastructures have been considered on the basis of their lifetime. The lifetime of wind and solar power sources is assumed to be 20 years, 40 years for coal, gas and oil-fired power plants, and 60 years for nuclear power plants. For each type of installation, the stock in tons of "Primary raw material in the infrastructure" is supplied by the flow in tons/year of "primary production", and the stock of "recycled metal in the infrastructure" is supplied by the "recycling" flow. The infrastructure manufactured at t_0 reaches its end of life in $t_0 + \Delta t$ where Δt = lifetime (20, 40 or 60 years). The stock "primary raw material in the in-

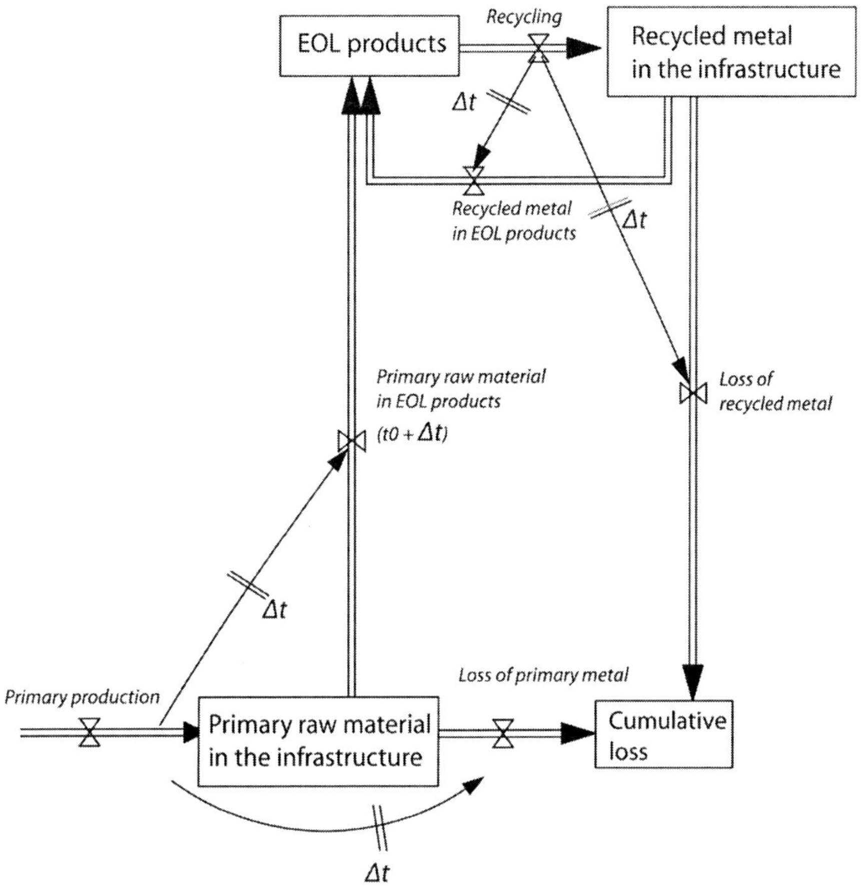

Fig. 2. – Schematic view of the model used to estimate the stocks and flows of matter during the construction and maintenance of the infrastructure of electricity generation [4].

frastructure" is then reduced by the outgoing flow of "primary raw material in EOL products", which represents the flow of primary raw materials from the working infrastructure to the stock of end-of-life products (EOL). This $t_0 + \Delta t$ flow is equal to the "primary production" flow at t_0 multiplied by a collection rate (CR). The rest of the material is lost at $t_0 + \Delta t$ through the flow "loss of primary metal" = "primary production" $\times(1 - \text{CR})$. The stock of "EOL products" is also supplied by the already-recycled metal flow entering a new recycling loop ("recycled metal in EOL products"). At $t_0 + \Delta t$, this "recycled metal in EOL products" flow is equal to the "recycling" flow at t_0 multiplied by CR. We have assumed that all the EOL products are immediately recycled, so that this stock is empty. This assumption implies that the collection rate (CR) used in the present study encompasses both the true collection and recycling rates of EOL products.

The model is run from 1900 to 2100, assuming that i) the installed capacities of

Table I. – *Energy consumed for the production of concrete, steel, aluminium and copper* [12], *and collection rates (CR) of metals from the infrastructure of energy or for all goods at the global scale.*

	2010	2030	2050	2050/2010
		Primary energy of production (MWh/t)		
Concrete	0.3	0.2	0.2	0.84
Steel	6.11	5.1	4.2	0.68
Al	23.6	20.8	18.1	0.77
Cu	8.3	8.3	8.3	1.00
		Recycling energy (MWh/t)		
Steel	1.9	1.5	1.1	0.57
Al	2.4	2.1	1.8	0.77
Cu	4.2	4.2	4.2	1.00
		CR of metals in the infrastructure (%)		
Steel	0.70	0.70	0.70	1.00
Al	0.70	0.70	0.70	1.00
Cu	0.70	0.70	0.70	1.00
		Global CR (%)		
Steel	0.65	0.67	0.68	0.73
Al	0.43	0.47	0.51	0.60
Cu	0.43	0.47	0.51	0.60

nuclear power stations increase linearly from 10 GW in 1950 to 609 GW in 2010, ii) the installed capacities of other fossil fuel power stations and hydropower plants increase exponentially at a constant rate of 3%/yr between 1900 and 2010, and iii) the installed capacities remain constant after 2050.

The amounts of material in the electricity-generation infrastructures are calculated using the material intensities in tons per MW installed capacity listed in table 5.1 of [4]. The energy required to produce the materials are calculated using the values in energy/mass from [12] for primary production and recycling, which are listed in MWh/t in table I. The collection rates (CR) of all metals from the electricity-generation infrastructure have been set at 70%.

3. – Concrete, steel, aluminium and copper requirements for the future electricity-generation infrastructure

For the three scenarios, the needs of copper for renewables represent more than 85% of the total needs (*i.e.* including the other sources of electricity generation). For the blue map scenario, the cumulative stock of primary copper in the infrastructure of electricity generation is about 39 Mt in 2050 (24 Mt primary and 15 Mt recycled, see the graph in the

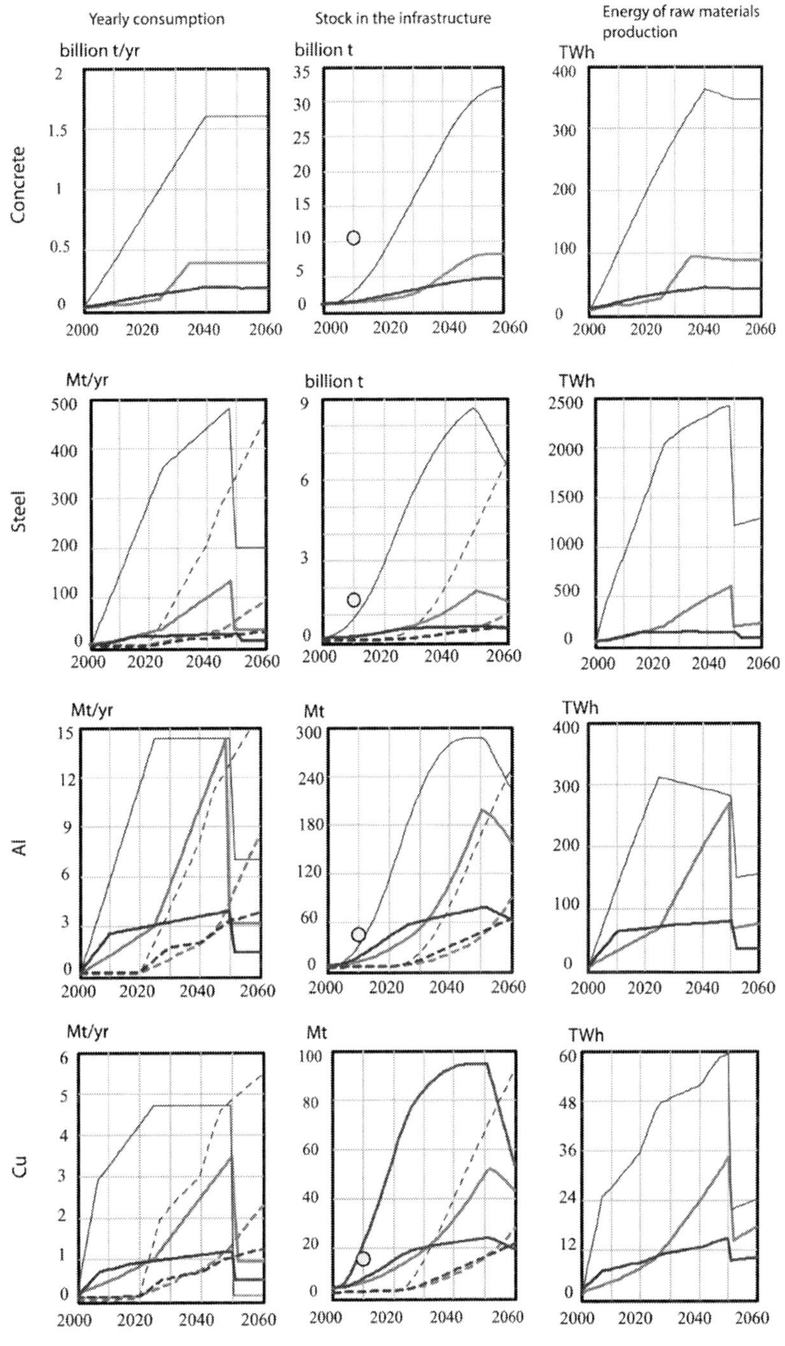

Fig. 3. – Annual consumption (first column) and cumulative requirements (second column) of primary materials (solid lines) and recycled materials (dotted lines) for the deployment of the infrastructure of electricity production. The third column shows the annual energy consumed for primary production and recycling. The red, green and blue lines show the GO, WWF and BM scenarios, respectively. Yellow dots: annual global production.

second column and last row of fig. 3). However, 30% of copper is lost during recycling, so that the true amount of copper to produce by 2050 is about 55 Mt. This estimated value is almost two times higher than that estimated by [5] for the same scenario (29 Mt). The cumulative amount of copper increases to 90 Mt for the WWF scenario (50 Mt primary, 22 Mt recycled and 18 Mt lost) and 205 Mt for the GO scenario (55 Mt primary, 90 Mt recycled and 60 Mt lost), in fair agreement with García-Olivares *et al.*'s [2] own estimates (183 Mt). Finally, the BM, WWF and GO scenarios need 2, 4 and 10 years of the present Cu global production, respectively, which will have to be allocated to the infrastructure of electricity until 2050. In the BM scenario, the annual production of copper assigned to the infrastructure of electricity generation peaks at 2.2 Mt/yr (1.2 Mt/yr of primary Cu) at the end of the deployment period. It drops to 1.6 Mt/yr (0.5 Mt/yr primary) after 2050, which is the flow of virgin Cu required to maintain the infrastructure and balance the loss of recycling. For the GO scenario, the yearly consumption reaches 9 Mt (4.5 Mt/yr primary) and it drops to 5 Mt/year (1 Mt/yr primary) after 2050.

Following our assumption that 70% of copper is recycled, from the infrastructure, primary copper consumption between 2030 and 2050 ranges from 1 to 4.5 Mt/yr for the three considered scenarios. This yearly consumption is equivalent to 5 to 20% of the present global production. García-Olivares *et al.* [2] estimated that the share of copper in the infrastructure of electricity generation represents 55% of the total amount required to achieve the full shift from fossil to renewable energy. The cumulative amount of primary copper to be produced until 2050 could be therefore as high as 270 Mt for the GO scenario and 145 Mt for the WWF scenario, which is equivalent to 7 to 13 times the present annual global production, and 20 to 40% of the known copper reserves. However, it is emphasized that a significant portion of the needed metals comes from recycling. The most exigent scenario (GO) requires 205 Mt copper, but if the infrastructure is recycled for the same use, "only" half of this amount is primary copper.

The cumulative production and consumption of steel and aluminium show trends similar to those of copper (fig. 3). About 500 Mt aluminium are necessary to achieve the GO scenario, 230 Mt for the WWF scenario and 120 Mt for the blue map scenario. These amounts represent 8, 4 and 2 years of the present global production, respectively. The yearly consumptions of primary Al and steel between 2030 and 2050 represent 5% (BM) to 25% (GO) of the present global productions. For the GO scenario, our estimated amounts of steel incorporated in the infrastructure (13000 Mt) are much larger than García-Olivares *et al.*'s own estimates (3800 Mt). The difference stems from the use of different steel contents per MW of installed CSP facilities and to a lesser extend from wind turbines. The steel content used in the present study for CSP technologies (563 t/MW) was estimated by [4] as an average of the literature data and it is similar to the average of the data reported by [5]. In contrast, García-Olivares *et al.* used data representative of Andasol 1 only (180 t/MW). Similarly, the steel content of wind turbines estimated by [4] (166 t/MW) considers a range of on-shore and off-shore installations, while García-Olivares *et al.* used the value of Vesta V-90 (3 MW) turbine only (132 t/MW). As for copper, a strong drop in primary metal consumption occurs after 2050, when the capacity of the installed infrastructure is assumed to remain constant.

The primary metal consumption after 2050 remains however significant (200 Mt/yr, see fig. 3), because the short lifetime of solar and wind infrastructures implies rapid renewal and high losses of material during collection and recycling of the end-of-life infrastructure. For the WWF and GO scenarios, the yearly recycling of metals from the infrastructure after 2050 oscillates around values equivalent to the present global recycling from all manufactured products. To achieve this goal, the present recycling capacities will have to be scaled up significantly.

4. – Generated and consumed energy

The consumed energy to produce the raw materials for the electricity generation infrastructure increases until 2050, it drops sharply after 2050 and remains almost constant until 2100 (fig. 3, last column). For the GO scenario, this energy rises from 63 TWh/yr in 2000 to 3050 TWh/yr in 2050, and it stabilizes at 1650 TWh/yr after 2050. The energy consumed to produce the materials of the infrastructure is thus multiplied by 50 in 50 years, and it is 25 times higher after 2050 than in 2000, even if the installed capacity remains constant. This is due to the fast replacement of wind and solar facilities, which are assumed to have a twenty-year lifetime. Nevertheless, the energy consumed to produce the materials represent only from 2 to 4.5% of the annual energy generated by the infrastructure between 2010 and 2050 and 1.5% after 2050. For the WWF scenario, the energy consumption increases from 63 to 950 TWh/yr between 2000 and 2050 (between 0.4% and 3% of the produced energy) and stabilizes at 400 TWh/yr after 2050. For the blue map scenario, the consumed energy increases to 250 TWh/yr in 2050 and remains below 1% of the produced energy. For the three scenarios, the energy used to produce the concrete, aluminium, copper and steel is less than 5% of the energy generated by the infrastructure. This fraction may appear small, but it omits the energy to produce other components than the generation system of the infrastructure, as well as the energy consumed by the numerous industrial processes that transform the raw materials into complex machines, the construction of the infrastructure, its maintenance and repair operations. Whatever the true amount of energy consumed to build the infrastructure of energy, our results show that greater amounts of energy are needed for the construction and maintenance of fully decarbonized electricity generation infrastructures than for an infrastructure maintaining a fraction of the electricity produced from fossil fuels. The shift to renewables will thus be a source of material and energy overconsumption. The WWF scenario assumes a significant reduction of global energy demand down to 260 EJ/yr (72000 TWh/yr) in 2050 from 393 EJ/yr in 2015 [11]. Is this reduction of final energy compatible with the increasing demand of energy at the global scale? This point is discussed below.

5. – Comparison with future global trends

The results presented above can be compared with the global production of concrete, steel, aluminium and copper as well as the global energy consumed for their supply. At this end, the model shown in fig. 1 was used and run for each commodity assuming an average lifetime for all goods produced globally. This allowed us to infer the evolution of primary and secondary productions, as well as the energy of production at the global scale. The lifetime of iron (steel), copper and aluminium at the global scale were assumed to be 30 years. The CR of steel was estimated in order to reproduce the 1900 to 2015 production data of crude steel reported by the USGS and the share of primary iron (pig iron) and global scrap steel consumption reported by the AME group [13]. Between 2007 and 2050, the primary production and recycling of steel were assumed to increase from 1300 Mt and 400 Mt, respectively, to 2200 Mt and 1000 Mt, respectively. To reproduce these trends, the collection rate of steel should increase from 0.65 in 2000 and to 0.68 in 2100 (table I). The collection rate of copper was constrained to 0.43 for the period 1930 to 2015 in order to reproduce the yearly amount of copper produced from old scrap since 1960 [14] and the amount of primary copper produced since 1900 reported by the USGS [15]. It was more difficult to find reliable data for aluminium and we assumed a collection rate similar to that of copper. The future primary copper production was assumed to follow the trend estimated by [16] and the future production of steel until 2050 corresponds to the base case scenario of [13], which is consistent with the BM scenario. For aluminium, we assumed a constant growth rate of production that reaches 150 Mt/year in 2050, in agreement with the estimates of [9] ranging from 142 to 190 Mt/yr. The growth rate of concrete production was assumed to increase to 7 Gt/year in 2050, which is slightly higher than the IEA estimates ranging between 5 and 6 Gt/year [9].

For the BM scenario, the proportions of metals for the infrastructure of electricity are 0.8 to 3% of the global production for steel, 1.5 to 6% of aluminium production and 2 to 5% of copper production. These proportions are much higher for the WWF scenario (up to 9.5% for steel, 15% for aluminium and 10% for copper between 2040 and 2050), and become huge for the GO scenario (9% for aluminium and 25% for steel and copper between 2025 and 2050). In the case of copper, the maximum consumption of metals by the infrastructure of electricity is reached at the time of the primary production peak forecasted by [16] and [17]. If these copper peaks occur, the implementation of the WWF and GO scenarios might be compromised by the availability of primary copper. In any case, a significant proportion of the global copper production should be reserved for the generation and use of electricity, and/or substitutes to copper will have to be found. The same trends are observed for the recycled portion. Between 2025 and 2075, 20 to 30% of steel, 10 to 20% of aluminium and 30 to 50% of copper recycled every year should be used for the infrastructure of electricity according to the GO scenario. It must be emphasized that a significant increase of recycled metal is not possible, because the amount of recycled metal is primarily constrained by the flow of metals to the EOL products stock. This flow corresponds to the total production 30 years earlier multiplied by the collecting rate. About 850 Mt of crude steel has been produced in 2000, while the expected demand in 2030 will be about 2300 Mt. The share of recycled steel in total production will thus be less than 35% in 2030. Actually, the true share of recycled steel in total production is calculated to decrease down to 25% between 2000 and 2010 and to remain stable until 2030. Steel recycling and the ratio recycled/total should then rapidly increase to reach 45% in 2050. Similar trends are observed for aluminium and copper. In the same time, the substantial increase of (recycled metal)/(total production) ratios after 2030 is accompanied by an equally significant decrease in raw materials energy consumption.

The huge needs in commodities for the GO scenario result from the enormous renewable production of electricity, which reaches 109000 TWh/yr in 2050. In comparison, electricity productions in the WWF and BM scenarios are 29000 TWh/yr and 36500 TWh/yr, respectively. The WWF scenario also assumes a decline in final global energy from 102000 TWh/yr in 2010 [11] to 72000 TWh/yr in 2050. For the energy efficiency improvements considered in table I, the energy consumed by the global production of cement, aluminium and steel is 10700 TWh/yr in 2010 and 17200 TWh/yr in 2050. It follows that the share of global final energy used to produce these commodities increases from 10% to 16% between 2010 and 2050 in the BM and GO scenario. It increases from 10% to 24% over the same period of time in the WWF scenario. If we assume that the industry worldwide will consume one third of the final energy in 2050 (as in 2015), 70% of the industrial consumption of energy should be assigned to the production of cement, steel and aluminium for the WWF scenario (fig. 4). This is quite unrealistic.

Apart from the raw materials availability, the WWF scenario thus raises the problem of available energy for the future global production of mineral commodities. To cope with the problem, the authors of the WWF scenarios assumed a 50 to 60% reduction of energy consumed for the production of cement, steel and aluminium. Part of this reduction can be achieved through an increase of the ratio recycling/primary production of metals, because recycling is much less energy demanding than primary production (see table I). However, as stated above, this ratio will remain unchanged until 2030, because the long lifetime of metals in finished goods implies that their available amounts for recycling are at best 50 to 80% of their primary production 30 to 50 years earlier. In a period of strong growth of metal consumption, the available amounts of old scraps are therefore much less than total production. Another way to reduce the energy of production is to significantly improve the energy efficiency. However, a 50 to 60% reduction of embodied energy of metals production in the next 40 years will be very difficult to achieve. In the past 40 years, the global average energy requirements for production of pig iron from ore and aluminium smelting have been lowered by 25 to 30% only, and steel and aluminium are already close to 60% efficiency relative to their thermodynamic limits [18]. Further energy savings are more and more difficult when approaching these thermodynamic limits. For the 16 to 32% energy improvement considered in the present study (table I), the global final energy must remain above 110000 TWh/yr in 2050 in order to keep a constant proportion of energy consumed by the production of concrete, steel and aluminium below 15%. This value is 40% higher than that envisaged in the WWF scenario, which is extraordinary low compared to the majority of published values, which range between 110000 and 250000 TWh/yr. The final energy consumption assumed by WWF is only one third of the value of the reference case proposed by the International Energy Outlook 2016 (815 PJ/yr = 238800 TWh/yr).

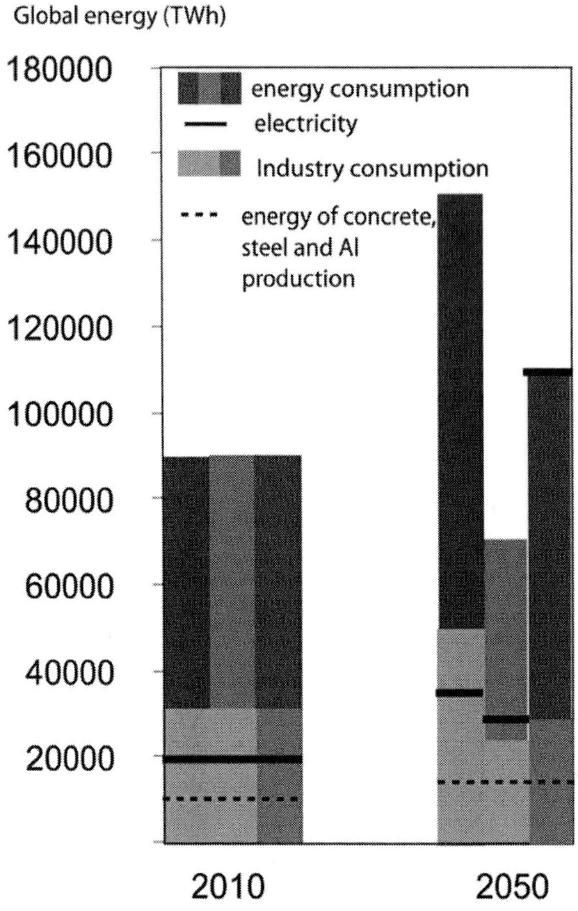

Fig. 4. – Global energy demand and electricity consumptions for the scenarios GO (red), WWF (green) and BM (blue). The industry consumption (black line) is estimated to be 30% of the global energy consumption. The energy consumed for the global production of concrete, steel and Al (dashed lines) is calculated from the relative proportions of primary and secondary production derived from dynamic modelling multiplied by the energies of production listed in table I.

Conclusion

Contrasting needs in raw materials and energy are estimated for the different energy scenarios used in the present study. This reflects the diversity of assumed energy mixes and installed capacities. The huge amount of raw materials required by the scenario GO makes it unrealistic to achieve in a short time. The WWF scenario is less exigent in terms of materials requirements, but the very low final energy production in 2050 is also unrealistic compared to the expected demand. Actually, the needs for the WWF scenario are similar to those of the GO scenario if the capacity to be installed is scaled up to the same value of the final energy. It must be emphasized that our estimates are fraught with large uncertainties because the variability of metal contents reported in the literature for the different kinds of electricity generation is equally large. This was illustrated by the differences discussed above for the steel content of CSP facilities, but the same uncertainties exist for all technologies and commodities. It is also possible that the energy efficiency of mineral resources production increases more than we have assumed in this study. Having said that, at least three qualitative conclusions can be drawn from the results of dynamic modelling:

1) With the current technologies, the transition to low-carbon energies implies a substantial increase in base materials consumption such as concrete, steel, aluminium and copper. This result is in line with previous estimates [1-8] and studies focusing on critical metals [19, 20]. The consumed raw materials and energy are maximum during the building of the new infrastructure of energy, but they remain significant after this period to maintain and renew the wind and solar infrastructures of with a short lifetime.

2) The results of modelling show that i) the total amount of metal to produce during the deployment of the infrastructure of energy is significantly higher than the cumulative amount stocked in the infrastructure. This is due to the losses during recycling. Assuming a twenty-year lifetime for the renewables infrastructure and $CR = 0.7$, about 30% of the total copper production is lost by 2050, and 50% is lost by 2100, which will have to be balanced by primary extraction; ii) a significant portion of metals in the infrastructure is supplied by the recycling of the infrastructure. Therefore, the cumulative amount of metals in the infrastructure cannot be interpreted in terms of primary resources consumption. Dynamic modelling is necessary to estimate the true amount of primary metal and other resources required to build the infrastructure. Since the relative proportions of recycled-to-virgin metals as well as the energy mixes both change with time, dynamic modelling should be coupled with ACV analyses to correctly estimate the impacts of the energy transition.

3) The needs in materials and energy will occur in a period of expected increase of primary metal consumption at the world scale and limited potential of recycling. The supply of metals from recycling is expected to increase but it will not become higher than primary supply before the full shift to renewable energy.

More work is necessary to reach a comprehensive estimate covering the needs in mineral resources and energy from other sectors of energy such as the storage, distribution and use of electricity for applications that traditionally used fossil fuels, (transport, heating). Cement, steel, aluminium and copper are hardly substitutable and bear the most important part of the energy consumed for the production of mineral resources. The supply and impacts of supply of these basic raw materials that do not present any sign of scarcity yet it is an issue as important as the supply of the so-called "critical elements" produced in much lower quantities and whose use is likely to evolve fast with technical innovation.

* * *

This work was supported by the project SURFER financed by "Agence de l'Environnement et de da Maitrise de l'Energie (ADEME).

REFERENCES

[1] Kleijn R., van der Voet E. and Kramer G. J. *et al.*, "Metal requirements of low-carbon power generation", *Energy*, **36** (2011) 5640.

[2] García-Olivares A., Ballabrera-Poy J. and García-Ladona E., "A global renewable mix with proven technologies and common materials", *Energy Policy*, **41** (2012) 561.

[3] Vidal O., Goffé B. and Arndt N., "Metals for a low-carbon society", *Nat. Geosci.*, **6** (2013) 894.

[4] Vidal O., *Mineral resources and energy* (ISTE and Elsevier edition) 2017.

[5] Hertwich E. G., Gibon T. and Bouman E. A., "Integrated life-cycle assessment of electricity-supply scenarios confirms global environmental benefit of low-carbon technologies", *Proc. Natl. Acad. Sci. U.S.A.*, **112** (2015) 6277.

[6] The World Bank: *The Growing Role of Minerals and Metals for a Low Carbon Future* (World Bank Publications) 2017.

[7] Falconer I., "Metals required for the UK's low carbon energy system: The case of copper usage in wind farms", Master's dissertation, University of Exeter, 2009.

[8] Kleijn R., "Materials and energy: a story of linkages, material requirements of new energy technologies, resource scarcity and interconnected material flows", PhD thesis, Leiden University, Netherlands, 2012.

[9] International Energy Agency (IEA), *Energy Technology Perspectives 2010: Scenarios and Strategies to 2050.* Paris, France, 2010b, `http://www.iea.org/publications/freepublications/publication/name,26100,en.html`.

[10] Ecofys-WWF, *The Energy Report 100% Renewable Energy by 2050* (World Wide Fund for Nature (WWF), Gland, Switzerland) 2012.

[11] IEA Sankey diagrams: `http://www.iea.org/Sankey/index.html#?c=World&s=Final%20consumption`.

[12] Birat J.-P., Chiappini M. and Ryman C. *et al.*, "Cooperation and competition among structural materials", *Rev. Métallur.*, **110** (2014) 97.

[13] AME: `http://www.ame.com.au/Website/FeatureArticleDetail.aspx?faId=12`.

[14] Glöser S., Soulier M. and Tercero Espinoza L. A., "Dynamic Analysis of Global Copper Flows. Global Stocks, Postconsumer Material Flows, Recycling Indicators, and Uncertainty Evaluation", *Environ. Sci. Technol.*, **47** (2013) 6564.

[15] Kelly T. D. and Matos G. R., *Historical statistics for mineral and material commodities in the United States*, U.S. Geological Survey Data Series **140** (USGS) open-life report 01-006, 2014 `https://minerals.usgs.gov/minerals/pubs/historical-statistics/`.

[16] Northey S., Mohr S. and Mudd G. M. *et al.*, "Modelling future copper ore grade decline based on a detailed assessment of copper resources and mining", *Resour. Conserv. Recycl.*, **83** (2014) 190.

[17] Sverdrup H. U., Ragnasdottir K. A. and Koca D., "On modelling the global copper mining rates, market supply, copperprice and the end of copper reserves", *Resour. Conserv. Recycl.*, **87** (2014) 158.

[18] GUTOWSKI T. G., SAHNI S. and ALLWOOD J. M., "The energy required to produce materials: constraints on energy-intensity improvements, parameters of demand", *Philos. Trans. R. Soc. A*, **371** (2013) 2012003 Available at: http://dx.doi.org/10.1098/rsta.2012.0003.
[19] ÖHRLUND I., "Science and Technology Options Assessment, Future Metal Demand from Photovoltaic Cells and Wind Turbines, Investigating the Potential Risk of Disabling a Shift to Renewable Energy" (Science and Technology Options Assessment (STOA), Brussels) 2012. Available at: http://go.nature.com/VUOs7V.
[20] MOSS R. L., TZIMAS E. and WILLIS P. *et al.*, "Critical metals in the path towards the decarbonisation of the EU energy sector, Assessing rare metals as supply-chain bottlenecks in low-carbon energy technologies", JERC Pub. No JRC65592, EUR 24884 EN, Joint Research Centre, 2013. Available at: http://publications.jrc.ec.europa.eu/repository/handle/111111111/22726.

AUTHOR INDEX

EDP Sciences – Web of Conferences
17, Avenue du Hoggar
Parc d'Activité de Courtabouf
BP 112
F-91944 Les Ulis Cedex A
France

ISBN 978-1-5108-7252-3

CURRAN ASSOCIATES INC.
proceedings.com